Chemistry of the Cell Interface

Contributors

HARRY DARROW BROWN

A. F. S. A. HABEEB

W. DROST-HANSEN

FRANK X. HASSELBERGER

CHEMISTRY OF
THE CELL INTERFACE

Part B

Edited by

HARRY DARROW BROWN

CANCER RESEARCH CENTER
AND
THE UNIVERSITY OF MISSOURI
COLUMBIA, MISSOURI

1971

ACADEMIC PRESS New York and London

CHEMISTRY

ACADEMIC PRESS, INC.
111 Fifth Avenue, New York, New York 10003

United Kingdom Edition published by
ACADEMIC PRESS, INC. (LONDON) LTD.
24/28 Oval Road, London NW1 7DD

LIBRARY OF CONGRESS CATALOG CARD NUMBER: 76-162934

PRINTED IN THE UNITED STATES OF AMERICA

Contents

Chapter VI. Structure and Properties of Water at Biological Interfaces
W. Drost-Hansen

Chapter VII. Matrix-Supported Enzymes
Harry Darrow Brown and Frank X. Hasselberger

Chapter VIII. Changes in Protein Conformation Associated with Chemical Modification
A. F. S. A. Habeeb

List of Contributors

Numbers in parentheses indicate the pages on which the authors' contributions begin.

HARRY DARROW BROWN (185), Cancer Research Center and The University of Missouri, Columbia, Missouri

A. F. S. A. HABEEB (259), The University of Alabama, Birmingham, Alabama

W. DROST-HANSEN (1), University of Miami, Coral Gables, Florida

FRANK X. HASSELBERGER (185),* Cancer Research Center, Columbia, Missouri

* Present address: The Missouri Academy of Science, Columbia, Missouri.

Preface

Biochemistry has developed from the *fiction* that components of a biological cell, separated from each other as "pure" entities, can interact in a way that is illustrative of the chemistry *in vivo*. Despite *prima facie* improbability, this approach has been successful. Many studies, however, have drawn attention to the fact that structure contributes to the nature of the chemical events. Because reactions cannot proceed independently in the same place at the same time, there is an inherent requirement for the existence of compartmentation. Morphologists using electron optics have vividly defined subcellular structure. It is now clear that many (and perhaps *most*) of the reactions occurring in cells are not solution reactions, and *none* can be equated with those which occur in solution under "ideal conditions." A present view is that biological reactions generally are interfacial phenomena in which one or all of the reactants are spatially restricted.

The contributions to this two-volume work are involved with the relationship of structure to biochemical reactions. They approach in several ways the nature of the generalities which have grown out of the scientific literature. Too, the authors bring understandings of interface chemistry to bear upon conditions within the biological cell. Hence, "Chemistry of the Cell Interface" is a consideration of reactions involving the cell's structured elements and of interfacial reaction systems which are extrapolations from the conventional methodology of solution biochemistry.

This treatise is comprised of two volumes: Part A (Chapters I–V) and Part B (Chapters VI–VIII). Chapters I through III deal with components of complex subcellular systems. In Van Bruggen's Chemistry of the Membrane (Chapter I), the various models have been related to the source data, and the implications of the acceptance of one model over another has been made overt. Thus, while recognizing that biomembranes are chemically as well as functionally diverse, the discussion does much to unify the cell membrane literature. Dieckert, in Chapter II, describes cell particles and the concepts which have developed from their study, thereby further elaborating on interrelationships inherent in subcellular organization. Consideration of enzymes associated with the membranous organelles (Brown and Chattopadhyay, Chapter III) focuses upon catalytic activity as a function of structural ties.

An *in vitro* interface relationships model, that in which the lipid membrane provides the structured environment, has been considered by Tien and James (Chapter IV) as an aspect of lipid–lipid, lipid–protein interactions. Laidler and Sundaram (Chapter V) deal with the interpretations of chemical phenomena in systems having restricted degrees of freedom, with emphasis on the reaction model.

In Part B, water's contribution to the reaction systems is elucidated (Drost-Hansen, Chapter VI). Perhaps in this literature, more strongly than elsewhere, it is apparent that the biological cell is not the solution system of convention. Hence it is essential that the nature of the cell's aqueous phases be understood if we are to correctly interpret the chemistry. The two concluding chapters deal with modified proteins as model reactants. Matrix-supported enzymes, the technology of the model, and the properties of enzymes bound to polymeric matrices have been contrasted with solution and membrane particle systems (Brown and Hasselberger, Chapter VII). The final words are about aspects of protein chemistry pertinent to the design of such interface experimental systems (Habeeb, Chapter VIII).

I wish to express my appreciation for the clerical, editorial, bibliographic, and graphic arts support which has been provided by associates at the Cancer Research Center, and to acknowledge, in particular, the help of Helen Estes, Cynthia Cunningham, Florence Brown, Libby Forbis, Mary Dorward, Robert Hahn, and Yvonne Chapman.

HARRY DARROW BROWN

Contents of Part A

CHAPTER VI

Structure and Properties of Water at Biological Interfaces*

W. DROST-HANSEN

* Contribution No. 1, LABORATORY FOR WATER RESEARCH, Department of Chemistry, University of Miami, Coral Gables, Florida.

1

I. Introduction

The unusual properties of water are directly traceable to the complex structure of this remarkable liquid. Unfortunately, the structure of water has, up to now, escaped a precise description—in fact, several mutually contradictory theories are currently discussed in the literature. Even less is known about the structure of water in moderately concentrated aqueous solutions or about water near interfaces. Vicinal water containing a relatively high concentration of both electrolyte and nonelectrolyte solutes is exactly the type of water which is encountered in biological systems—"the witches brew," according to Professor Henry Frank. Up to now, biochemistry and biophysics, in attempts to describe biological systems, have been fairly successful merely by ignoring the problem of the detailed structure of water in the systems under consideration. The purpose of the present article is to call attention to several unique facets of vicinal water structure in the hope that the recognition of these features may assist in elucidating the role of water in biological systems. Essentially, the plan is first to delineate some ideas about water in general and vicinal water in particular and, next, to present evidence for unusual temperature dependencies of many types of aqueous phenomena: it will be demonstrated that a large number of these phenomena may be explained quite readily in terms of changes in the underlying structure of the vicinal water. With this information in hand, the likely structural characteristics of vicinal water in biological systems becomes more easily understandable, and this is illustrated through a review of some specific properties of a number of biologically interesting systems. The value of this approach is that, however qualitative it may be, it does afford the opportunity to make rather general (qualitative) predictions which can be tested against experience, and, thus, "the exercise is removed from the realm of pure speculation to that of empirical science." The present chapter is not intended to serve as a review—certainly not a critical, carefully annotated review. Instead, it is an attempt to speculate along a line of reasoning (based on water structure) that has not previously been exploited to any large degree. Hence, the chapter should be read for its "inspirational value" rather than as a final or definitive treatment of the subject of the role of water structure near the cell surface. Unfortunately, as will become apparent, the treatment will leave something to be desired, as our understanding of water is still highly fragmentary and incomplete. This, of course, ensures many generations of biochemists, biophysicists, and biologists a fertile field for future *Forschung*.

One can hardly overstress the incomplete understanding which presently prevails regarding the structure of water in general and in biological systems in particular. The essence of the present study is to advocate that future work in the fields of biophysics and biology should recognize from the onset the possibility that the vicinal water of living systems (and, indeed, possibly all the water of the cell) exists in a structurally modified form (compared to bulk water). Although basically not an unreasonable assumption, this approach is made exceedingly difficult by lack of the appropriate physical chemistry of aqueous systems in which the aqueous phase does not possess the structure and properties ordinarily discussed. On the other hand, it is undoubtedly better to seek in the darkness for that which is missing than to search under the light of oversimplified traditional conceptual (and mathematical) models and approaches where it is obvious that the answer is not to be found.

A thorough, general survey of the problem of water near the cell surface should start with a comprehensive review of water structure. However, the subject of water structure has been reviewed by many authors, and the reader is referred to the available survey articles and monographs; particular attention is called to the monographs by Kavanau (1964), Samoilov (1965), Luck (1964), and Eisenberg and Kauzmann (1969). See also the extensive writings by Frank (and co-workers) (1945, 1957, 1958, 1959a,b, 1963, 1965a,b, 1966, 1967). At this time a number of excellent general tools for water structure studies are rapidly becoming available, and these will also be useful for the study of water in biological systems—nuclear magnetic resonance (NMR) and neutron inelastic scattering are typical examples. Unfortunately, even these tools have their restrictions, ranging from instrumental to theoretical (for interpretation of data), or are otherwise limited, for instance, in their geometric resolution of the area or volume to be examined. They far surpass, however, many of the tools which have been available in the past, not to mention that even those tools were frequently not used; indeed, the entire problem has been rather neglected by most researchers. Furthermore, we often seem to encounter tremendous inhomogeneity in our overall approach to problems. Thus, our understanding of the structure of water is qualitative, fragmentary, and incomplete, whereas an eloquent mathematical treatment is developing for the description of irreversible thermodynamics which is achieving great sophistication in its application to biological membrane phenomena. Yet, the advances are often made, not because of, but in spite of, our current understanding of the very medium in which the processes and phenomena of interest occur. In quantitative biology (as well as in a great number of other fields of research), numerical agreement between calculated and observed values for a single parameter (based on a simple, specific mechanism) is often taken as "proof" of a particular mechanism with little regard as to the ability of the underlying theory to explain—even qualitatively—one single additional parameter or set of phenomenological observations. To the present author, an attempt to achieve great theoretical sophistication in, for instance, the analysis of membrane functioning or enzyme reactions seems currently an inordinately misdirected effort because of the uncertainty regarding the influence of the very structure of the phases in which the processes take place.

II. Overview of Structure of Water

A. Structure of Bulk Water

1. Safford's Survey of Water Structure Models

As indicated in the introduction to this chapter, the problem of the structure of water (in bulk) remains unsolved. In fact, several contradictory and mutually exclusive theories are currently discussed in the literature. A number of these theories are outlined briefly in Section II,A,2. At this point we present, in tabular form (Table I), an overview of current water structure theories, as prepared by Safford and Leung (1971). (The original report by Safford is recommended for more detailed information, and, specifically, for a careful analysis of the use of neutron inelastic scattering as a tool for determining water structure.)

2. Comments on Specific Models

The question of the possible existence, in liquid water, of discrete structural entities of geometric, identifiable characteristics is one of the most fundamental current problems regarding water structure. The existence of such structural entities has often been proposed, but the alternative view (i.e., that, on the average, there is no order in water, neither long-range nor short-range order) has been maintained for a number of years, particularly by some spectroscopists. The notion that structurally identifiable units are present in liquid water has been advocated for a long time; thus, mixture models have been envisioned, employing (in older papers), for instance, the existence of dimers, tetramers, hexamers, octomers, and other "polymers" of water (conceptually, the notion of large polymers resembles the more recent idea of clusters). In the 1930s and 1940s, "cybotactic swarms" were envisioned by Stewart (1931) and by Frenkel (1955). In each swarm, the molecules were all presumed to possess one or more physical properties in common (such as dipole orientation) and were considered to contain as many as 10,000 water molecules. Presently, liquid crystals (nonaqueous systems) are probably the closest known to cybotactic swarms. However, even before the turn of the century, it was suggested that other types of structured elements might exist in water; as an example, in one of these mixture models the presence of "microcrystalline chunks" of Ice–Ih* lattices was assumed. The presence of ice-like elements, in this view of the liquid structure, were seen merely as the result of the (incomplete) thermal breakdown (upon melting) of the ice lattice. In view of the enormous amount of information available regarding water and aqueous solutions (often data of extreme precision), it is truly remarkable that the problem of a continuum versus a mixture model continues to be the subject of active discussion in the literature. Currently, the majority of investigators apparently favor the mixture model (to which the present author also subscribes). In connection with the mixture models, we call attention in particular to some aspects which have been stressed by Frank (see, for instance, Frank, 1970) and co-workers. The first of these is in the fact that regardless of the specific nature of the structural units, they probably have the attributes of "flickering clusters." This means that the lifetimes (stabilities) of the structured units are quite short in terms of the times studied by most ordinary analytical tools, but are still long compared to individual, molecular vibrations. A reasonable estimate for the lifetime of any structural arrangement in liquid bulk water might be 10^{-11} second (within a factor of $10^{\pm 1}$). Thus, in approximately 10^{-11} second the structure entity may break down due to an unfavorable thermal fluctuation which disrupts the local order. However, the unit elements, in a similar interval of time, will regroup or reform in part (or in connection with molecules from other, disrupted structures) to form new, short-lived, structured elements. Another feature of some, but not all, mixture models is the possibility that a certain degree of cooperativeness may occur due to the nature of hydrogen bonding. This subject is also one of active discussion in the literature; Del Bene and Pople (1969) in particular are studying this problem through purely quantum

* Ordinary, hexagonal ice is referred to here as Ice-Ih.

TABLE I

MODELS OF WATER STRUCTURE[a]

Description and principal features of the model	References	Comments
I. Continuum Models		
Tetrahedral—four-coordinated hydrogen-bonded local ordering. No local domains having structural differences. Bonds become bent at melting point	J. D. Bernal and R. H. Fowler, *J. Chem. Phys.* **1**, 515 (1933)	Based on X-ray data
Upon melting, bonds become relatively flexible and may bend continuously, resulting in rotational distortions	J. Lennard-Jones and J. A. Pople, *Proc. Roy. Soc., Ser. A* **295**, 155 (1951)	Based on measurements of dipole moments
No separate vapor-like and structured regions, but rather a continuum without definitive structure	T. F. Wall and D. F. Hornig, *J. Chem. Phys.* **43**, 2079 (1965)	Based upon the widths and shapes of observed Raman lines. The observed linewidths and the width of maxima in the X-ray radial distribution function are correlated with continuous variations in the O—O distances of bonded molecules
II. Specific Structure Models		
Water has distorted or expanded Ice-I structure in which defects may occur and in which a definite number of H_2O molecules may pass through faces of surrounding tetrahedra and take up interstitial positions. A molecule which has relaxed from the framework occupies a shallow potential in a void and is partially hydrophobized by the high symmetry of the field in the void. The interaction between framework and void must be weak in order that the framework itself does not collapse	O. Ya. Samoilov, "Structure of Aqueous Electrolyte Solutions and the Hydration of Ions." Consultants Bureau, New York, 1965; E. Forslind, *Acta Polytech.* **115**, 9 (1952); Yu. V. Gurikov, *J. Struct. Chem.* (*USSR*) **4**, 763 (1965)	Based in part upon X-ray radial distribution functions. Fisher and Andrianova, and Gurikov have calculated such quantities as mean coordination number, fluctuation in the coordination number, isothermal compressibility, entropy, and free energy for this model. They find agreement with experiments on mean coordination number and a low percentage of molecules in the voids. The agreement with entropy and fluctuation in the coordination number is found to be poor unless an excess of about 3% of the molecules can exist in the framework. In addition, they point out that the lack of knowledge on the long-range order of the system and its influence on the entropy can give rise to significant error
It has been shown that water has a quasi-tetrahedral short-range structure and that the radial distribution curve is not compatible with an octahedral water structure. The short-range ordering in water is described in terms of a blurred-out Ice-I structure in which molecules have been displaced from the structure by thermal excitation and have occupied interstitial positions. Each framework oxygen atom has three neighbors at 2.94 Å and one at 2.77 Å, and each interstitial has neighbors at 2.94, 3.30, 3.40, and 3.92 Å. The interstitials show a larger temperature coefficient associated with their longer neighbor distances. At 25°C, 50% of the framework cavities are filled	J. Morgan and B. E. Warren, *J. Chem. Phys.* **6**, 666 (1938); G. W. Brady and W. J. Ramanow, *ibid.* **32**, 306 (1960); A. H. Narten, M. D. Danford, and H. A. Levy, *Discuss. Faraday Soc.* **43**, 97 (1967)	This model was based upon fits to observed X-ray radial distribution functions. Such a fit is necessary, but alone does not constitute a sufficient test for the model's validity

The associated clusters in water are clathrate structures similar to those of the gas hydrates. Twenty of the water molecules lie at the corners of a labile pentagonal dodecahedron with an unbonded molecule at the center of the dodecahedron	L. Pauling, *Science* **134**, 15 (1964)	The radial distribution function calculated for this model is inconsistent with that measured for water [see M. D. Danford and H. A. Levy, *J. Amer. Chem. Soc.* **84**, 3965 (1962)]
A hydrogen-bonded framework or Pauling clathrate structure is assumed with interstitial molecules to calculate the statistical, mechanical and thermodynamic properties of water. The framework may relax or "flicker" yielding a third state of H_2O's which are unbonded and have not yet re-formed a cluster	H. S. Frank and A. S. Quist, *J. Chem. Phys.* **34**, 604 (1961)	Yields a satisfactory representation of the PVT properties of water over a limited range of pressure and temperature. It is suggested that the inclusion of the third state would yield a better fit of the heat capacity of water and partial molal properties of nonpolar solutes
An equilibrium is assumed to exist between species having densities similar to Ice-I and Ice-III. Both fluidized vacancies and monomers exist on melting, and the monomers may pack into voids in the Ice-I-like units	M. S. Jhon, J. Grosh, T. Ree, and H. Eyring, *J. Chem. Phys.* **44**, 1469 (1966)	Good agreement is obtained for molar volumes, the vapor pressure below the boiling point, the specific heat, and the pressure dependence of the viscosity
Open- and close-packed structures formed of "puckered hexagonal rings" coexist in equilibrium. In particular, the rings are not viewed as static structures at a given temperature, but as representative of localized short-time interactions. A given water molecule is not always in a ring and is unbonded or monomer-like only when changing from one state to another. Thus, this model characterizes the monomers in terms of a residence or relaxation time for the structure, and rearrangement takes place by jumps of individual molecules	C. M. Davis, Jr. and A. T. Litovitz, *J. Chem. Phys.* **42**, 2563 (1965)	Accounts for the radial distribution curve out to 4 Å, the thermal expansion of water between 0° and 100°C, the relaxation portion of the isothermal compressibility, and the specific heat. The fraction of hydrogen bonds was estimated by Raman spectroscopy

III. Cluster Models

The existence of clusters containing 10,000 water molecules was postulated	G. W. Stewart, *Phys. Rev.* **37**, 9 (1931)	Based on X-ray diffraction only
Frank and Wen proposed that the formation of hydrogen bonds in the liquid was a cooperative phenomenon and that short-lived (10^{-10}–10^{-11} sec) ice-like clusters of varying extent are mixing and exchanging with nonbonded molecules. No specific structure is assigned to the clusters except that H_2O molecules in the interior be four-coordinated	H. S. Frank and W. Y. Wen, *Discuss. Faraday Soc.* **24**, 133 (1957)	These arguments were based upon the partially covalent cluster of the hydrogen-bonded molecules and supported by data on densities, relaxation times, structural changes in solution of nonpolar solutes, and thermodynamic parameters
Nemethy and Scheraga have done a statistical thermodynamical calculation for the flickering cluster model. The compact clusters are made up of four distinct species corresponding to molecules with one, two, three, and four hydrogen bonds and a sharp energy level is assigned to each species. A fifth species is the unbonded monomers which is in equilibrium with the clusters. No long-range ordering is specified and irregular arrangements in the clusters are allowed	G. Nemethy and H. A. Scheraga, *J. Chem. Phys.* **36**, 3382 (1962)	Agreement is obtained with calculated values of free energy enthalpy, and entropy, but poorer agreement with the heat capacity. This theory was also able to account for first and second nearest-neighbor maxima in the X-ray radial distribution curves. Nemethy and Scheraga have pointed out that variations as large as 50 cm^{-1} in the frequencies used for the partition functions produce almost negligible changes in the calculated thermodynamic functions. The frequencies considered correspond to torsional oscillation and hindered translation of H_2O molecules bonded to structured units in the liquid and occur below 500 cm^{-1}. However, shifts of 50 cm^{-1} in these frequencies are not negligible and can correspond to significant changes in the bonding and the lattice geometry for H_2O molecules

TABLE I (Continued)

Description and principal features of the model	References	Comments
Buijs and Choppin obtained near infrared results that, in general, were in agreement with the theory of Nemethy and Scheraga. However, they considered only three species with zero, one, or two of the OH groups of an H_2O molecule bonded. Best agreement on the relative numbers of hydrogen-bonded water molecules was obtained at low temperature	D. Buijs and G. R. Choppin, *J. Chem. Phys.* **39**, 2035 (1963)	Both the theory and the infrared results have been subject to recent questions [see D. F. Hornig, *J. Chem. Phys.* **40**, 3119 (1964)]. D. D. Boettger, D. D. Harders, and D. D. Luck [*J. Phys. Chem.* **71**, 459 (1967)] have pointed out that the populations obtained by Buijs and Choppin for the species are not unique. Further, Stevenson [D. P. Stevenson, *J. Phys. Chem.* **69**, 2145 (1965)] has argued that the concentration of non-hydrogen-bonded water molecules between 0° and 100°C, as predicted by models of Nemethy and Scheraga and by Buijs and Choppin, are two orders of magnitude too high. He concludes that the number of monomers in liquid water between 0° and 100°C is less than 1% of the molecules in the liquid. More recent studies by Griffith and Scheraga (J. H. Griffith and H. A. Scheraga, *150th Meet. Amer. Chem. Soc., 1965, Abstract I-43*) give lower concentrations of monomers than in previous mixture models, but still well above those estimated by Stevenson
Vand and Senior argued that better agreement is obtained between the results of Buijs and Choppin and theory if the sharp energy levels for each species, as assumed by Buijs and Choppin, are replaced by broad energy bands. As these bands overlap, a continuous distribution of molecular states is approached. Thus, the energy state of a species varies with the coordination. Branched chains of molecules exist and can be either free or attached to a cluster. In the limit of this picture, water would be viewed as a loosely bound solid. Vand and Senior have shown, furthermore, that the particular model they chose is not unique in being able to explain the thermodynamic data, but a model based upon one species with an energy band distribution that also fits the thermodynamic data	V. Vand and W. A. Senior, *J. Chem. Phys.* **48**, 1869 (1965), W. A. Senior and V. Vand, *ibid.* p. 1873; V. Vand and W. A. Senior, *ibid.* p. 1878	This model yields good agreement with the Helmholtz free energy, the internal energy, and the specific heats
Walrafen, from Raman measurements, has proposed a model of water similar in certain features to those of Nemethy and Scheraga and of Vand and Senior. Intermolecular librational modes corresponding to tetrahedral five molecule units having C_{2v} symmetry are observed and are in equilibrium with unbonded species which contribute little, if at all, to the spectra. With increasing temperature, these tetrahedral species may distort and break down, giving an increase in species of a lower degree of coordinations	G. E. Walrafen, *J. Chem. Phys.* **40**, 3249 (1964); **44**, 1546 (1966)	The data of Walrafen appears in contradiction to that of Wall and Hornig. The broad Raman lines the widths of which Wall and Hornig ascribed to a continuous variation in O–O distances, appear partially resolved into components in Walrafen's spectra. From the changes in the line intensities with temperature, he obtains reasonable estimates of enthalpy, entropy, and fair agreement with the heat capacity of water

[a] Table prepared by Safford and Leung (1971) and reprinted here by permission of the authors.

mechanical calculations (see also Hankins *et al.*, 1970). The results to date suggest strongly that stabilization of hydrogen bonds may occur where cyclic arrangements of chains of water molecules are possible. We finally stress that some mixture models allow for the existence of discrete sites or voids in the structure. Later in this chapter we return to specific possible structured elements in water, and particularly in water near interfaces. The structured elements include ice (that is, the structural elements similar to ordinary hexagonal ice, i.e., Ice–Ih), high-pressure ice polymorphs, clathrate hydrates, and clusters (in the Nemethy–Scheraga sense). For a review of ice polymorphism, see the excellent report by Kamb (1968) and the report by von Hipple and Farrell (1971); the general field of clathrates has been surveyed by van der Waals and Platteuw (1959) and by Gawalek (1969).

We can summarize a number of significant features. The results of certain theoretical calculations notwithstanding, at present, it appears that the evidence for a mixture model for water structure is quite substantial— water is probably best described in terms of a mixture of structured elements and, possibly, monomers (or other relatively low molecular weight associated species). The major likely candidates for the structured elements are Ice–Ih (somewhat unlikely), the high-pressure ice polymorphs, clathrate cage structures, or various types of clusters.

B. WATER IN ELECTROLYTE SOLUTIONS

1. *Safford's Survey of Water–Ion Interactions and Electrolyte Solution Structure*

The reader is referred to Tables II and III (from Safford and Leung, 1971) for a short survey of recent work on the subject. Further references can be found in the books by Kavanau (1964), Samoilov (1965), and review articles by the present author (Drost-Hansen 1967a) and by Hertz (1970). For an excellent review of hydration of ions, see the article by Desnoyers and Jolicoeur (1969). Very readable accounts of molal volume aspects of electrolyte solutions are presented by Millero (1970, 1971).

2. *Solute–Solvent Interactions*

It appears that most solutes in water affect the structure in the vicinity of the solute molecule (or ion) and several different types of structural arrangements may result. Some solutes (especially electrolytes) are known as "structure breakers," whereas others are "structure makers." One of the difficulties in this connection is that of semantics. Thus, a structure maker

TABLE II

MODELS OF WATER–ION INTERACTIONS[a]

Description and principal features of the model	References	Comments
Water molecules in primary hydration layer subject to strong centric-symmetric forces and are ordered. Beyond this region is a zone of disordered water structure which, at larger distances from the ion, blends into the undisturbed water structure	H. S. Frank and W. T. Wen, *Discuss. Faraday Soc.* **24**, 133 (1957)	This model was based upon corroborative evidence from measurements on heat capacities, dielectric relaxation, diffusion of H_2O in salt solutions, ionic mobility, entropies of solution, and viscosity
Water forms frozen patches or microscopic icebergs about solute molecules. The hydration of ions—their ability to break the water structure and the sizes and bonding of their hydration layers—depends both on size and on the charge of the solute molecules	H. S. Frank and M. W. Evans, *J. Chem. Phys.* **13**, 507 (1945)	Based on observed entropy measurements and the large negative partial molal heat capacities of ions—ions such as Al^{3+}, Mg^{2+}, Li^+, and F^-—can build icebergs about them and decrease the fluidity. In particular ions of high charge, such as La^{3+} and Eu^{3+}, may give rise to a "superlattice" ordering of H_2O molecules with sufficient quasi-solid-like ordering to support lattice vibrations
Ions may either increase or decrease the activation energies of H_2O molecules in their vicinity and act, respectively, as positive or negative hydrators. Ions such as Mg^{2+}, Ca^{2+}, Li^+, and Na^+ show positive hydration, whereas ions such as K^+, Cs^+, Cl^-, Br^-, and I^- show negative hydration. This hydration is assumed to occur in the region of the ion and largely determines the kinetics of the solution. There is also a more distant region where the water is mainly influenced by the ionic field	O. Ya. Samoilov and T. A. Nosnova, *J. Struct. Chem. (USSR)* **6**, 767 (1965); O. Ya. Samoilov, "Structure of Aqueous Electrolyte Solutions and the Hydration of Ions." Consultants Bureau, New York, 1965	Samoilov argues that, in dilute aqueous solutions, ions interact with H_2O molecules so as to yield the minimum modification of the solvent structure. X-Ray data are cited to indicate that regions having structures similar to pure water can coexist with hydrated ions. At higher concentrations, water molecules coordinate to ions so as to yield a similar local ordering to that of solid salt hydrates
Ions are shown to fall into two classes depending on the parameter $\sqrt{Z}/\bar{r}$, where Z is the charge and $\bar{r}$ is the distance from the ion to the dipole center of the first layer of H_2O molecules about the ions. Ions with $\sqrt{Z}/\bar{r}$ greater than about 0.3 (i.e., Li^+, Mg^{2+} La^{3+}) are surrounded by an ice structure varying from 3 to 10 water molecules	E. Glueckauf, *Trans. Faraday Soc.* **61**, 914 (1965)	Data on molar volumes have been analyzed to draw conclusions on the structure of water about ions and the iceberg-building tendency of ions
Small ions (i.e., Li^+ and Na^+) may fit into "holes" in the water structure without breaking, but only slightly stretching and bending bonds. Larger ions, such as K^+, would not fit into the structure and would break it down	F. Vaslow, *J. Phys. Chem.* **70**, 2286 (1966)	Electrostatic energies of alkali ions in the field of a water calculated as a function of the angle of the ions with the dipole axis of the water molecule. A series expansion was used for the potential due to the ionic charge, together with quadrupole moments of the water molecule
Dissolved electrolytes produce distortions in open structure of water lattice. The solute-solvent interaction in general does not give rise to the formation of a distinct coordination sphere of H_2O about the ion. Rather, the ion is integrated into the water lattice by ordinary chemical bonds of variable strength and duration	E. Bergqvist and E. Forslind, *Acta Chem. Scand.* **16**, 2069 (1962)	Based on concentration dependence of proton magnetic resonance chemical shifts

[a] Table prepared by Safford and Leung (1971) and reprinted here by permission of the authors.

does not necessarily mean a solute that enhances the intrinsic (or latently present) structural characteristics of pure water, but, rather, merely indicates the enhancement of some type of "lattice rigidity" over that present in the bulk solution. On the other hand, a structure breaker is a solute which disorders the (local) intrinsic water structure. It should be mentioned that there is some evidence that, even in rather concentrated solutions, there appear to persist structural elements characteristic of pure bulk water (see, for instance, Safford, 1966; Drost-Hansen, 1967b). The suggestion of continued existence of undisturbed water structure elements in strong electrolyte solutions (say, 0.5 M) is, indeed, surprising. In such solutions of a strong 1-1 electrolyte, the ratio of water molecules to individual ions is approximately 1:50. The distance between ions, measured in terms of diameters of the solvent molecules, is only 3 or 4. Thus, it would be expected that all of the water molecules would be under the strong centrosymmetric force fields of the ions, and that no undisturbed elements of the original water structure would remain (see, however, Vaslow, 1963).

C. WATER IN NONELECTROLYTE SOLUTIONS

1. General Comments

The physical chemistry of aqueous solutions of low molecular weight nonelectrolytes has been studied far less frequently than that of aqueous electrolyte solutions. With regards to biological systems, it is even more unfortunate that practically no studies have been made on electrolytes in aqueous nonelectrolyte solutions. We return briefly to that specific problem later in this chapter. Much of the available information on aqueous nonelectrolyte solutions has come from the Russian school of authors (see, for instance, Samoilov, 1965; Mikhailov, 1968; Krestov, 1969, and Yastremskii, 1963), from England (Franks, 1967; Symons and Blandamer, 1968; Blandamer et al., 1969a,b), and from the studies by Ben-Naim (1969). Discussions of the water–nonelectrolyte mixed solvent systems can be found in the proceedings from two recent Symposia (Franks, 1967; Covington and Jones, 1968). We summarize below a few of the more important facets of these studies. Again, no general agreement as to interpretation of experimental data exists regarding structural features of an aqueous nonelectrolyte solution. However, a number of pertinent generalizations can be made.

2. Aqueous Alcohol Solutions

Franks and Ives (1966) have presented a lucid review of the structural properties of alcohol–water mixtures. Aqueous solutions of the lower

TABLE III

PARTIAL SUMMARY OF EXPERIMENTAL RESULTS ON THE STRUCTURE OF IONIC SOLUTIONS[a]

Technique–solutes	Ref.[b]	Information obtained
X-Ray		
$Ag(NO_3)$, $Pb(NO_3)_2$	(a)	Th^+ and UO_2^{2+} have a regular or "super" arrangement in the liquid. The ions are surrounded only by H_2O molecules. Ag, Pb, and Ba nitrates show a large fraction of disassociated (gas-like) molecules. I^-, Br^-, and Rb^+ show no arrangement, but only a broad interference between heavy ion and surrounding molecule
KCl, NaCl, LiCl	(b)	At low temperatures these solutions showed the second maximum in the radial distribution curve of pure water which is interpreted to show a nonhomogeneous structure at low temperatures, with water-rich regions
LiCl, LiBr, RbCl	(c)	The 4.0 and 2.58 M LiBr and RbBr retained the principal diffraction maximum of water. In more concentrated solution (i.e., 13 M LiCl), the structure resembles that of the corresponding hydrated salt
KOH, KCl	(d)	For KOH, the K^+ substitutes an H_2O in the quasi-tetrahedral water structure, whereas the OH occupies an interstitial position. For KCl, the Cl^- breaks down H_2O structure by distorting the tetrahedral coordination and squeezing out H_2O molecules
LiCl	(e)	The water structure is broken by Li^+ ions. Hydrated Li^+ ions then pack around Cl^- ions giving a hydration number of 8 to 9. The hydrated Li^+ ions have four water molecules in a tetrahedral configuration about the ion
$ErCl_3$ and ErI_3	(f)	The H_2O molecules are firmly held in octahedral arrangement around the Er^{3+} ions. There is evidence for an ice-like ordering of H_2O molecules resulting from the higher degree of orientation about the cation
	(g)	For lithium and sodium halide solutions, 7–9 water molecules occupy the first hydration layer of the halide ions. Their number increases slightly with ion size. Second and third hydration layers are correlated with the anions. The region of influence of Li^+ and Na^+ is in general smaller than that of the anions and corresponds to first and second hydration layers. The halide ions lie along the OH axes, whereas cations lie on the dipolar axes of the primary waters. Considerable ion–ion contact was detected in the cesium salt solutions, but not in the others
Raman		
Concentrated (>4N) solutions of Li, Ca, Al, Cr, and Th nitrates were compared to their corresponding solid salt hydrates	(h)	From a correspondence of frequencies in the region 700–1600 cm^{-1} between the solutions and the solid hydrates, it was concluded that the relative placement of the ions in the concentrated solutions is characteristic of the solid hydrate
Solutions of the nitrates, sulfates and perchlorates of Cu, Zn, Hg, In, Mg, Tl and Ga	(i)	Lines in the 360–400 cm^{-1} region are assigned to metal–oxygen stretching frequency of cation–water complexes. The increased binding of the hydration sheath is correlated to increased ionic charge. There is considerable electron sharing in the metal–oxygen bond
	(j)	A similar conclusion was reached from IR studies for metal–oxygen bonding in solid hydrates
LiCl, NaCl, KCl, NH_4Cl, LiBr, KBr, NH_4Br, KNO_3, $Ca^+(NO_3)_2$, Li_2SO_4	(k)	Strongly hydrated units exist in electrolyte solutions. Librational frequencies of H_2O molecules in the primary hydration layer of the ions are observed around 900 and 400 cm^{-1}. Reasonable agreement is obtained for $O\cdots HOH \cdots Cl$ and $O\cdots HOH \cdots$ librations in solutions and those reported for solid hydrates
Concentrated solutions of $In_2(SO_4)_3$, $In(ClO_4)_3$, $In(NO_3)_3$	(l)	Changes in the region below 500 cm^{-1} indicated formation of large, highly ordered clusters of H_2O molecules centered on In^{3+} ions and extending several water molecules in depth. Solutions become more ordered with decreasing temperature. NO_3^- and SO_4^{2-} tended to displace H_2O molecules from the hydration sphere of the cations at $+25°C$

Electronic Spectra		
$CoCl_2$, $CoBr_2$, CoI_2 and $NiCl_2$, $NiSO_4$, $Ni(NO_3)_2$	(m)	The observed frequencies below 500 cm⁻¹, characteristic of a cation complex, appear within 30 cm⁻¹ of the corresponding solid hydrate in each case
$ErCl_3$, $Er(NO_3)_3$	(n, o)	In the solutions, thermally activated lattice vibrations were seen which closely paralleled those of the solid hydrate. The structures in the spectra of the solution were more diffuse than in the solid spectra. They became more diffuse with decreasing concentration. The cation appears surrounded by a quasi-solidlike patch which can support lattice vibrations. Frank and Evans have argued that a similar superlattice may be associated with La^{3+} ions in solution
Isotopic Mobility		
Concentrated $LiNO_3$	(p)	The solution contains aggregates with molecular orientations similar to the crystal lattice of the solid hydrate
Solubility		
KSCN	(q)	The nearly linear SCN^- to an extent occupies channels in the water structure
Nuclear Magnetic Resonance (NMR)		
Indium halides	(r)	Evidence exists for $In(H_2O)_6^{3+}$ complexes in solution
General review of data for many salts	(s)	In general, the hydrated ion destroys the structure of water and forms complexes of type $Me(H_2O)_6$. However, ions may enhance strength of hydrogen bonds of H_2O molecules beyond the first hydration layer due to polarization. Vibrations of the complex are in general not harmonic
1-1 Electrolytes	(t)	The ion–water complex is treated as a molecular species and effective hydration numbers are calculated. A decrease in the effective hydration number occurs with increasing ionic radius. Among halide ions, it is suggested that only the F^- forms a hydrate structure. The larger halide ions break down the water structure. A structure-making effect is suggested for Li^+
NMR O^{17} Absorption		
Aqueous solutions of H^+, Li^+, Be^{2+}, Mg^{2+}, Ba^{2+}, Sn^{2+}, Hg^{2+}, Ga^{3+}, Bi^{3+}	(u)	Exchange times for H_2O molecules between the hydration shell of an ion and the solvent were obtained. For Al^{3+}, Be^{2+} and Ga^{3+}, the time exceeds 10⁻⁴ second, whereas for all others it is less than 10⁻⁴ second
Quadrupole Relaxation		
KCl, CsCl, NaCl, LiCl, $MgCl_2$, $AlCl_3$	(v)	K^+ and Cs^+ increase the rotational freedom of H_2O molecules in the hydration sphere of the ion. In contrast, Na^+, Li^+, Mg^{2+}, and Al^{3+} reduce it
Proton Relaxation		
Alkali halide solutions	(w)	The configuration of H_2O molecules is more stable about Li^+, Na^+, or F^- ions than for pure water. For other ions, it is less stable. The degree of stability decreases with increasing ionic radius
Theory		
Alkali–metal cations and halide anions	(x)	Estimates of interaction energies of ions with their nearest H_2O molecules have been obtained using LCAO-MO theory. The change in the energy of electrons on hydration decreases in the sequence $Li^+ > Na^+ > K^+ > Rb^+ > Cs^+$. The energy changes are smaller for anions than for cations. The translational mobility of H_2O molecules close to the ion should increase in going from Li^+ to Cs^+. The Raman studies summarized above also indicate metal–oxygen electron sharing in aqueous solutions. Infrared studies yield similar results for the solid salt hydrates

[a] Table prepared by Safford and Leung (1971) and reprinted here by permission of the authors.

aliphatic alcohols appear to be the only systems for which reasonable estimates can be made of the free energy, enthalpy, and entropy of solutions at infinite dilution. Practically no systematic information is available which covers a wide range of concentrations, and most of the available data were obtained only at one (or at best, a few) temperatures. The study of aqueous alcohol solutions involves the separation of solute–solvent effects from solute–solute and solvent–solvent effects and attempts to understand the alcohols in terms of the interaction with the solvent of the functional group (the hydroxyl group) and the (nonpolar) hydrocarbon part of the molecule. In the studies of solutes at infinite dilution, it is, of course, the solvent–solute interaction in which we are interested. This interaction will naturally depend on the water structure, that is, the solvent–solvent interaction. However, the available experimental data often pertain to a wide concentration range of the solute, but not necessarily at very high dilutions. This necessitates considering also the solute–solute interactions, and, because the measurements are frequently not carried out to sufficiently high dilutions, the extrapolation to infinite dilution often becomes tenuous and sometimes impossible. The studies by Franks (1966, 1968) are especially interesting in connection with the alcohols; he has also reviewed more generally the state of organic nonelectrolytes in water. The reader is referred specifically to the recent article on effects of solutes on the hydrogen bonding in water (Franks, 1968).

A great deal of information has been learned about the behavior of nonelectrolytes in water from studies of molal volumes. Figure 1 shows a characteristic set of curves of partial molal volumes for ethanol, dioxane, and hydrogen peroxide. The pronounced minimum for alcohol near 0.1 mole fraction is practically interesting, and Fig. 2 shows some additional data for ethanol and t-butanol at various temperatures. The behavior is obviously highly complex and no doubt reflects properties intrinsic not only to the water structure but also to the solute which, in turn, interacts with the water through the functional group as well as thorugh the nonpolar hydrocarbon part. Indeed, (a) the water structure may contract, (b) the solute may fit into already existing voids in the water, (c) the solute may induce voids in the water structure, or (d) the solute may experience a confining effect due to the water, reducing possible (larger-volume) conformations of the nonpolar group. Franks and Ives (1966) and, more recently, Franks (1968) have discussed the possibilities in great detail. We quote Franks and Ives:

> It is therefore proposed, for purposes of present discussion, that the effect need not depend exclusively on use by solute molecules of the pre-existing cavities natural to pure water (although this may be preferred), nor to the formation

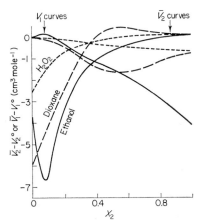

FIG. 1. Partial molar volumes for the ethanol–, dioxane–, and hydrogen peroxide–water systems, respectively, at 0°, 25°, and 0°C. (Franks and Ives, 1966, reproduced with permission from the Chemical Society, London.)

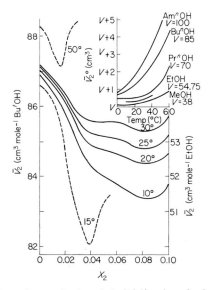

FIG. 2. Partial molar volume of ethanol (solid lines) and t-butanol (dotted lines) in aqueous solution. Top diagram: partial molar volumes at infinite dilution for various alcohols, as function of temperatures. (Franks and Ives, 1966, with permission from the Chemical Society, London.)

of cavities peculiar to the structures of a limited number of stable gas–hydrates. Instead, it is envisaged that, in virtue of its versatility in three-dimensional hydrogen bonding, water may have an intrinsic cavity-stabilising function that meets, in greater or lesser degree, the stearic requirements of any solute molecule, perhaps better for a spherical molecule than for one of another shape. In effect, a solute molecule will control and protect the hydrogen-bonded packing of water molecules in its vicinity, replacing by its own volume one or more of the natural cavities that would otherwise have been present, with appropriate overall economy of space.

In passing, it should be observed that anomalies have been reported by numerous investigators for properties of aqueous alcohol solutions in the concentration range of about 0.05 to 0.1 mole fraction. Volumetric, ultrasonic absorption, and spectral data are available to demonstrate such anomalies. However, recently Lang and Zana (1970) have studied the ultrasonic absorption of nonaqueous alcohol mixtures. In all cases studied, the excess absorption, when plotted against mole fraction, displayed sharp maxima (for concentrations less than 0.1 mole fraction). This undoubtedly is related to dimerization (or higher degrees of association) of hydrogen-bonded aggregates of the alcohol in the organic solvent. Although the interpretation which is usually placed on the explanation of similar experiments in aqueous solutions is most likely essentially correct (i.e., the water structure is, indeed, profoundly changed), future work must also more carefully allow for the self-association of the alcohols in the aqueous systems.

Robertson and Sugamori (1969) have studied the kinetics and mechanisms of hydrolysis reactions and, in particular, the energy and heat capacity of activation for solvolysis of t-butyl chloride in alcohol–water mixtures. As would be expected from previous studies by Arnett and co-workers and from the studies by Symons, Blandamer, and co-workers, dramatic changes were found to occur in the structure of the aqueous solvent as a function of the alcohol concentration and this is reflected in the kinetic parameters. The paper by Robertson and Sugamori (1969) is of interest because of the light it throws on the question of enhanced structuring in aqueous nonelectrolyte solutions for low concentrations of the nonelectrolyte (0–0.15 mole fraction). However, it is perhaps equally significant that Robertson and Sugamori recognized the tremendous advantage of extremely careful measurements which permit evaluation not only of the apparent enthalpies of activation and apparent entropies of activation but, more importantly, of the temperature dependencies of the apparent enthalpy of activation—in other words, the apparent specific heat capacity. That this approach is presently possible is largely due to Robertson and co-workers who have long demonstrated an impressive ability to perform exceedingly precise measurements. A similar accomplishment (to which we return below) has been made by Goring and co-workers (see Ramiah and Goring, 1965) in their studies (for instance, volumetric measurements) of water–polyhydric alcohol, including cellulose interactions.

3. *Classifications of Solutes*

Table IV is an attempt by Franks (1968) to classify solutes depending on the relative magnitude of the excess mixing functions. From this table, it is seen that various solutes of interest to biochemical systems may have opposite effects on the water structure.

The typical aqueous mixtures are those which predominantly seem to enhance water structure (act as structure makers). A tempting possibility for the interpretation of such structure stabilization is a model of water that allows for the discrete existence of sites or voids into which the nonpolar moieties may fit or in which the nonpolar part of the solutes

TABLE IV

CLASSIFICATION OF SOLUTES ACCORDING TO THE RELATIVE
MAGNITUDE OF EXCESS MIXING FUNCTIONS

Typical nonaqueous	Typical aqueous
$|\Delta H^E| > T|\Delta S^E|$	$|\Delta H^E| < T|\Delta S^E|$
Hydrogen peroxide	(Hydrocarbons)
Nitriles	Alcohols
Dimethyl sulfoxide	Amines
Amides	Ketones
Urea	Glycols
Glycerol	Ethylene oxide
Glycerides (?)	Polyoxyethylene derivatives
Polyhydroxy compounds	Pyridine bases
Carbohydrates	Hexamethylene tetramine (?)
Polyamino compounds	Ethers
	Dioxane (?)

may be locally stabilized or enhanced. A particularly useful parameter in this connection is obviously the partial molal volume. It is, for instance, well known (see the preceding section) that the molal volume of alcohols in water shows a pronounced minimum at some low mole fraction, corresponding approximately to 1 solute molecule per 20 water molecules.

The typical nonaqueous solutes will, in general, tend to disrupt the water structure, and an inspection of Table IV will show that this may well be due to the presence of multiple functional groups per molecular moiety, which may interact destructively with the intrinsic water structure. The problem of the structure of such solutions is extremely difficult, and, in fact, rather contradictory conclusions have been reached on the basis of different experimental techniques. This is important particularly to the biologist in connection with solutes such as urea and the glycerides. Table V shows the current status of interpretation of the effects on water structure of dioxane in dilute solutions. This table clearly illustrates the diversity of conclusions reached by employing different techniques.

Part of the Russian school of physical chemists appears to favor primarily an interpretation of nonpolar solutes in water as promoting structure through clathrate cage formation. On the other hand, the school, led by Samoilov (1965), is built essentially on an interstitial, ice-like model of water. The reader is referred to the papers by the Russian authors for de-

TABLE V

INTERACTIONS BETWEEN DIOXANE (D) AND WATER IN DILUTE
AQUEOUS SOLUTIONS

Experimental method	Type of interactions proposed	Reference
Raman scattering	Hydrogen bonds between D and H_2O	Rezeav and Shchepanyak (1965)
Dielectric relaxation	D Promotes water structure	Haggis et al. (1952)
Dielectric relaxation	D Breaks water structure	Clemmett et al. (1964)
X-Ray diffraction	D Breaks H_2O structure	Cennamo and Tartaglione (1959)
^1H Chemical shifts	$D \cdot H_2O$ and $D \cdot 2H_2O$	Muller and Simon (1967)
Ultrasound absorption	$D \cdot 2H_2O$ and $(D \cdot H_2O)_2$	Hammes and Knoche (1966)
Density, viscosity	$D \cdot 4H_2O$ and $D \cdot 2H_2O$	Tsypin and Trifonov (1958)
Density	$D \cdot 3H_2O$ and $D \cdot 2H_2O$	Schott (1966)
Conductance	$D \cdot 2H_2O$ only	Trifonov and Tsypin (1959)
Freezing points and enthalpy of mixing	$D \cdot 6H_2O$, $D \cdot 3H_2O$, $D \cdot 2H_2O$	Goates and Sullivan (1958)
^1H Chemical shifts	$D \cdot 2H_2O$	Hall and Frost (1966)
^1H Nuclear magnetic resonance relaxation	D Promotes structure in water	Clemmett (1967)

tails. It should be pointed out that the Russian school of physical chemists has probably provided more data than any other school on the properties of electrolytes in aqueous nonelectrolyte mixtures. The reader is also referred to the writings of Ben-Naim (1965, 1967, 1968, 1969), Arnett and McKelvey (1965, 1966), Arnett (1967), Arnett and co-workers (1969), Wetlaufer et al. (1964b), and Frank and Franks (1968). Finally, attention is called to the extensive studies by Symons and Blandamer (1968), Blandamer et al. (1969a,b, 1970) and especially Blandamer and Fox (1970). These authors have contributed greatly to the elucidation of aqueous nonelectrolyte studies in water (for instance, through very extensive studies on aqueous solutions of tertiary butanol).

4. Discrete Size Effects

Before leaving the subject of nonelectrolytes in water, attention is called to possible discrete size effects of organic nonelectrolytes on the structure of water. Figure 3 shows the mole fraction solubility and the free energy of solution of the normal paraffins as a function of carbon number. This illustration is taken from a note by Franks (1966) and clearly suggests that the size of the solute affects the thermodynamic properties. This is undoubtedly of signal importance in understanding aqueous solutions of biologically important materials such as proteins and enzymes with nonpolar side chains of the sizes illustrated in Fig. 3.

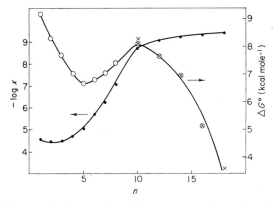

FIG. 3. Solubility (mole fraction) and free energy of solution for the normal paraffins in water as function of carbon number. (Franks, 1966, with permission from the publisher of Nature.)

An interesting study of the discrete effects of carbon chain length was reported by Clifford et al. (1965) in an NMR and Raman study of water in colloidal systems. These authors measured the initial slopes of the intensity of absorption at 3450 cm^{-1} Raman band against concentration as a function of chain length. The "hydrocarbon" was a sodium alkyl sulfate. Between C-4 and C-6, a very notable, abrupt change occurred in the observed curves, suggesting discrete effects of solutes below a certain critical size. Even more impressive are the frequency shifts in the Raman spectrum. The authors note that the Raman measurements show the occurrence of two different types of behaviors: for C-2 and C-4 sulfates, the behavior is notably different than for C-6 and C-8. The conclusion is that C-6 and C-8 sulfates create a more powerful H-bonded solution than water alone. As a result, the authors suggest that the Raman results are consistent with the NMR chemical shift data if one assumes an increase in the degree of covalency of the hydrogen bonding. The authors also observe that the range between C-4 and C-6—the point where the abrupt changes are observed—corresponds to the carbon chain lengths where the transition to micelle-forming soaps occur.

D. Some Specific Structural Models

1. *Hildebrand on "Icebergs"*

Not everyone agrees with the interpretation of water structure in terms of a clathrate or latent clathrate-like stability for water, nor is there unanimous agreement that some nonpolar solutes, such as methane, induce clathrate cagelike structures in their immediate vicinity. Few people have challenged the concept of clathrate stabilization in solution more succinctly and eloquently than Hildebrand (1969). This can be well illustrated by the following quotation from his paper concerning the relative diffusivities of methane in water compared to carbon tetrachloride:

> Methane diffuses in water at 25°C, 3/5ths as fast as it does in carbon tetrachloride. Since diffusivity depends mainly upon temperature, viscosity of the solvent, and molecular cross-sections of the diffusant, and since viscosities of H_2O and CCl_4 at 25°C are almost identical, one may infer that molecules of CCl_4 in H_2O are not imprisoned in "icebergs," and are retarded only by hydrogen bonds, not by encounters with "ice-like" aggregates.

Hildebrand's notion becomes particularly intriguing when compared to some recent neutron inelastic scattering experiments by Franks *et al.* (1971). The studies of dilute solutions of *t*-butanol in water have almost conclusively proven that the solute does, indeed, enhance the water structure. Whether or not the data can best be interpreted by invoking a clathrate hydrate structure remains to be seen. It is possible that the solute merely stabilizes (in time) whatever particular intrinsic water structure is present. Franks and co-workers introduce the term "glassbergs" to connote that the structural entities are not stabilized Ice–Ih-like elements, but in a sense, perhaps, have rather "amorphous" structural arrangements. Stability then means merely that the time for exchange between neighboring positions in this "vitreous lattice" has been lengthened considerably. Franks and co-workers emphatically point out that the induced water structures bear no resemblance to (ordinary) ice. This should be compared to the unpublished X-ray scattering data by Lipscomb, quoted by Lumry and Rajender (1971), which suggest that the water associated with proteins may show Ice–Ih-like characteristics.

The existence of the crystalline solid clathrate hydrates has provided some of the most "inspirational evidence" for the possible occurrence of clathrate-like entities in aqueous solutions—and possibly in pure water itself. This approach has been particularly advocated by Glew and co-workers (1968), who have studied the properties of aqueous solutions of acetone, ethylene oxide, propylene oxide, tetrahydrofuran, dioxane, and t-butyl alcohol. Anomalous properties of these solutions occur for solute concentrations between 3 to 6 mole percent. For all six solutes studied, notable downward field shifts were obtained in the NMR studies of the water, suggesting a stronger water–water hydrogen bonding. The water-to-solute ratios giving rise to maximum effects correspond to coordination numbers of the solutes of 24 to 28 water molecules similar to the structure-II clathrate hydrates. See also Section IV,C,4.

2. NMR Information on "Icebergs"

As mentioned in the introduction, NMR is playing an ever-increasing role in the study of the structure of water and aqueous solutions. To this field, Hertz and co-workers (see Hertz, 1970) have contributed notably. Recently, Hertz et al. (1969) have studied the tetraalkylammonium ions in aqueous solutions. These ions have commanded particular interest as they provide a water–nonpolar hydrocarbon interface. Thus, clathrate hydrates of the tetraalkylammonium halides are well known, and it would be expected that all such ions would be capable of engaging in clathrate-like formation in solution. Yet, from the study of the self-diffusion coefficient of the water in solutions of aqueous tetraalkylammonium chlorides, it was found that there was no evidence for rigid, long-lived hydration shells. Similar results were obtained with t-butanol, tetrahydrofuran, and acetone. In view of the extensive and very convincing studies by Symons, Blandamer and co-workers, and Arnett and co-workers in this country, it is difficult to resolve the conflict that by one measuring technique, no large hydration effects are obtained, whereas other techniques seem to give results which can be interpreted only in terms of rather extensive hydration effects.

3. Clathrates versus Ice Structure

As stressed before, it is remarkable that it has not been possible to prove or disprove firmly any of the various theories proposed for water structure. In particular, it would seem possible that discriminating experiments should be able to determine, for instance, whether or not the structure of bulk liquid water is best described in terms of a clathrate-like mixture model. The clathrate cage model is one of two different mixture models which provides for separate discrete voids or sites in a pseudo-lattice (short-ranged order) in water. The other model for water which allows for

discrete sites is the one assuming the existence of elements of Ice–Ih like-
ness between adjacent hexagonal rings, forming a closed, hexagonal,
cylindrical microprism; this model implies that discrete and rather large
sites may occur. This view has been advocated and exploited by Samoilov
(1965). Thus, experimental data tending to suggest discrete effects of
solute size upon the properties of aqueous solutions might be taken as
evidence for either a clathrate-like or an ice-like model. To distinguish
between these two possibilities, however, it would be natural and necessary
to look for other characteristics of the respective structured elements,
such as the fivefold symmetry occurring in the individual sides of many
polyhedra of the clathrates (for instance, in the pentagonal dodecahedron)
compared to the hexagonal symmetry of the basal plane configuration in
the hexagonal Ice–Ih lattice. Experimentally, a search for symmetry elements is not easy.

If anomalous, abrupt changes occur in solution behavior as a function of
the size of the solute molecule, this may perhaps be taken as indication of a
"fit" of the solute into a discrete void, but it must, of course, also be kept
in mind that the effects observed must be distinguished from specific
hydration effects, for instance, due to functional groups.

Helium apparently does not form a clathrate hydrate—undoubtedly
because it is small enough to permit this atom to diffuse freely through
the confining walls of a host lattice. On the other hand, whereas methane,
ethane, propane, and butane form clathrate hydrates quite readily, pentane does not appear to form such compounds. This may be interpreted as
due to the ability of the classically recognized clathrate cages to accommodate molecules as large as a four-membered (iso-C_4H_{10}) hydrocarbon
chain, but not a five-carbon atom chain.

III. Structure of Water near Interfaces

A. Traditional View

In the field of water structure, as in many other intellectual endeavors,
some virtue has been ascribed to the adherence to the principle of maximum intellectual economy (*simplex sigillum veri*). This has led to the
construction of many theories of interfacial phenomena, based entirely on
the concept that the structure of water remains unchanged from the bulk
up to say, one, or at most, two molecular diameters from the surface. In
Section C below is summarized some of the available evidence which suggests that this assumption is untenable, even for such simple systems as

air–water, mercury–water, or "general solid"–water interface. Instead, there appears to be strong evidence for notable structural rearrangements of water adjacent to almost any interface be it a solid–water, water–immiscible liquid, or air–water interface.

B. General Approach to Structural Ordering near Interfaces

Most previous authors have not challenged the traditional view of aqueous interfacial structures; that is, they have ignored the possible existence of structural ordering in water near an interface. Among those who have taken issue with this oversimplified view are notably Derjaguin (1965) in Russia and Henniker (1949), Low (1961), and the present author (1969b) in this country. The traditional evidence for structuring in liquids, particularly for water adjacent to an interface, has come from measurements of viscosity of liquids in narrow pores, anomalous diffusion coefficients or energies of activation for ionic conduction in capillaries, etc. Recently, the present author has added to this type of evidence an additional, independent set of arguments for demonstrating the probable structuring of water near interfaces.

C. Evidence for Vicinal Ordering Based on Thermal Anomalies

1. *Illustrative Examples*

The new evidence is based on the existence of thermal anomalies in the properties of vicinal water. It appears that these anomalies can best be interpreted as manifestations of cooperative phenomena (order–disorder phenomena), requiring the stabilization of structures which form cooperative, large-scale entities among the molecules involved. The cooperativeness is reflected in the abrupt temperature responses of these layers. Some typical examples will demonstrate this phenomenon.

a. Disjoining Pressures. Figure 4, taken from a study by Peschel and Adlfinger (1967), is an example; in this illustration the disjoining pressure of water between two quartz plates (separated 100 Å) is plotted as a function of temperature. The disjoining pressure is a measure of the repulsive forces between two surfaces and obviously depends on the structure and general properties of the intervening liquid layer. It is seen that the disjoining pressure does not vary in a simple, "regular" manner with temperature; on the contrary, notable anomalies are observed. It is of interest to call attention to the fact that the temperatures at which maxima are encountered in the disjoining pressure are close to 15°, 30°, and 45°C. As

24

W. DROST-HANSEN

will be discussed below, these temperatures do, indeed, have great importance for biological systems. We postpone at this time a discussion of the possible implications of this and proceed to discuss other examples of such thermal anomalies.

b. Low-Frequency Mechanical Damping. Figure 5 shows the half-life of mechanical vibrations of an oscillating quartz capillary filled with water. The data are from a study by Forslind (1966). It is seen that the half-life of the vibrations (of a capillary tube filled with water and oscillating in a vacuum) goes through a notable maximum near 30°C. Similar results have been obtained by Kerr (1970). In this connection, attention is

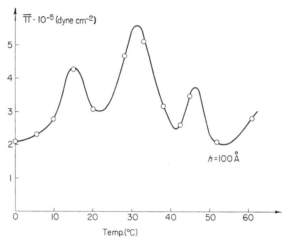

Fig. 4. Disjoining pressure of water between quartz plates: separation between plates, 100 Å. (Peschel and Adlfinger, 1967.)

called to the previously discussed values for the (apparent) entropies of surface formation of pure water (Drost-Hansen, 1965b). In the latter case we obviously deal with an interface which is vastly different from the quartz–water interface; however, the entropy of surface formation was shown to exhibit a notable maximum also near 30°C!

We are here dealing with three different, specific examples, all showing the existence of extrema near 30°C in the properties of vicinal water. The explanation of these anomalies may be relatively simple: below 30°, one particular structure of water is stabilized by the proximity to a surface or an interface (regardless of the nature of the interface), whereas somewhere above this temperature, a different structure prevails (i.e., a different structure is energetically favored). Around, say, 29–32°C, neither of these structures predominates, i.e., they lack stability. A transitional region exists, most likely characterized by enhanced randomness due either to increased numbers of "monomeric" water molecules or, at least, to smaller structural entities. Thus, if this hypothesis is correct, we should expect increased disjoining pressure as the result of increased kinetic energy of a larger number of smaller, discrete, kinetic entities (due to the lower degree of structuring—possibly the enhanced concentration of monomers). Furthermore, we should expect the mechanical coupling to the wall of the vibrating quartz capillary to be greatly reduced because of the smaller, vicinal, anchored structural entities and, hence, expect the half-life of the vibrations to be notably increased, as is, indeed, observed. Finally, if more disorder prevails at this temperature, one must expect higher

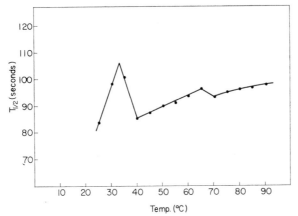

Fɪɢ. 5. Viscous damping of water in vibrating hairpin capillary. (Data by Forslind, 1966.)

entropy of surface formation and this has been reported earlier (see Drost-Hansen, 1965b; also Bordi and Vannel, 1958, 1962; Cini et al., 1969; but compare Gittens, 1969).

c. *Surface Viscosity.* As a final example of the occurrence of a thermal anomaly in a relatively well-defined aqueous (nonliving) system, we mention some recent data by Peschel and Adlfinger (1967).

Using an experimental approach somewhat resembling that previously used for the study of the disjoining pressure, Peschel and Adlfinger measured the apparent viscosity of water between two closely spaced quartz plates. Their results are shown in Fig. 6. It is seen here that anomalies are observed in the viscosity of water near interfaces (determined for different plate separations) in the vicinity of 15°, 32°, and 45°C. Indeed, the present example, as well as the study of the diffusion constant for thiourea in water by Dreyer et al. (1969), discussed below, clearly indicate the occurrence of structural transitions in vicinal water affecting different transport properties.

d. Diffusion Data and Activation Parameters. Elsewhere in this chapter we mentioned that Hildebrand, on the basis of measurements of the diffusion constant for methane in water and carbon tetrachloride, concluded that there was no evidence for iceberg formation or other structuring of the solvent around the diffusing solute molecules. The measurements by Hildebrand refer only to one temperature; thus, the energy of activation for the diffusion process is not known. It is certainly conceivable that the process of initiating the displacement of the solute molecule—resulting in

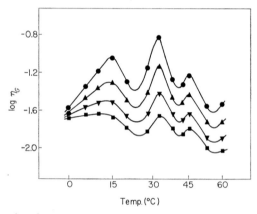

Temp.(°C)

Fig. 6. Surface viscosity as a function of temperature for various separations between two adjacent quartz plates. Plate separations: 300 (●), 500 (▲), 700 (▼), and 900 Å (■). (Data by Peschel and Adlfinger, 1969.)

the diffusion process—hinges on the (rapid) breakdown of one hydrogen bond only in a structured cooperative entity. The subsequent events might well be the fast breakdown of the entire structured entity and the overall rate and its temperature dependence would, therefore, be dependent only on the one individual molecular happening—the initial breakdown of a single H-bond.

The statement made by Hildebrand regarding the diffusion mechanism in water obviously applies only to a bulk phase. Recently, very strong evidence has been obtained for anomalous diffusivities in water near interfaces. Dreyer *et al.* (1969) have measured the diffusion coefficient of thiourea in water over the temperature range from 2° to 65°C using a novel pulsation diffusion method. In this method, the system under study (i.e., the interface across which the diffusion takes place) is continually deformed and enlarged to enhance greatly the net rate with which the diffusant is transferred in the concentration gradient. The results of this study are shown in Fig. 7. It is seen that anomalies occur near 17°, 28°, 43° and 60°C; that is, near the temperature where the thermal anomalies have been observed in general for vicinal water. It is obvious from this graph that four thermal anomalies are manifested in the data and that in this case it is useless to try a simple Arrhenius- or Eyring-type rate equation for the analysis of the activation parameters involved. We stress that the measurements are not representative of bulk water, but rather of a solution adjacent to an interface—the capillary wall (capillary diameter ~80

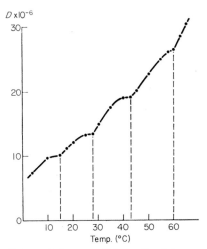

FIG. 7. Diffusion coefficient of thiourea (0.3 gm/liter) in water. Abscissa—temperature; ordinate—observed diffusion coefficient. (Dreyer *et al.*, 1969, with permission from the publisher of *Naturwissenschaften.*)

μ). The authors correctly point to the importance of this phenomenon for the understanding of the properties and structure of water adjacent to membranes or macromolecules. (However, in the opinion of the present author, but contrary to that of Dreyer *et al.*, the results are not necessarily related to the anomalous water reported by Derjaguin.) Since the observed anomalous diffusion coefficients do, indeed, represent an interfacial phenomenon in an aqueous system, similar anomalies may well be expected to occur in many other systems including and, in particular in biological systems.

Elsewhere in this chapter, we shall point to the use and misuses of rate data in biochemical and biological studies. The data by Dreyer *et al.* strongly suggest that caution is necessary in the analysis of rate data on interfacial systems. The studies by Good (1960, 1961a–d, 1967) and Coldman and co-workers (1969a,b; Coldman and Good, 1967, 1968a,b, 1969; see below) appear to be reasonable attempts to use an Eyring rate expression for analyzing molecular aspects pertinent to the nature of the activated complex. Even so, in one of the graphs shown by Good (1960), it is also seen that the rate data themselves, in an Arrhenius plot, do not necessarily follow a straight line very precisely. On the contrary, there are systematic deviations, and, although it could be argued that these deviations individually might be within the experimental error, it is also possible that this is another instance where trends in the data cannot be neglected. As an example of an overt abuse of an Arrhenius rate expression, we discuss separately the study by S. M. Johnson and Bangham (1969) where it is obvious that very persistent, notable trends in the data have been ignored and obviate any detailed interpretation of the data, arrived at by averaging over a wide range of temperatures.

For many years, advances in the understanding of thermal anomalies in the properties of water were hampered by the erroneous notion that the anomalies were manifestations of sudden changes in the properties of bulk liquid water in general. Very careful measurements, for instance, by Senghaphan et al. (1969), Cini et al. (1969), Rushe and Good (1966), and Korson et al. (1969) proved that thermal anomalies are absent in the bulk properties of pure water. However, thermal anomalies do exist, but, as emphasized specifically by the present author (Drost-Hansen, 1969a), the anomalies are manifestations of changes in the properties of water near interfaces. Thus, those experiments which may conceivably be influenced by the proximity to a surface, such as viscosity measurements in narrow capillaries, are likely to show the existence of the unusual temperature dependencies, but bulk properties are undoubtedly not involved, at least not for pure water (for solutions, see discussions by the present author elsewhere 1967a). However, there is little doubt now that the thermal anomalies are, indeed, manifested in the properties of water near interfaces—the vicinal water. The origin, as discussed in the present chapter, is probably related to higher-order phase transitions. It is emphasized again that (in the absence of large solutes in appreciable concentrations) the phenomenon is probably restricted to water near interfaces, but such water may, indeed, be extensively structured.

e. *Influence of the Substrate on Vicinal Structure.* (*The "Paradoxical Effect"*) With regard to water near interfaces, it is important to distinguish between the different effects of various types of solids, just as the nature of hydration of solutes in aqueous solution depends on the nature of the solute. Structurally different types of arrangements are expected adjacent to nonpolar-, polar-, or ionic-type interface. The important aspect is the fact that in all cases, structural changes are apparently induced and the structured zones may extend over many molecular diameters from the surface. It should be stressed also that these effects appear to be superimposed on the discrete separate effects of charged double layers (and other interfacial phenomena traditionally accounted for in colloid chemistry).

There exists presently a paradoxical state of affairs with respect to the occurrence of anomalies in the surface properties of aqueous systems. It is quite certain that anomalies occur in the temperature dependence of many properties of vicinal water, undoubtedly related to structural anomalies; however, these anomalies are observed in properties of water adjacent to highly dissimilar interfaces. Thus, thermal anomalies are found in the properties of water near the air–water interface, the mercury–water interface, the decane–water interface, many types of silica–water interfaces, the lead iodide–water interface, and the polyvinyl toluene–water interface, as well as the interface between water and many other substances. However, notwithstanding the enormous diversity of the chemical nature of the substrates, the anomalies are usually observed to occur at or near the same temperatures for all the systems under consideration. That the effect is, indeed, a surface effect rather than a bulk effect appears certain from the observation that the larger the surface-to-volume ratio of the system, the more pronounced are the observed anomalies. It seems paradoxical that the specific nature of the nonaqueous part of the interface should play practically no role in determining the temperatures of the thermal transitions. The only explanation which can be offered at this time is to assume that, at least for water near solids, the effect of the interface on the structure of the water is primarily to act as a "momentum sink." Thus, structures which are only latently present in bulk water may become stabilized near the interface; in other words, the interface may act as a momentum sink for thermal fluctuations which in the bulk would have led to the disruption of the structured entity. There are some indications, however, that specific, if minor, influences of the substrate are superimposed on the general tendency for structure stabilization near any water–solid interface. It would, indeed, be hard to see how the water immediately adjacent to, say, a lead iodide crystal (with its ion–water dipole interactions) could be identical to the nature of the water adjacent to a lipid or a hydrocarbon.

In connection with the "paradoxical effect" some recent NMR results by Woessner (1971) are of interest. For about a decade, Woessner (1966) and Woessner and co-workers (1963, 1968; 1969a,b; 1970a,b) have studied water near interfaces, particularly of minerals (quartz and clays). For such materials, Woessner has concluded that the water is preferentially oriented immediately adjacent to these surfaces—or within a few molecular layers from these interfaces. I have also discussed interactions at quartz and clay surfaces (Drost-Hansen, 1969b), but the conclusions by Woessner differ somewhat from mine, at least regarding the extent of the possible structuring of the vicinal water. However, it is of interest that Woessner noted some unifying features for the orientation of water near a variety of interfaces. The measure of orientation employed by Woessner is the quantity

$$(3 \cos^2\theta - 1)_{av}$$

where θ is the angle between the molecular axis and the direction of the magnetic field. The systems studied by Woessner are such that this quantity is non-zero for sufficiently long times. Woessner observed that the ratio of values for θ differs between D_2O and H_2O, but that the ratio is "very nearly the same for sodium hectorite clay, collagens, Li-DNA, and, within a large experimental error, for Rayon." This result appears to agree with the conclusion from the study of the thermal properties, namely that the anomalies occur in vicinal water at almost identical temperatures, regardless of the specific nature of the material in contact with the water (the "paradoxical effect"). Woessner also points out that the proton T_1 minimum for water, adsorbed on montmorillonite, occurs at about the same temperature as it does for proteins. These results, therefore, appear to show that the presence of a surface is generally more important than its specific nature in determining some of the dynamic and structural aspects of the water at or near the interface (Woessner, 1971, personal communication).

Some anomalous temperature dependencies in the proton transverse relaxation times of water adsorbed on silica gel are shown in Fig. 2 and Fig. 6 from one of the papers by Woessner (1963). Without the notion that thermal anomalies exist in vicinal water, the observed results might easily have been attributed merely to experimental uncertainty and "noise." However, while the two examples from the study by Woessner certainly do not prove the existence of the structural transitions, the idea of thermal anomalies does offer a logical explanation for the observed results.

D. SOME SPECIFIC STRUCTURAL MODELS OF WATER NEAR INTERFACES

We have discussed some of the evidence for the existence of thermal anomalies in the properties of water near interfaces. This evidence suggests the occurrence of higher-order phase transitions which, in turn, are taken as evidence for the existence of large, ordered, cooperative arrangements of water molecules. It remains to speculate on the structure and extent of such ordered groupings of water molecules. The reader is referred to Drost-Hansen (1969b) for a more detailed set of speculations; only a few paragraphs regarding vicinal water structure are quoted here:

The present author believes that the model for the structure of water which is most likely to prove correct is one involving the existence of discrete, structural elements. Most likely candidates as structural units are those associated with a mixture model containing structured clusters such as clathrate cages or high-pressure ice polymorphs in equilibrium with monomeric molecules. Here, as always, the reader is warned against the danger of generalizing the notion of structural entities to mean "discrete, permanent, microcrystalline chunks" of one form of lattice or other. Recently, Eisenberg and Kauzmann (1969) have given a good discussion of the very meaning of the term "structure" as it applies to liquid water. Suffice it to say that, both on general grounds, as well as in view of the structural implications of this paper, we shall adopt a conceptual model of water involving structured elements and discuss the likely occurrence of specific structured entities adjacent to various interfaces. Thus, notwithstanding the recent criticism of the utility of the concept of water structure in general (Holtzer and Emerson, 1969) the use of structural models appears highly practical for the discussion at hand. Again, it must be emphasized that the structural entities envisioned are not "stable, permanent, microcrystalline chunks"; rather, they are distorted appearances of elements with identifiable symmetries or other structural features, possessing the characteristics of flickering clusters. In other words, they may suffer frequent disruptions due to thermal motions, but when recreated elsewhere in the liquid volume under consideration, they again present a preponderance of one structurally characteristic element or another; for instance, the occurrence of pentagonal rings or similar geometric, identifiable features. Undoubtedly, the essential point in this discussion is the temporal stability of such identifiable characteristics. When discussing structure near an interface, the implication is only that the structure under consideration may have a lifetime which is notably longer than the corresponding lifetime of a structured entity in the bulk of the liquid. It remains to be determined whether this means an increase in the relaxation time by an order of magnitude (say, from 10^{-11} to 10^{-10} second), or from an increase corresponding to the increase in (dielectric) relaxation time on going from water to ice. Undoubtedly, future refinements of dielectric and NMR studies will throw considerable light on this question.

We finally attempt a "molecular" interpretation in terms of some conceivable vicinal structures. Figure 8* shows a possible interfacial structure for water near a polar surface. The water molecules near the solid are oriented by dipole-dipole interactions. These interactions are not likely to propagate over very many molecular diameters. Sufficiently far removed from the surface are the structures in bulk water. In Figure 8 these are indicated by pentagonal circuits and partial pentagonal outlines. These outlines are intended to convey only the presence of geometrically identifiable structural entities (but not necessarily structures related to pentagonal dodecahedra or other clathrate-like units) be they clusters, clathrate cage-like or high-pressure ice polymorphs. It

---

* Figure numbers have been changed to conform to the numbering in the present chapter.

is suggested that the structured entities are in equilibrium with monomeric water molecules (indicated by arrows) and possibly present voids or "sites" which may or may not be occupied by individual water molecules (the Pauling and the Frank–Quist model for water). The intermediate zone is the disordered transition between the ordered vicinal water structure and the differently structured bulk. Similar conceptual models can easily be constructed for the other cases discussed in the preceding paragraphs; one such case is shown in Figure 9.*

It is seen that only limited progress can be expected regarding details of vicinal water structure until a general theory of bulk water structure has been

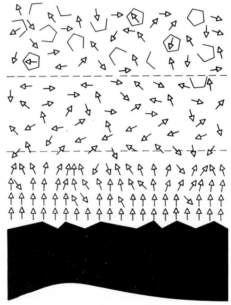

Fig. 8. Highly schematized model of water adjacent to a polar (or ionic) solid surface. Ion–dipole orientation in the vicinal layer. Note possible occurrence of disordered transition zone between ordered, oriented dipoles and bulk water. (Reproduced with permission from *Industrial & Engineering Chemistry*.)

reached. However, certain features of water near interfaces may still become apparent before a more complete understanding of bulk water is achieved.

IV. Structural Aspects of Water in Biological Systems

A. STRUCTURAL ROLE OF WATER IN GENERAL

1. *Szent-Györgyi on Vicinal Water*

Historically, Szent-Györgyi was probably the first to lend stature to the notion of ordered water structures as important elements in biological functioning. Earlier, Jacobson (1953, 1955) had been concerned with the possibility of ordered water ("liquid ice") in biological systems, but no one showed greater insight than Szent-Györgyi when he stated (1957):

> We can thus suppose water structures to be built around dissolved molecules, structures which may have a different crystalline structure according to the polar or non-polar nature of the atomic groups on that molecule and the mutual distance of these groups in relation to the lattice constants of the different possible water crystals. It is believable that different spacing pro-

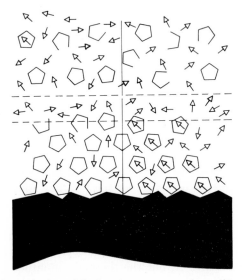

Fig. 9. Highly schematized model of water near nonpolar solid surface. Note the possible enhanced stabilization of clathrate-like entities. (Reproduced with permission from *Industrial & Engineering Chemistry*.)

motes different crystal forms or, if unfavorable, inhibits order and lattice formation. Possibilities are rich, relations complex.

Without ignoring the notable contributions of individual researchers (such as Jacobson and others), it is probably fair to say that in the period since World War II, the most inspired attack on the problem of water in biological systems was that led by Szent-Györgyi. Indeed, it became a great deal easier to pursue the idea of ordered water structures in biological systems under the banner of Szent-Györgyi's writings. Much has been added to the relatively simple, qualitative speculations of Szent-Györgyi— especially by Klotz, Lumry, Fernandez-Moran, Glassel, Warner, Hazelwood, and Ling. The writings by these authors have now become so numerous and detailed that the reader is directly referred to the more important, individual contributions of these authors for details. In the following paragraphs, however, I have summarized some of the major facets of these studies.

Szent-Györgyi (1971) has recently speculated on the role of water structure in an article aptly entitled "Biology and Pathology of Water." The unique ability of water to form an impressive array of different structural arrangements has been stressed in this chapter, together with the use of such possible arrangements to accommodate the different environments possible in biological systems. Szent-Györgyi stresses essentially the same idea when he observes "water can thus play an important role not only in structure and physical properties, but also in function and in pathology. The wonderful many-sided reactivity of water, the ease with which it can form bonds of low energy, and the ease with which these bonds can be broken, may take us a long way toward the understanding of the wonderful subtlety of life and its reactions."

2. Dynamic Aspects; Flickering Clusters

In discussing structural aspects and ordering effects near interfaces, it is easy to lose sight of the truly kinetic nature of all molecular systems. Indeed, it is important to remember the "traffic" approaching and across an interface—not only of the solutes which may permeate the cell wall, but particularly of the solvent molecules. In the case of greatest imaginable stability of a possible (if unlikely) ordered, vicinal system, consider ordinary ice (i.e., the Ice–Ih lattice). The separation between neighboring water molecules is 2.76 Å, but the thermal amplitude just below 0°C is 0.25 Å or approximately one-eleventh of the equilibrium distance! At the other end of the spectrum, compare, for instance, the traffic across an air–water interface of an area equal to the average area covered by one water molecule: the number of collisions per second is of the order of one hundred thousand (arriving at and being reflected or exchanged across the interface). This rapidly fluctuating environment does, however, still allow for the existence of structures and one may (and, indeed, one must) retain the concept of structures near interfaces; these structures probably differ from those which predominate in the bulk solution. However, we stress again that the situation is an exceedingly dynamic one with the vicinal structures suffering many types of distortions and fluctuations, consistent with the notions implied in the Frank model of the "flickering clusters." [In this connection, see the discussion of the concept of structure in the text by Eisenberg and Kauzmann (1969) and also the paper by Drost-Hansen, (1969b).]

3. Sizes of Clusters

Berendsen (1966) has made some simple estimates, based on fluctuation theory, of the probability of clusters of various sizes in water (at 0°C). The results are shown in Table VI. From the data in this table, Berendsen concludes that the probability is only of the order of one in thousands that clusters in the size range from 50 to 100 molecules will occur; thus, Berendsen concludes that the concentration of such clusters is probably insignificant. However, as mentioned elsewhere in the present chapter, the effect of the interface is likely to stabilize structural arrangement in the liquid through the influence of the substrate, acting as a momentum sink for thermal fluctuations which might otherwise disrupt the lattice stability. [In this connection, see also the brief discussion of the stabilization of cytotactic swarms in the monograph by Frenkel (1955).]

TABLE VI

PROBABILITY OF "ICE CLUSTER" FORMATION IN WATER AT 0°C

Number of molecules N	Density			Enthalpy			
	Standard deviation (gm/cm³)	Ratio deviation to SD	Proba-bility (%)	Standard deviation (kcal/mole)	Deviation $N \times 1.44$ (kcal/mole)	Ratio deviation to SD	Proba-bility (%)
10	0.0815	1.0	32	10.5	14.4	1.4	16
20	0.0576	1.5	14	14.8	28.8	2.0	5
50	0.0365	2.3	2.2	23.4	72	3.1	0.2
100	0.0258	3.2	0.14	33	144	4.4	0.001

4. Hechter's Survey—The Best of Two Worlds

In this section we give a brief (and slightly dated) review of the progress along the lines presented by Hechter (1965) at the symposium on "Forms of Water in Biologic Systems". However, before summarizing Hechter's analysis, it should be noted that the advance made by Hechter is essentially in the nature of a compromise. Hechter proposed that it is necessary for the understanding of water in biological systems to operate conceptually with both (a) specific processes, such as in membranes (and in particular, with the role of water in an individual, specific, and, hopefully, relatively simple membrane transport process) and (b) a "general structuring" of all (or most of) the water in the interior of the cell. Thus, it is implied that all (or nearly all) the water of the cell is influenced by the proximity to an interface and that this water is more highly ordered than bulk water. As indicated, superimposed on this idea is the adherence to the possible existence of specific active transport processes across the cell membrane. The idea that one may combine elements from either of the two (often sharply differing and at times, opposing views), is not acquired cheaply! As stressed elsewhere in this chapter, an allowance for an additional degree of freedom in model making must be paid for by an unusual degree of accomplishment by the proposed theory.

Hechter's basic view is that the "monolithic dichotomy of either/or" must be replaced by the conceptually far more difficult "also/and" pluralism. The either/or dichotomy refers to the membrane thesis of transport for cellular control as opposed to the "holist" theory which invokes the more-or-less complete intracellular ordering of the water of the cells and its ability to determine and effect the necessary thermodynamic and kinetic properties. Biochemically, the choice is to select between accepting an active transport, driven by metabolic energy through various transmembrane pumps, or simply to invoke changes in activities of solutes due to different aqueous environments. The latter would require far less metabolic energy; the energy requirements would now be primarily those needed initially to order the entire system under consideration, but would require very little energy expenditure for the continued maintenance of that state. In other words, the energy is not directly coupled to the transmembrane solute flow.

Having delineated the opposing views of the two major attitudes toward cellular functioning, Hechter makes the reasonable suggestion that facets of both approaches may be required for the complete description and understanding of cellular physiology. It is obviously difficult to argue against such a conciliatory attitude, but as stressed above, acceptance of a "combination approach" requires that the success of the approach be more complete than either of the individual avenues of approach.

The present author is, in a sense, willing to "pay the price" for the combination (unified) approach. He proposes specifically to pay in the currency of the ability, *sui generis*, of water to exist in vastly different physical forms under conditions that are physically almost identical. We mention briefly in this chapter the remarkably small energy differences between the high-pressure ice polymorphs, as one example of the versatility of water structuring, and added to this, the different types of clathrate cage structures available.

Another possibility has been advocated by Hazlewood (1971) who suggests that active transport may operate where ion transfer occurs from low to high concentration between two *bulk* phases, while transfer from the bulk outside solution to the intracellular protoplasma may be affected by the structuring of the water inside the cell. See the discussion in Section V,G.

We may conceptually combine (and likely confound) the approach to the understanding of water structures in biological systems by considering a two-dimensional array of possibilities. Consider laterally displayed the possible major types of water structures (ice polymorphs, clathrates, etc.) which may be stable near an interface (even if not necessarily thermodynamically stable as bulk phases), while considering the various cellular environments vertically displayed, as envisioned by Hechter (1965):

...(a) water *within* the membrane structure; (b) one or more layers of water directly in contact with extended relatively immobile macromolecular surfaces, be it at the surface of membranes or highly ordered structural components of certain cell types (e.g., the contractile system in muscle fibers), or filaments of DNA in the chromosomes of the nuclei; (c) water *between* closely paired systems of unit membranes, as in the aqueous region between the interior and outer membranes of the mitochondria, or in the cysternae of the endoplasmic reticulum; (d) water within interior chambers of an organelle, like the mitochondrion or between the structural nucleoprotein components of the nucleus; (e) water in the hyaloplasm between the various organelles in the cell.

Consider further that this array should probably be extended in more than three dimensions, by adding, as independent variables, temperature, pressure, and the nature and concentration of solutes. Perhaps rather than despair over the fairly modest progress which has been made up to the present, one might be encouraged by the fact that—in the face of such enormous possible complexity—any degree of understanding has been achieved at all! Add to these problems the possibility that the systems may not be in thermodynamic equilibrium, but, at best, in a steady-state condition or merely in a diuturnal metastable condition; it is possible also that hysteresis effects (thermal memory effects) occur (see Section VI,J), as well as, almost certainly, sheer rate-dependent phenomena.

For the other half of the combination approach, one must adopt the notion that active transport processes also play an important role in determining biological functioning. Here, the essential feature is that such processes are driven through metabolic energy. Note, however, that this does not exclude a secondary and possibly important role of vicinal water in this nonholist part of the combination approach—as stressed in Sections IV,A,9 and V,B,6,e, water structure undoubtedly plays a crucial role in membrane functioning because of the mutual interaction between the stabilized water structures (near the interface) and the substrate itself (the membrane).

5. Ling's Theory

The studies by Ling clearly exemplify the holist attitude toward water structure in biological systems. Ling's theory is a more general theory based on an "association–induction" hypothesis. This has been particularly elaborated on in the monograph by Ling (1962) and summarized in his contribution to the symposium "Forms of Water in Biologic Systems" (1965), from which we quote:

> The association–induction hypothesis which deals with a broader topic agrees in essence with Troschin's sorption theory concerning ionic and non-electrolyte distribution problems, although the two theories were developed independently. The association–induction hypothesis offers, however, specific molecular interpretation of the differences in solubility properties of the cell water in terms of restricted rotation of poly-atomic non-electrolytes and de facto poly-atomic hydrated ions and of differences in the H-bond formed in the protoplasmic system. The theory also stresses that the living protoplasm and hence, protoplasmic water, does not exist in one single physical state, but as a rule, exists reversibly in more than one metastable cooperative state in the course of its normal physiological activity. Anticipating the evidence to be presented, we may state that it is our purpose in this paper to demonstrate that all or nearly all water molecules in a living cell can be considered to exist as polarized multi-layers oriented on the surfaces of cell proteins.

6. Derjaguin's Structured Water

In connection with water in biological systems, Derjaguin (1965) reviewed some of the anomalous properties of water near interfaces in the symposium "The State and Movement of Water in Living Organisms." Derjaguin mentioned the measurements made by Metsik and Aidanova (1966) of the thermal conductivity of parallel stacks of mica sheets with layers of water of various thicknesses. Derjaguin points out that the average conductivity of these water layers increases as the thickness decreases, reaching a value "several dozen times higher" than the bulk conductivity for thicknesses of the order of 0.1 μ. Derjaguin also points to the dielectric studies by Zhilenkov (1963), in which it was found that the dielectric constant for the second layer of water on silica gels is only 8 to 10 and this value remains unchanged upon further addition of water up to as many as 10 or 20 layers. In other words, the water appears to have lost its orientational ability, thus, reducing the net dielectric permeability.

7. NMR Results

Notable support for the holist theory has been obtained by Fritz and Swift (1967). These authors studied the state of water in the polarized sciatic nerve of the frog using an NMR technique. The results from this study suggested that "very marked changes in the state of the intracellular water accompanied nerve polarization. If this change does occur in untreated nerves it is of great potential significance in the theory of neural phenomena...." The authors call attention to the nearly simultaneous article by D. Chapman and McLauchlan (1967) who studied the NMR spectra for various orientations in the magnetic field of the rabbit sciatic nerve. These authors found clear indications of two types of proton environments consistent with the findings of Fritz and Swift.

An early study of water in biological systems by NMR techniques was reported by Odeblad (1959). Assuming that various possible disturbing influences are not operating, Odeblad contends that "there may be a mechanism that makes possible the arrangement of the order of, for example, 10^2 layers of water molecules on the cell surfaces."

Recently, Damadian (1971) published an interesting paper on the application of nuclear magnetic resonance to biological problems. He demonstrated that the proton relaxation times in malignant tumors are notably different from the values obtained for normal tissues, thus suggesting an increase in the motional freedom of tissue water. This is in agreement with the observations and speculations presented in Section VI,H,2, in which the effectiveness of the hyperthermia therapy of cancer is related to the probable greater disorder of the water in malignant cells.

Other NMR studies on biological systems were reported in a symposium "Magnetic Resonance in Biological Systems," edited by Ehrenberg and co-workers (1967). A more detailed discussion will be presented in Section V,F.

8. Limitations on Role of Vicinal Water Structure

For an excellent introductory review of the more classic aspects of water in biological systems, see, for instance, the small monograph by Dick (1966). References to more complete treatments can be found in this book. Dick mentions briefly the possibilty that the structure of water in cell systems may be different from that of bulk water. However, not everyone has recognized the probable occurrence of phase transitions in vicinal water. Even those in other fields who have contributed notably have occasionally overlooked details of some importance—at least so far as surface and interfacial effects are concerned. Thus, Cole had shrewdly observed that "if we fail to look carefully and worry effectively about the exceptions, we may too long postpone the appearance of some radical, undreamed of, unifying concept." However, Cole (1968) appears to overlook the likely existence of abrupt changes in vicinal water (near the transition temperature) as reflected in dielectric properties. Cole reports that "there have not been many or extensive measurements of temperature coefficients of membrane capacities, probably because they have seemed rather dull and not very important," and notes further "it seems safe to assume that within the range of 0 to 40°C, there is no phase change such as

those which produce dramatic effects in many dielectrics." In view of what
is reported in the present chapter, this certainly seems to be a notable
oversimplification or omission. However, some additional comments are
appropriate here. There is little doubt that water of hydration may exist
adjacent to many biologically interesting interfaces, ranging from the cell
surface (membrane interfaces in general) to the "interface" between the
water and the dissolved macromolecules (such as the proteins and enzymes). To demonstrate unequivocally the existence of such hydration
structures (ordered structures of vicinal water) is by no means simple.
Very few direct approaches are possible (see, however, discussion of recent
NMR results above). More often, our evidence is only indirect and sometimes tenuous. In some cases the thermal anomalies are probably good
indicators of the effects of cooperative phase changes of the structured,
vicinal water; at the same time it must be kept in mind that the effects of
temperature may not always produce anomalous behavior (a "kink") as
the details of vicinal water structure do not play a predominating role in
all reactions or equilibrium properties of the biological system under consideration. Certainly, some changes will be relatively independent of whatever the attendant water structure may be. At the same time it is undoubtedly also worth while to pay careful attention to temperature effects
in biological systems and this is only possible through very careful studies,
that is, studies of high precision, carried out at closely spaced temperature
intervals. The effects of the thermal anomalies are often seen superimposed on those general processes which are partially understood in terms
of our general insight into the structural and kinetic aspects of molecular
biology. A large amount of such evidence is available but has not been
exploited. Again, it is important to recognize that the interaction between
the water and the macromolecules is a mutual interaction, that is, different
water structures are stable in different temperature regions, and these hydration structures, in turn, influence and modify the underlying substrate—be it, for instance, a membrane or a protein. We stress again that
the temperatures at which the structural changes occur in the vicinal water
are very likely "invariants," that is, relatively independent of such variables as, say, electrolyte and nonelectrolyte concentration and (perhaps to
a lesser extent) pressure. On the other hand, the conformation of the
macromolecules is obviously very sensitive to environmental changes in
the form of electrolyte concentration of hydrostatic pressure. Thus, again
we emphasize that the effects of the structural changes in vicinal water
are sometimes merely superimposed upon the behavior of the biological
macromolecules per se, rather than determined through the overall temperature–solute–pressure dependencies.

9. Mutual Effects—Structure of Substrates and Solvent

It is important to stress the fact that the macromolecules in biological systems significantly influence and probably order the structure of the water (at least the vicinal water) in the cell. However, this water, in turn, imposes some restrictions and conditions on the properties and structure of the substrates. Indeed, this is one of the reasons why notable effects may be observed in biological systems at discrete temperatures—the proteins (and other macromolecules) in the system will be influenced by any drastic structural change in the water of hydration. Similar arguments must apply to the water vicinal to a membrane and in the membrane pores. Thus, order–disorder transitions in vicinal water introduce an additional element of cooperative behavior of the macromolecules. Cooperative phenomena are the type required in order to have all-or-none responses, i.e., triggering phenomena. For an interesting example of exceedingly abrupt temperature effects in biological systems, see, for instance, Section V, E where some neurophysiological responses to temperature are briefly outlined; see also the examples of electric potential responses of marine algae as described by Drost-Hansen and Thorhaug (1967) and the studies of Thorhaug quoted by Drost-Hansen (1969c).

10. Review by Tait and Franks

For a general survey and introduction to the problems of water in biological systems, see the article by Tait and Franks (1971). These authors eloquently point to many unsolved problems regarding water in biological systems. Throughout the article the authors describe the urgent need for more detailed studies, but note the apparent futility of the typical X-ray diffraction studies (which have incidentally come mainly from England). Such studies have indeed led to "exciting discoveries of biopolymer structures," but appear not to be well suited to a study of the dynamic behavior or other characterization of the aqueous component of the macromolecular systems. Tait and Franks end their article on a rather discouraging note regarding the study of water in biological systems: "It is not anticipated that much progress can be expected in this area, while the intermolecular nature of bulk water itself is still a matter of lively controversy." The article by Tait and Franks should be studied carefully by anyone interested in the topic of water in biological systems. A few specific points will be noted here, while other points will be discussed elsewhere in this chapter. The authors emphasize the studies by Warner (1965) and Berendsen (1967) on the "lattice fit" between water molecules and the unique 4.8 Å spacing of oxygen atoms in many bio-

logically active molecules, observing that water "can discriminate between molecules as similar as α- and β-methyl pyranosides, which differ
only in the position (axial against equatorial) of one hydroxy group".
They also note that the effect of carbohydrates are of a short-range nature as opposed to the structuring that results in the case of hydrophobic
solutes.

B. Bound Water

1. *Szent-Györgyi on Semantics*

A matter of semantics might be mentioned here briefly. Biologists and
biochemists have long realized the existence of bound water. Bound water
has always been a rather ill-defined term, probably owing its origin to
water not readily removed by "reasonably mild drying action." However,
it may possibly be necessary to distinguish bound water from the type of
water with which we are primarily concerned in this chapter, namely,
structurally modified water. Szent-Györgyi has suggested the distinction
between bound water and oriented water in terms of energetics. He
(Szent-Györgyi, 1957) states:

> The formation of such water structures should not be confused with the old
> idea of "bound water." "Binding" involves rather the idea of energy than that
> of structure. "Binding" means a certain force, energy needed to remove a
> molecule from its site. Such "bound" molecules, having their dipole forces en-
> gaged, are also unfit to serve as solvents for other molecules. Such a binding is
> especially strong around free charges, as those of ions. The order thus pro-
> duced is "short-range order" the number of more firmly held layers of mole-
> cules being very small, 1–2. Contrary to this the building of lattices means
> "long-range order" in which the single molecules collaborate collectively.

Indeed, we owe Szent-Györgyi a great deal of gratitude for his contributions to the general question of the role of water in biological systems as
exemplified in his small book "Bioenergetics" from which the above quote
was taken. However, it should be obvious from the present discussion that
the distinction between bound water and the long-range ordered or structured water may not be as simple and clear cut as implied in the statement
by Szent-Györgyi.

2. *Nonsolvent Water*

Methods for studying water in biological systems, and particularly
those for determining the amounts of bound water, were reviewed in a paper by Higasi (1955). The older methods were primarily based on measurements of water loss upon dehydration; dielectric measurements; and

water which does not possess the "normal" solvent properties of water. The idea of nonsolvent water implied that water which was sufficiently intimately tied to the underlying substrate would not be able to act as ordinary water with its usual solvent properties. Conceptually, this is a reasonable, if somewhat crude approach, and only because of the availability of more refined methods of physical measurements, such as dielectric studies, NMR spectroscopy, and other techniques, has the approach via nonsolvent water slowly been abandoned. However, attempts are still being made to discuss the apparent nonsolvent character of water, and recently de Bruijne and van Steveninck have applied the same approach (1970). These authors studied the permeation into the eventual equilibrium (or at least steady-state) concentrations of nonelectrolytes in yeast cells.

Among the experimental approaches which have played a role in the understanding of the problem of the structure of water in biological systems are dielectric studies. However, in this field great problems are encountered even in very simple, physicochemical systems. These problems are in part experimental (electrode polarization) and in part conceptual and fundamental owing to the difficulties of measuring dielectric properties of a very "leaky condenser" which, furthermore, is heterogeneous, thus giving rise to various charge separation and accumulation phenomena (Maxwell–Wagner polarization). Regarding the utility of dielectric studies as a means of studying water near interfaces, the reader is referred to the article by the present author on the nature of water near solid interfaces and, particularly, to the references therein. With regard to water in biological systems, see, for instance, the articles by Schwan (1965) and the monograph by Cole (1968). See also Section V, D, 2 for an interesting study of dielectric properties of a lipid membrane.

C. POSSIBLE WATER STRUCTURES NEAR BIOLOGICAL INTERFACES

1. Stabilizing Effect of an Interface

Before proceeding, attention is called to the fact that, although in the past the existence of ordered water structure at interfaces has often been doubted, it is certainly important to recognize that there is no lack of available "structured" species of water which can be invoked for such vicinal stabilization. Thus, we have mentioned (and will discuss in more detail below) the ice polymorphs and the clathrate hydrates (both the simple and the mixed clathrate hydrates). In addition, one can certainly envision water structure stabilized by dipole–dipole interactions without conforming to either of the above-mentioned categories of stable structures. Finally, a solid interface may act as a momentum sink serving to stabilize what might possibly be only intrinsically latent (metastable) structures in the bulk by removing some of the thermal fluctuations which would disrupt the structured units in the bulk phase. We also call attention to the fact that electric double layers certainly play a large role in the stability of colloid systems; it is the contention here, however, that such electric double layers may exist independently of the aqueous structures imposed for the reasons discussed above (rather than that they involve the primary mechanism giving rise to stability).

Perhaps the ultimate in "long-range effects" at an interface was discussed by Schulman and Teorell (1938). These authors measured the amount of water carried from one compartment to another in a trough by a moving monolayer of oleic acid. The thickness of the water layer, moving with the monolayer, was found to be of the order of 30 microns! Exactly how much of this can be explained in terms of some "classic hydrodynamic" effects and how much is due to vicinal water structuring must await a more detailed analysis.

Subsequent to the study by Schulman and Teorell (1938) Pak and Gershfeld (1967) carried out somewhat similar type measurements and observed the effect of steroids on the thickness of the water layers carried along with a monolayer. These authors conclude "if the monolayer is taken as a model for a biological receptor site located at an interface, the possibility must be considered that the steroid hormones exert their influence indirectly on the receptor site by altering the aqueous environment." This suggestion is similar to the idea mentioned briefly in Section IV,6,C: namely, that some of the biological effects of drugs may be due to types of interactions other than direct chemical bond formation between the solute molecule and the receptor site. See also the discussion of the paper by DeHaven and Shapiro (1968).

2. Solution Aspects of Cell Fluids

It is important to note that in biological systems we are dealing with rather concentrated solutions. Biological fluids are approximately 0.1 M in electrolytes and quite concentrated in nonelectrolytes (about 20% of the net weight is proteins). It is almost certain that many of the difficulties which have been encountered in general physiology and related areas stem from the fact that the aqueous phases have often been treated as mere electrolyte solutions without proper regard to the nonelectrolytes present. Particularly, the Russian school of physical chemists has demonstrated the profound structural changes that take place in water and aqueous electrolyte solutions upon adding nonelectrolytes. Various specific aspects of this problem have also been surveyed recently by Franks (1968), by Franks and Ives (1966), and by Symons, Blandamer, and co-workers (for a review of the papers by the later authors, see the recent review by Blandamer and Fox, 1970). The systems which have been studied most intensively are aqueous solutions of the lower aliphatic alcohols and, to a somewhat lesser extent, of organic solutes such as acetone, dioxane, urea, and reducing sugars. It is interesting that somewhat less attention has been paid to solutions of amino acids as far as the influence on structural properties of solution is concerned. Solutes such as the alkaloids, clathrate hydrate formers, and hormones will be discussed separately, but briefly in the present chapter. Finally, a vast literature exists on the polypeptides and proteins, but these systems are poor as model systems insofar as vicinal water structures are concerned because of the intrinsic complications which derive from the larger size of the solutes, the simultaneous presence of nonpolar, polar, and ionic groups, and conformational changes.

3. "Ice-Likeness" of Vicinal Water

As discussed in various papers by the present author, it appears likely that Ice–Ih-like structures may not play an important role in the structure of bulk (pure) water nor may such Ice–Ih-like structured units be present in ordinary solutions. However, due to the uncertainty regarding the structure of water, the preceding statement is merely a tenable hypothesis. The absence of elements of Ice–Ih in bulk water obviously does not preclude the possibility that Ice–Ih might be stabilized by proximity to a surface. This idea has been advocated by those who have envisioned nucleating agents for supercooled water (for instance, in supercooled clouds), as promoting the growth of ice by epitaxy from oriented Ice–Ih-like layers; that is, induced solidification caused by the proximity to a suitable substrate with a lattice configuration in the basal plane similar to that of ordinary ice. This problem is far from resolved and a possible alternative to the simple notion of epitaxy has been presented by the present author. The suggestion is almost exactly contrary to the "conventional wisdom." In this new (very tentative) hypothesis it is proposed that nucleating agents are capable of inducing a rigid structure of water adjacent to the substrate, but this structure is significantly different from the bulk structure. In the transition zone between the two aqueous structures (the stabilized, vicinal structures and the bulk structure) may possibly exist a layer of enhanced disorder—the disordered layer in the the three-layer model of vicinal water (see Section III, D). It is certainly conceivable that it is this disordered water which facilitates nucleation, i.e., permits the easy rearrangement of the water molecules (by random, thermal fluctuation) into nuclei for the ordinary Ice–Ih lattice.

In molecular biology, considerable attention has been paid to the possibility that Ice–Ih-likeness may play an important role. One of the first to advocate this idea was Jacobson (1953) who proposed that the spacing of turns in helical proteins might be stabilized by an ice-like structure of the vicinal water (water of hydration) of the proteins. This basic idea has been further pursued by Warner (see the symposium on "Forms of Water in Biologic Systems," 1965).

An impressive example of the utility of hexagonal ice-likeness as an element in the stabilization of biologically interesting macromolecules is the tobacco mosaic virus (TMV) protein. Warner has shown with models how hexagonal conformation can be built from subunits and such units serve as building blocks for the morphogically hexagonal TMV rods; however, HOH bond angles and hydrogen bond angles are not yet sufficiently fully determined to make the proposed model more than an interesting possibility. Recall, for instance, that the difference in free energy between Ice–Ih and Ice–II is only 19 cal/mole. Ice II is not hexagonal, but rhombohedral; Ice–Ih is merely stabilized by the positional disordering of the protons. In passing, we might note also that, although many of the high-pressure ice polymorphs do, indeed, require inordinate (and unphysiological) pressures, the energies of transformation from one to another of the polymorphs is quite low, ranging from 19 cal/mole (Ice–Ih to Ice II) to approximately 550 cal/mole with most of the transitions around 200 cal/mole. Obviously, in the range of temperature of physiological interest, 200 cal/mole is only one-third of kT—in other words, relatively small compared to thermal energy fluctuations. This, indeed, makes it unlikely that Ice–Ih should be the only preferred or particularly stable form of all the ice polymorphs.

4. Clathrate Structures

The strongest impetus to study clathrate hydrates in connection with biological systems was the nearly simultaneous and independent studies by Pauling (1961) and by S. L. Miller (1961) on "hydrate microcrystals" as an important element of the mechanism of anesthesia.

Clathrate hydrates have been known for more than 150 years; Faraday prepared and described a chlorine hydrate (stable at temperatures below 28.7°C; dissociation pressure, 252 mm). The name "clathrate compound," however, is far more recent—it appears to have been introduced by Powell (1948). Over the past two decades, clathrate compounds have come to play an important role in technology as well as in biology.

The crystalline clathrate hydrates are solid solutions of a minor component in a water lattice. The essential feature is the existence of a "host lattice," which by itself is thermodynamically unstable, but which becomes stabilized by the presence of a "guest molecule": the water lattice becomes stabilized upon forming cages around the guest molecules. Usually, though not invariably, the solute molecules act as guests in the host lattice and are small molecules without a large dipole moment. (Molecules with large dipole moment generally tend to interact strongly with specific water molecules, thus destroying the symmetry required for the stabilization of the host network.) The interactions between the guest molecules and the host lattice are of the van der Waals-type forces, and no chemical bonding (generally) is incurred. The solutes that do not form clathrates are those with high charge density (small ions), molecules with large dipole moments, or molecules that are either too small (helium) or too large. However, it is important to recall that, although a number of substances do, indeed, tend to form clathrate hydrates with water (for instance, argon, krypton, xenon, methane, ethane, propane, chlorine, methyl chloride, methyl bromide, bromine, or ethyl chloride), no direct evidence is available as yet to demonstrate that an aqueous solution of these compounds in water induces clathrate-like entities in the solution. Glew (1962a,b), among others, has advocated on an indirect basis, that such structuring of the solution undoubtedly takes place and this view seems to have been adopted by such authors as Franks, Blandamer, Symons, and others, as well as by the present author.

In connection with the clathrate hydrates, particular interest attaches to the so-called "mixed clathrate hydrates." These are clathrates in which there are two types of guest molecules. Since clathrate hydrates in general may occur in several forms, consisting of polyhedra of different sizes, different-sized cavities are available to serve as host voids. Thus, typical examples are double hydrates containing Ar, Kr, or Xe with either CH_2Cl_2, $CHCl_3$, or CCl_4 or SF_6 with H_2S.

The existence of such mixed clathrates may be particularly important in connection with biological systems. In the cell, the water adjacent to a protein molecule or a cell membrane may be stabilized in part by some of the solutes in solution with the help of the smaller, nonpolar side chains of the proteins (see Klotz, 1958). Thus, it is of interest to speculate that nitrogen, which forms a hydrate only at physiologically "extreme" pressures (160 atm), may possibly be able at moderate pressures to form a clathrate hydrate vicinal to a protein molecule or a membrane surface by the formation of mixed clathrates (through the effects of the smaller, nonpolar groups mentioned above). This type of reasoning is of importance in connection with the problem of solubility of hydrocarbons in protein solutions or deep-diving physiology where "nitrogen narcosis" has been observed (see Section VI, E, 3).

For a review of anesthetic effects in terms of clathrate hydrate formation, the reader is referred to the original articles by Pauling and by Miller and, more recently, to the papers by Catchpool (1966) and the discussion in Section IV, C, 4.

In the pentagonal dodecahedra, a Type I clathrate, the polyhedra are regular twelve-sided structures with pentagonal faces. However, these cannot fill space (as required in order to form an extended three-dimensional crystalline solid hydrate). The cavities between adjacent dodecahedra are formed by polyhedra with fourteen sides; twelve are pentagonal, and two are hexagonal. The smallest of the units in the unit of structure contains 46 water molecules, a total of six, larger, fourteen-sided cavities, and two twelve-sided cavities. Another type of clathrate is the so-called Type II (or 17-Å type) consisting of 136 water molecules per unit cube with sixteen dodecahedra and eight hexakaidecahedral cavities. We point to these crystallographic facts to stress the likely existence of different types of clathrate cages in our total inventory of possible stabilized water structures which can be induced near solid surfaces. Note that these clathrate hydrates have near tetrahedral arrangement of the water molecules, but fivefold symmetry (pentagonal faces). Bernal (1960) has asserted that fivefold symmetry may be more commonly encountered in nature than normally assumed—a fact which Berendsen has also emphasized from his studies on water near various protein molecules, particularly, collagen and silk fiber.

If and where extensive networks are formed of clathrate hydrates, it is undoubtedly worth noting that the bond angle in the clathrate hydrate is 108° compared to the 109°28′ encountered in the tetrahedral arrangements necessary to make hexagonal structures. The details of the potential energy curve as a function of the HOH bond angle remain uncertain. However, it would be remarkable if the free water molecule bond angle (of approximately 105°) could be opened up through a rehybridization by the mere presence of neighboring water molecules to give exactly the tetrahedral bond angle of the hexagonal structure, 109°28′. Hence, whether or not the potential energy as a function of the bond angle shows a deep or a relatively flat minimum, it is certainly energetically more favorable to require only to open the bond angle from the vapor phase bond angle to 108° rather than to 109°28′.

An example of a clathrate hydrate in which one part of the guest molecule is also part of the host lattice is hexamethylene tetramine, studied by Mak (1965). In this compound, the water cage is somewhat different from the other clathrates with slightly puckered, six-member rings of water molecules. Other interesting hydrate forms, which may play a role in connection with water in biological systems, are the hydrates of ether, ethylene oxide, and acetone. These will be discussed in Section V,B,2.

5. Enhanced Reaction Rates

In connection with the problem of ordered structures, attention is called to the fact that proton mobility is very high in ice. A proton is readily passed from one water molecule to another, facilitated by the rotation of the water molecules in the lattice. Similarly, facilitated conduction may possibly take place in other types of "rigid water." This probably plays a role in various biological systems, although at the present time, the evidence for this conjecture is rather weak. It has been alleged that certain reactions of biochemical interest progress faster in ice than in solution; in other words, reaction rates in some aqueous systems have shown an increase in rate upon freezing. In terms of a facilitated proton mobility, this proposition does, indeed, seem plausible. It is not yet certain that experimental artifacts could not play a significant disturbing role in these cases. Thus, the freezing of ice leads to the exclusion of most electrolytes (and nonelectrolytes) which may subsequently concentrate in small "brine pockets." Although the temperature is lowered, the reactions may proceed more rapidly merely because of the vastly increased concentrations. This field, however, deserves considerable further study. For enzyme reactions in frozen systems, see the article by Bruice and Butler (1965) dealing with two specific examples of relatively simple reactions in ice formed by the freezing of various aqueous solutions of the reactants.

In connection with the effect of ordered structure on proton mobility, see the article by Privalov and the present author's comments on this notion (Drost-Hansen, 1967b).

6. Effects at Low Concentrations

One of the obvious problems in research on the effects of various pharmacological agents on cellular functioning is the problem of the activity and concentration of the solute. As discussed elsewhere in this chapter, there are reasons to believe that the activity of the solute may be different from that which a classic physicochemical study would suggest on the basis of equilibrium measurements on the separate bulk phases. Vicinal water may be so highly organized that the ordinary calculations of ionic activities are inapplicable, or at best, unknown. The other disturbing factor is obvious—the concentrations reported are generally the concentrations calculated on some volume, or mass, bulk basis. Such concentration units may have little relation to physiological effects; in the classic study by Franks and Ives (1960) on the interfacial tension between water and hexane, notable increases in interfacial tension were observed for methyl and ethyl alcohols in concentrations of less than 10^{-10} M. This simple example clearly shows that even for an "unsophisticated" system, a concentration of bulk solute so low as to leave completely unaffected the overall bulk structure may play an enormous role in interfacial phenomena. Obviously, the problem has not escaped the attention of biologists, to wit, the concern of the toxicologist who may be dealing with lethal effects of selected toxins in concentrations of less than 1 part in 10^9 (calculated on a per weight basis of body fluids). However, only rarely has the possibility been considered that these effects may reflect phenomena that are manifestations of induced water structure changes rather than specific, chemical bond-type interactions between functional groups of biochemically important substrates and the toxic compound.

7. Relative Size Effects

As discussed above, it is likely that different types of structural entities may be stabilized in water adjacent to a solid surface, depending upon the nature of the solid. It is probable that a nonpolar, low dielectric material (such as a monolayer on an air–water interface or hydrocarbon–water interface) is able to stabilize structured units similar to or identical with clathrate hydrate structures. On the other hand, in the case of strongly polar or ionic surfaces, such as cellulose, quartz, or lead iodide, other types of interactions between the water molecules and the solid substrate should lead to other types of stabilized, vicinal structures.* In these cases, dipole–dipole and ion–dipole interactions will play the dominant role. In either case it is possible, and in some cases very likely, that a disordered zone may exist between these vicinal, ordered structures and the bulk-like structure. At this time the suggested existence of a disordered zone remains a mere hypothesis based on indirect evidence and physical intuition rather than on concrete, experimental evidence.

Relative sizes very probably play a prominent role in biological systems; thus, it is possible that in small cells all the water present is "near" a cell surface, and, in addition, there is likely to be a sufficient concentration of macromolecules to bring all the water under the influence of some type of "surface entity," thus giving rise to a predominant water structure throughout the entire cell. At the same time, recall that in relatively concentrated solutions of both electrolytes and nonelectrolytes, there appear to be elements of water structure which retain the structural attributes of the original ordinary, bulk water (unaffected by the proximity to the

* See, however, discussion of the paradox of surface influences (Section III,C,1, e).

solute) as suggested by the neutron inelastic scattering results by Safford (1966; see, also, Drost-Hansen, 1967a). However, the smaller the cell, the less bulk structure would be expected to exist (as the ordered structures would tend to exclude more of the bulk-like regions and, in some cases, possibly also the disordered region should such exist). No wonder, then,

TABLE VII

FIT OF NATURAL MOLECULES TO WATER LATTICES

Molecule	Repeat in direction of best fit (A)	Number of water repeats	Deviation from fit (%) based on	
			4.74 A	4.52 A
Collagen	28.6 Axial	6	+1	+5
Deoxyribonucleic acid	34 Axial	7	+2	+7
Feather keratin	23.6 Axial (quasi-repeat[a])	5	0	+4
Also observed by X-ray diffraction	18.9 Axial	4	0	+4
Tobacco mosaic virus	23 Axial (quasi-repeat[a])	5	−3	+2
Cross-β-protein	4.65 Axial	1	−2	+3
β-Protein				
parallel-chain pleated sheet	4.73 Perpendicular to fiber	1	0	+5
antiparallel-chain pleated sheet	9.46 Perpendicular to fiber	2	0	+5
chitin	4.69	1	−1	+4
apatite	9.43	2	0	+5
gramicidin S				
etamycin		1	≈+1	≈+6
actinomycin	4.8 (Model)			
Circulin		(Fit in hexagonal pattern)		
Staphlomycin				
repeat in myelin	4.7		≈−1	≈+4

[a] Not precisely in axial direction.

that biological systems have escaped any profound or accurate quantitative description up to the present.

8. Berendsen on Vicinal Water Structures

Berendsen has considered repeat lattice distances for various molecules of biological significance and the number of water molecules required to repeat the patterns in the direction of best fit. Table VII from one of Berendsen's articles (1966) shows the various repeat numbers proposed and their degree of fit. It is worth noting that Berendsen observes that the fit of the water to the fibrous molecules to be barely on the border of significance because five to seven water-repeat distances are involved.

Berendsen makes an interesting observation regarding the surface entropy of various structures. He concludes that water structure near a (flat) hydrophobic surface is characterized by a relatively large positive entropy. However, in the case of the clathrate hydrates, a stabilization and decrease in entropy is observed, and he, thus, proposes that the entropy change upon cavity formation in water depends on the radius of curvature, suggesting a changing of the sign of the entropy at around 3.5 Å.

Berendsen (1967) has further commented on the model proposed for (vicinal) water by Warner, particularly in connection with the stabilization of the TMV protein subunits. Berendsen notes:

> It is not at all sure, and not even very likely, that Warner's models represent protein conformation in solution or even in crystals, but the proposed models are very interesting with respect to biological activity. The marginal stability of each of the possible conformations of proteins relative to each other makes the actual structure critically dependent on the nature of the surroundings. The occurrence of a "hydrophobic" side make Warner's structure very acceptable in conjunction with hydrophobic surfaces, as lipid layers.

As stressed throughout this chapter, there is little doubt that the substrate influences the structure and properies of the vicinal water and that the converse holds true also, namely, that the particular water structure, intrinsically stable within a certain temperature interval, influences the structure of the underlying macromolecular substrate or membrane.

Berendsen (1966, 1967; also see Berendsen and Migchelsen, 1966) has long been advocating the idea of orientation of water molecules around the collagen molecule and contributed significantly to this through NMR studies. Figure 10A shows some of the results obtained by Berendsen and Migchelsen (1966). The line width (in reciprocal milliseconds) for the outer peaks of the proton resonance in a collagen sample (45 gm water/100 gm collagen) is plotted as a function of the reciprocal absolute temperature. The general shape, with the broad minimum near 0°C, is that expected as the result of the superposition of two straight lines in the composite relaxation process. Figure 10B shows the data points obtained by Berendsen and Migchelsen, replotted, and the curve used by these authors has been omitted. It is now seen that an additional definite trend exists in the data. For temperatures above 16°C (for the reciprocal of the absolute temperature less than 3.47×10^{-3}), all data points (except one) fall on a relatively straight line with far smaller slope than the limiting slope (line E). It is obviously not possible to delineate specifically what this anomaly implies for the structural arrangement of the water adjacent to the collagen strand. The coincidence between the observed temperature of this change

and the thermal anomaly in the vicinity of 13° to 16°C strongly suggests
that the anomaly owes its origin to a structural change in at least part of

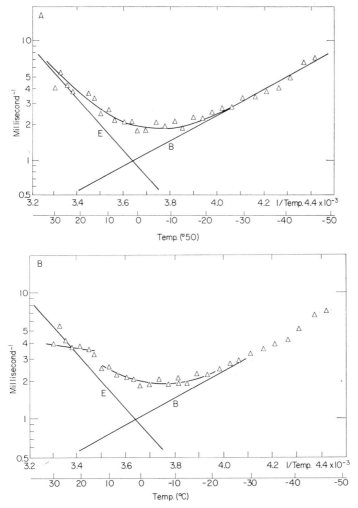

FIG. 10. (A) Line width (in reciprocal milliseconds) for the outer peaks of the
proton resonance spectrum in a collagen sample. (Berendsen and Migchelsen, 1966,
with permission from the publisher.) (B) Proton resonance line width in collagen
sample. Same data as in Fig. 10A but with curve redrawn by the present author.

the water associated with the hydration of the collagen. It is worth noting
that, upon careful inspection, a considerable number of such thermal
anomalies can be seen in other published data for collagen properties.

Among more recent studies of water in biological systems, special attention should be paid to the proceedings of a symposium held in Russia (Kayushin, 1969).

9. *Positronium, Anomalous Water, and Other Exotic Aspects*

Among the more unusual techniques which have been applied to the problem of the possible existence of structured water in biological cells is a recent report by Gustafson (1970) who studied the positron annihilation radiation from the water in biological cells. Gustafson used this approach based on the observation that in ordinary ice the rate of positronium formation is far higher than in liquid water. Thus, it was to be expected that if the water in a cell were essentially ice-like, a notable increase in positronium formation from such cells would be observed. The results obtained by Gustafson were negative in the sense that there was no significant increase in the positronium formation rate in the cells studied (abdominal muscle from white rats); thus, the results obtained suggest that the water in the biological cell is not in an ice-like state (at $+4°C$). Gustafson points out, however, that the results cannot exclude the possibility that other structured forms of water exists. Water structures with densities greater than that of water may still be present but might not result in enhanced positronium formation rates. Gustafson considers the possibility that anomalous water ("polywater," "Derjaguin water") might play a role in the cell, but the present author is certainly more than willing to adopt Gustafson's general feeling that "practically nothing is known about the behavior of positrons in polywater." In fact, the likelihood of the existence of anomalous water in biological cells is vanishingly small (a possible, but improbable exception—active silicosis).

We might draw an analogy (or, perhaps more precisely, a parable) regarding anomalous water in biological systems. Just as it is known that xenon is a potent general anesthetic, exerting its effect without forming any chemical bonds, it is also known that chemical compounds of xenon do exist. Similarly, anomalous water may exist (rather than being an experimental artifact, such as a silica gel). However, anomalous water appears only to form on freshly prepared glass or silica surfaces, preferably from a subsaturated vapor phase. This suggests that the formation requires the presence of a very high-energy surface site—a very unlikely part of a biological system. Hence, it is reasonable to assume that should anomalous water truly represent an intrinsic new form of water, it has probably as little bearing on biological systems as does the ability of xenon to form genuine chemical compounds have on its ability to induce narcosis.

10. Note

We conclude this section by noting that Clifford and Pethica (1968) have discussed the nature of water near a variety of interfaces. Table VIII from their paper shows the types of water–solute interactions considered and the corresponding sizes of the water domains—essentially the amount of water present compared to the amount of solute or matrix. This table does not give any quantitative indication of the extent of water structuring, but it does provide a rather inclusive survey of the types of

TABLE VIII

OVERVIEW OF WATER STRUCTURES IN SURFACE SYSTEMS

| Type of system | Size of water domain | | |
	I Essentially as single molecules	II Small groups of molecules	III Large enough for bulk water structure
(A) Water–solid particle	Water physisorbed on solid surfaces less than 2 monolayers or water in micropores	Water in the smaller "intermediate pores" or in concentrated colloidal dispersions	Water in dilute dispersions of particles or in dilute colloidal dispersions
(B) Water–macro-molecule	Water adsorbed in small amounts on macromolecules	Systems containing similar amounts of water and macromolecules	Dilute solutions of macromolecules
(C) Water–oil emulsions	Dilute solutions of water in oil	Water in oil emulsions (very small droplets)	Water in oil emulsions (large droplets) and oil in water emulsions
(D) Water–soap systems	Hydrate water in solid soaps	Water in concentrated micellar systems and in reverse phase micelles	Dilute solution of single molecules of micelles in water
(E) Small molecules	Dilute solutions of water	Solutions of intermediate concentration	Dilute solutions in water

systems most likely to be encountered in biological systems (especially if one substitutes the phrase "water–membrane" for "water–solid particle." The conceptual approach represented by Table VIII regarding water structing and order as related to solutes in general is similar to that of this chapter.

V. Possible "Sites of Action" of Water Structure Effects

A. INTRODUCTION

There is little doubt that the structure of vicinal water may undergo abrupt changes at or near a number of discrete temperatures, primarily in the ranges between 13° to 16°C, 30° to 32°C, 44° to 46°C, and 60° to 62°C. All of these temperatures are in the range of physiological interest. In view of the fact that all biological systems possess a large surface-to-volume ratio, it is not surprising then that some biological systems reveal abrupt changes at or near these temperatures. The question arises, At what specific point in the total biological system is the site (or sites) of action of the water structure effects? It will undoubtedly be some time before this question will receive a final answer; however, in this section attention is directed to a few of the probable major sources of influence of water structure changes. Among the possible sites are macromolecules in the biological systems (polypeptides, proteins, and enzymes) and lipids, membranes, nerves, and muscle. The effects of water structure on metabolic processes are in all probability primarily through the activity and functioning of adenosine triphosphate (ATP) and enzymes. Electrolytes and gas exchange rates are probably affected through the influence of the water structure of membranes. Consciousness and other nervous activity are affected through the structural changes of the water in or near the nerve fibers and synapses—a problem in lipid–water and membrane interactions. Only relatively crude speculations as to details are possible at this time. On the other hand, as illustrated in the following section, the evidence is overwhelming that the structural role of water in biological systems does, indeed, play an important role in the overall response of the cellular processes to temperature.

B. Solutes

1. Hydrophobic Bonding

To explain the interactions of many biologically interesting molecules, as well as to describe the behavior of even simple alkanes in solution, the concept of hydrophobic bonding has found considerable utility. Objections have been raised, yet, it appears that the concept is useful and has at least operational significance. Hydrophobic bonding is essentially the stabilization (of the water structure) which occurs as two nonpolar entities in solution approach each other. In this process, the intervening water structure experiences an increase in nearest-neighbor coordination number. The classic treatment of hydrophobic bonding is the one by Nemethy and Scheraga (1962a,b,c; see also Nemethy, 1967), but other quantitative and semiquantitative treatments have been attempted. See also the extensive studies by Ben-Naim (1965, 1969a,b, 1971a,b,c). The hydrophobic interaction does not result from unfavorable energetics for the nonpolar group in contact with the water, but, instead, from the ordering experienced by the water molecules, lowering the entropy. Much evidence suggests that the effective (average) number of H-bonds (per water molecule) is increased for the water adjacent to (and between) nonpolar moieties; the effect is due to the possibility of increasing the number of attractive intermolecular contacts (due to van der Waals forces). Other interpretations of hydrophobic bonding have been proposed by different authors. Thus, Franks (1970) notes that:

> Small angle x-ray scattering on some aqueous solutions shows sharp maxima in the scattering intensity at certain concentrations, e.g., mole fraction 0.12 in the case of t-butanol. It is also known that this peak intensity increases with rising temperature so that we have the indication of a lower critical demixing behavior. It is also known that, in the system water–tetrahydrofuran, demixing actually occurs at 72°. It is conceivable, therefore, that this lower critical demixing phenomenon can be identified with the hydrophobic bonding effect observed in dilute aqueous solutions.

From purely geometric considerations of hydration structures around the nonpolar solutes, Stillinger (1970) has tentatively proposed a water stabilization effect due to a preponderance of "eclipsed orientations" of hydrogen–oxygen lone pairs in the vicinal water structures (without invoking any specific geometric structures for this vicinal water, such as clathrates). It can be seen that, although no definite unique model is yet available, it may still be useful to employ the concept in discussing the behavior and structural properties of aqueous solutions of nonpolar solutes.

2. Organic Hydrates

Before discussing the hydration of macromolecules (of biological interest) in solutions, attention is called to the structure and properties of organic hydrates. Although crystalline inorganic hydrates have been the subject of much study, this topic is of only limited interest to the biophysicists. In electrolyte solutions the hydrations of the ions, particularly in the primary hydration shell, is dominated completely by ion–dipole interactions, and the structure of the pure solvent itself plays but a minor and indirect role. However, in the case of the organic hydrates the situation is sometimes considerably different. Recently, Jeffrey (1969) has reviewed the question of water structure in crystalline solid organic hydrates. Jeffrey has classified organic hydrates and other water structures according to the degree to which the ordering is determined by the water–water interaction. First, the epitomy of control by water–water interactions, is, of course, the various forms of ice (Ice-I and the high-pressure ice polymorphs). Second, Jeffrey considers clathrate hydrates to be essentially of two types—the simple gas hydrates with nonbonded guest molecules stabilizing the voids of the host lattices and the peralkylammonium salt hydrates where water and anions form closely related hydrogen-bonded structures with the cations occupying the polyhedral voids. Third, Jeffrey introduces the term "semiclathrate hydrates" in which the water–host structure has polyhedral clathrate voids, occupied by hydrogen-bonded alkylamine molecules. In addition to these three groups, water structures may be three dimensional, two dimensional, or one dimensional and correspondingly form hydrogen-bonded nets, sheets, columns, ribbons, or chains. These structures contain no identifiable clathrate cagelike voids, and the main characteristic is the strong interaction between the functional group of the organic molecule and the water molecule. Finally, Jeffrey briefly considers hydrates with isolated water molecules. For the latter, it suffices to note that there is a very large number of ways in which water molecules can adapt themselves to fit the lattices in hydrates of organic compounds or inorganic salts.

Of interest to biological systems are those hydrates where the ratio of water molecules to solute molecules in the solid state is large. One may reasonably expect that the major contribution to the lattice stability derives from the water–water interaction. Were it not for the formation of voids which leads to a low density relative to ice, the water in the structures of the gas hydrates would be energetically competitive with ice!

Jeffrey has also pointed out that in the solid crystalline hydrates the ratio of solute to water is an order of magnitude higher than the solubility of the solute in liquid water! Of particular interest to biological systems are the alkylamine hydrates. The studies of these materials were begun toward the end of the last century by Pickering, but little attention has been paid to his very extensive studies. In these hydrates a large number of different ways of hydrogen bonding occurs between the functional groups, in competition with the tendency for clathrate hydration of the hydrocarbon (nonpolar) groups. The alkylamine hydrates have hydration numbers (as reported by Pickering) ranging from 0.5 molecules of water per organic molecule to 36 or 37. The list of polyhedra present in some of the alkylamine hydrates is most impressive; only a few of these are irregular polyhedra. Many eight-, eleven-, twelve, fourteen-, fifteen-, sixteen-, seventeen-, eighteen-, and twenty-six-sided polyhedra are reported. Thus, there is no shortage of crystalline equivalent arrangements of water into lattices which could conceivably occur in the water vicinal to the side chains of the proteins of biological interest. It is also worth noting that the host lattice in some of these structures may accommodate larger molecules. The reason that pentane, for instance, does not form a clathrate hydrate, whereas some of the larger amines do, is simply due to the additional component of lattice energy in the amine hydrate as the result of dipole interactions between the amines and water–host lattice.

Jeffrey has also pointed out that with one exception (hexamethylene tetraamine hexahydrate) all clathrates and semiclathrate hydrates are polyhedral. This should be considered in connection with the recent theoretical calculations by Del Bene and Pople (1969) which suggest considerable hydrogen bond stabilization through resonance effect in cyclic structures of water molecules, including tetramers, pentamers, and hexamers.

TABLE IX

Hydration Properties of Organic Molecules in the Crystalline State

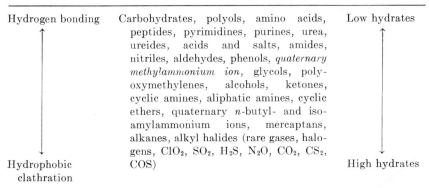

| Hydrogen bonding ↑ | Carbohydrates, polyols, amino acids, peptides, pyrimidines, purines, urea, ureides, acids and salts, amides, nitriles, aldehydes, phenols, *quaternary methylammonium ion*, glycols, polyoxymethylenes, alcohols, ketones, cyclic amines, aliphatic amines, cyclic ethers, quaternary *n*-butyl- and isoamylammonium ions, mercaptans, alkanes, alkyl halides (rare gases, halogens, ClO_2, SO_2, H_2S, N_2O, CO_2, CS_2, COS) | Low hydrates ↑ |
| Hydrophobic clathration | | High hydrates |

Returning briefly to the general problem of hydrates, Jeffrey (1969) has presented a table in which the participation of the water ranges from low hydrates, determined by direct hydrogen bonding between solute and the water structure, to the high hydrates characterized by hydrophobic clathrates. Table IX shows the results presented by Jeffrey. Finally, we quote Jeffrey on the multitude of possible interactions between amines and the water lattice:

In summary, then, the amine molecules have been observed to interact with surrounding water framework in the following ways: (1) forming no hydrogen bonds with the water structure, as in $16(CH_3)_3CHN_2 \cdot 156H_2O$; (2) forming one donor and one acceptor hydrogen bond which bridges across the water oxygen vertices at opposite sides of a void, as in $12(CH_3CH_2)_2NH \cdot 104H_2O$; (3) forming two donor hydrogen bonds bridging across two adjacent oxygens which would form an edge in a regular gas hydrate type of polyhedron, as in $10(CH_3)_2CHNH_2 \cdot 80H_2O$; (4) forming two acceptor hydrogen bonds from a bridging water oxygen, as in $4(CH_3)_3N \cdot 41H_2O$; (5) replacing a water oxygen vertex and bridging across a void by a hydrogen-bonded dimer of two amine molecules, as in $16CH_3CH_2CH_2NH_2 \cdot 104H_2O$.

There is good reason to believe that further studies of the remaining hydrates reported by Pickering will each reveal a new water framework suitably engineered to fit the particular stereochemistry of the alkylamine.

3. Gels

Gels play an important role in biological systems, but the role of water structure in gel formation is only poorly understood. Essentially, the problem is that gels can be formed at very low concentrations of solutes with viscosities which are several orders of magnitude larger than the viscosity of the pure solvent. On the one hand, it appears that in some gels rigidity (or at least, greatly increased viscosity) is achieved for solute concentrations so low that the viscosity increase cannot merely be the result of polymer "entanglement." This, then, would suggest a structural role of vicinal water, stabilized by the proximity to the gel-forming solute molecules. On the other hand, there is evidence that the microscopic viscosity in such systems is only slightly influenced by the presence of the gel-forming solute. Thus, ionic conductivities in gel systems appear almost idential to those in bulk solutions of the pure electrolytes (in contrast to what would have been expected if Walden's rule were obeyed), and ion diffusivities also appear relatively unaffected. Furthermore, the proton relaxation times appear almost unaffected; mobilities, diffusion coefficients, and relaxation times differ only by factors of about 2 for viscosity changes up to 10^5 to 10^6. Obviously, in some cases, including in particular some of biological interest, sufficient gel-forming material is present to impart overall lattice rigidity due to simple polymer entanglement. However, systems have been described (such as the cetyltrimethylammonium bromide–β-naphthol complex) for which the viscosity is highly shear rate-dependent in concentrations of solute of less than 0.02% (corresponding to a ratio of water molecules to solute of approximately 1:50,000). As mentioned by Kruyt (1949), the copper salt of cetylphenylether sulfonic acid imparts a detectable elasticity to water at a concentration of only 0.0002%. In this case the average separation between individual solute molecules is approximately 200 water molecules—indeed long-range ordering. See also the discussion by Glasel on macroscopic viscosity and NMR relaxation in protein solutions (Section V,B,5), both of which again confirm that little, if any, relation exists between local and macroscopic viscosity.

Cerbon (1967) has studied immobilized water in lipid systems using an NMR technique. Vastly different effects were observed for the influence of the relative viscosity on the proton transverse relaxation time (T_2, seconds) in the case of dextran, yeast, ribonucleic acid (RNA), and an aqueous lipid system. Thus, the increases in relative viscosity decreased the relaxation time in the RNA solutions markedly, whereas dextran (commercially used as an effective water thickener) had only a minimal influence on the relaxation times.

4. Peptides and Nucleotides

The most obvious link between water, on the one hand, and biological systems, on the other, is via the role of water in determining protein properties. Indeed, "polypeptides plus water equal proteins," and it appears that only through the action of water do proteins and lipids hang together in lipoproteins, etc. Great advances have been made in the understanding of the structure of RNA, deoxyribonucleic acid (DNA), and other nucleotides; yet, hardly anyone has begun to comprehend, or even describe, the role that the water must play in the stability of these macromolecules and in the functioning of these biomolecules. As an example, it is known that DNA requires 30% water to stabilize the double helix; yet, details of the structural functioning of this water remain almost completely obscure. Among the many other biologically important compounds in which the role of water is still poorly understood are also the polysaccharides and mucopolysaccharides (which play an important role in gel formation).

5. Glasel's Studies

Glasel (1970a,b) has recently studied some fundamental questions regarding the role of the water structure in connection with the problem of conformational changes of biologically interesting molecules (also see Glasel, 1968). In the first of the papers by Glasel, a number of relatively small molecular weight solutes was studied, including dimethylsulfoxide (DMSO), t-butanol, and urea, as well as some quaternary ammonium compounds and some nonionic surfactants. Particular interest attaches to the study of urea as a denaturant of proteins since urea has long been studied in considerable detail. Glasel writes:

> For relatively large τ_c this indicates that the number of waters hydrated is small and hence the interaction with water is not a long range phenomenon.*
> For urea, the interaction at *any* concentration is weaker than for any of the other molecules studied, and is the same order of magnitude as for simple electrolytes. Therefore, explanations of the denaturing properties of urea must lie elsewhere than in a change in the structure of the solvent. That is, an explanation in terms of the stability of the denaturant–polymer–hydration complexes must be sought.

* τ_c = rotational reorientation time.

This statement deserves considerable amplification. It should be noted first, however, that the denaturing effects of urea are not perhaps as astounding as may be implied by Glasel. In fact, it is customary in practical protein chemistry to use very high concentrations of urea solutions for denaturing the proteins (in the concentration range of 2 to 8 M). This is undoubtedly important as it clearly suggests that the denaturation does not depend on the individual single effects of the denaturant on a specific singular binding site in the protein. However, the fact that the urea does not appear to be very highly hydrated is consistent with the findings by Frank and Franks (1968); indeed, the urea appears not to interact with the water structure at all, but behaves merely as a "perfect solute." It is proposed here that, although urea may fit into the (bulk) water structure without notably affecting this structure (neither through reorganization of the solvent nor through extensive, direct hydration), the urea may not fit at all in the structure of the vicinal water associated with the macromolecules! Thus, the suggestion by Glasel that "an explanation in terms of the stability of the denaturant–polymer–hydration complexes must be sought" may well be related to an inability of the urea as a solute to fit into the rather unusual solvent, consisting of the differently structured, vicinal water.

From the studies on the NMR relaxation rates, Glasel makes a number of general conclusions, of which we quote a few:

> 1. Organic molecules and cations can best be described in solution by association of water molecules with their surfaces, and not by long-range effects on water structure.
> 2. The association of water molecules with organic solutes of the type studied here is not especially strong, compared with ionic effects of the same concentration. The observed effects on magnetic resonance relaxation in water are dominated by the longer times of reorientation of the larger solute molecules. Thus, the magnitude of the observed effect is very large in micelle suspensions of surfactants.

In addition, Glasel notes,

> 3. DMSO–H_2O solutions are composed of mixtures of molecular complexes in the ratio 1:2 and 1:3.

Finally,

> 4. The biophysical activity of these compounds in promoting denaturation depends upon the competing interaction with polymers.

Glasel proceeds to propose a simple means of hydration of the polymer, the counterions, and the solute. This study by Glasel was followed by a second contribution in which the deuteron spin relaxation rates were studied on a number of heavy water solutions of some macromolecules of gen-

TABLE X

Molecular Structures and Abbreviations of Polymers Used in Study by Glasel

Poly (vinyloxazolidi-
none methyl) (PVO)

(I)

Poly (L-lysine)

(V)

Poly (vinylpyrroli-
done) (PVP)

(II)

Poly (adenylic acid)

(VI)

Poly (methacrylic acid)

(III)

Poly (L-glutamic acid)

(IV)

Poly (uridylic acid)

(VII)

eral interest in connection with biopolymers. These polymers (see Table X) were chosen as representative of various elements of typical macromolecules, having a number of properties in common with the larger biopolymers including cloud points at high temperatures and/or high pH and phase behavior similar to the proteins (induced by the presence of urea and ammonium sulfate, etc.).

The results of these studies are rather remarkable. One of the most important observations is undoubtedly that the polymers that interact relatively strongly with water are those for which the dominant interactions take place between the unpaired species of the polymer and the water. Poly(methacrylic acid) is mentioned as a case in point. Glasel notes that phenomenologically, the following functional groups do not form strong interactions with water:

$$\begin{matrix} \diagdown \\ \diagup \end{matrix} C = O \qquad \begin{matrix} \diagdown \\ \diagup \end{matrix} NH \qquad \overset{\displaystyle O}{\underset{\displaystyle \parallel}{-C}}-O^-M^+ \qquad -C-NH_3^+X^-$$

whereas

$$\overset{\displaystyle O}{\underset{\displaystyle \parallel}{-C}}-OH \qquad -C-NH_2$$

do form strong interactions with water. Furthermore, strong interactions between water and polymers occur only in those charged polymers where there is total or at least partial intramolecular or intermolecular neutralization, i.e., when the counterion effects are eliminated through the polymer–polymer interactions. Finally, no interaction with water occurs in polymers where the geometric (helix-coil) fluctuations of intramolecular structure are large and with characteristic times of the order of a millisecond.

Two important statements from Glasel's study (1970b) should be noted and a comment will be made on these in connection with the occurrence of thermal anomalies in aqueous solutions of biologically interesting maccromolecules. Glasel states:

> Rule 1: evidence for this is based on the behavior of polymers I, II, III (basic), V (acidic) and VII. This and Rule 2 point out clearly the importance of counterions in shielding polymers from water interaction. When counterion shielding is present, conformational changes in the polymer do not effect water–polymer interactions because there are no interactions. This is illustrated by the behavior of poly(methacrylic acid) as shown in Figure 5. Similar conclusions on the basis of partial molal volume changes have been made for this polymer.[19] The results obtained from Figure 6 indicate *specific* interaction of water molecules with at least this one polymer. For poly-U, where the pyrimidine pK is of the order of 9.5, neither the charged nor the uncharged form shows any interaction with water. This and the similar absence of interaction for the acid form of poly(L-lysine) indicate the weakness of imide–water interaction. In addition, all of the polymers studied having simple carbonyl functions display no observable interaction. To summarize: in order to display interaction, there must be a proton donor without counterion shielding.

Rule 2: The coverage of Rule 1 is supported by the observations on III (acidic), IV, V, and VI. All of these have labile proton donor groups which may hydrogen bond to water in their uncharged form.

Finally, we quote:

Evidence as obtained from high resolution NMR experiments on polymers undergoing helix-coil transitions (26–28) suggests that such fluctuations exist and that their lifetimes are 10^{-3}–10^{-2} seconds. The experiments described here indicate that these fluctuations destroy the interaction of water with poly-A and poly(L-glutamic acid). The interaction of water with polymers is, in those polymers where it exists, remarkably stable as long as the topology of the polymer is stable. In the case of poly(L-lysine) the effect is not noticed because there is no interaction with the charged form and, hence, no mimima can be observed.

In his conclusions, Glasel notes that not only is there, contrary to expectation, no evidence for the interaction of water with imide or carbonyl functional groups, but, furthermore, no hydrophobic interactions are found where expected. Also, in those cases where functional groups could hydrogen bond to water, it appears that such bonding is very sensitive to counterion shielding.*

Thermal anomalies are sometimes observed in physical properties of polypeptides (and many other biopolymers) in aqueous solutions, while at other times anomalies appear to be completely absent. This observation is not the result of experimental uncertainty; instead, in view of the findings by Glasel, it appears more likely that some biopolymers are, indeed, interacting with and inducing notably different and extensive vicinal water structures, whereas in other cases, no such ordering occurs. Thus, we should expect only those processes and phenomena that are governed by the water-interacting polymers to exhibit the thermal anomalies. Hence, a possible rationale may be sought along these lines to explain the fact that some biophysical and biological systems show pronounced thermal anomalies, whereas other systems equally clearly demonstrate the absence of any thermal transitions.

6. Proteins

a. Two-State Processes. As mentioned elsewhere in this chapter, polypeptides, enzymes, and proteins in general are obviously among the most important biologically interesting macromolecules. However, this topic

* It is unfortunate that in the otherwise impressive study by Glasel, a temperature of 31° ± 1°C was employed throughout. Considering the amount of evidence to indicate a structural change may take place at around this temperature, some of Glasel's conclusions may need to be rechecked at different temperatures, or preferably, over an extended temperature range.

can only be given a most cursory treatment here; the reason for this is obviously the immense complexity of the problem. In general, changes in temperature will have obvious consequences for the state of aggregation of any macromolecule, especially those with the complexity of enzymes and other proteins. The amino acid sequences of the proteins play an enormous role in determining the conformation of these molecules. In a sense, it is only secondarily that the solution structure plays a role; yet, it is beyond question that the secondary and tertiary hydration structures do significantly interact with the proteins and, thus, jointly determine thermal stability properties. Variations in ionic strength obviously will be reflected in the properties of the protein through the influence on the ionizable groups of the protein. However, as stated before, although the proteins themselves in part influence the structure of the vicinal water, this water, in turn, impresses constraints and conditions for stability upon the protein. Because of the complexities involved, the proteins are poor model systems with which to probe the role of water in biological systems.

The nature of proteins is such that over narrow temperature intervals rather abrupt conformational changes may occur. These have been described as all-or-none processes or two-state processes in the cases where essentially only two possible states occur—namely, the initial and the resultant state (corresponding to a higher temperature). The transitions, however, are not truly discontinuous; they are indeed, abrupt and thus somewhat resemble all-or-none processes. However, superimposed upon these transitions, all-or-none transitions in the water structure must also be taken into account. Recall in this connection that thermal anomalies are frequently observed in water structure vicinal to vastly different types of water–solid interfaces, be they hydrophobic (nonpolar), hydrophilic, or ionic surfaces. Thus, in studies of temperature effects on proteins, one must be equally concerned with the detailed structural aspects of water.

b. *General Reviews.* One of the classic papers on protein hydration and protein properties is the contribution by Klotz (1958). There is little doubt that this paper must be considered among the first serious attempts to investigate the extensive hydration of proteins and the functional role of this hydration phenomenon; this places the paper in the class of the contributions by the early pioneers Jacobson and Szent-Györgyi. Lumry and Biltonen (1969) have reviewed various aspects of protein conformation, including water structure effects, in an article "Thermodynamic and Kinetic Aspects of Protein Conformation in Relation to Physiological Functions." The authors have presented an extensive review of protein conformation; they summarize other earlier contributions and specifically

implicate the structure of water in the stability and nature of the conformation of the proteins.

Privalov, of the Institute of Proteins, Academy of Sciences of the USSR,
has also long been concerned with the problem of the structure and properties of water in biological systems. Some interesting suggestions were
made earlier by Privalov (1958), whereas another study of the role of
water structure in thermal denaturation of macromolecules appeared in
a symposium, "Water and Biological Systems," edited by Kayushin
(1969). Privalov makes the important point that it is specifically necessary to take into account the superimposed effects of the tertiary hydration structures of the proteins in considering thermal denaturation of proteins. The reader is referred for details to the rather brief note by Privalov.
Sidorova (see, for instance, Kayushin, 1969) and co-workers have written
extensively on the state of water in biological systems. Sidorova has, in
particular, applied infrared techniques to study the changes in water
structure of macromolecules as a function of temperature.

A slightly dated review of the structure of proteins was presented by
Richards (1963). The review still deserves careful reading, particularly
the discussion on apolar bonds (i.e., hydrophobic interactions). Specifically, Richards reviewed the controversy between the ideas of Kauzmann
and those of Klotz. The present author sympathizes with the excruciat-
ingly honest observation by Richards to the effect that "On reading the
papers and listening to talks by these various people, this reviewer finds
himself in the embarrassing position of being convinced by each one in
turn. He would like to suggest that perhaps there is a measure of truth
all around."

Von Hippel and Schleich (1969) have presented an extensive review of
the effects of various salts on the structure and stability of biologically
interesting macromolecules in solution. This study is, without a doubt, one
of the most extensive on the problem. The authors have paid particular
attention to the effects on water structure by the presence of the electrolytes; as such, the chapter by von Hippel and Schleich represents a readable summary. However, it is unfortunate that it is yet too early to attempt to correlate the effects of electrolytes on water structure with the
effects of ions on the water structure near most biopolymers. Only the last
section of their extensive chapter even attempts to approach the problem.
Von Hippel and Schleich note that the ions will affect the water structure,
as will the nonpolar groups of the macromolecules; in addition, the "local
water" (undoubtedly the water referred to in this chapter as vicinal water)
will also be perturbed by the ions present, and this, in turn, will influence
or determine the extent to which the water can become organized by the

exposed nonpolar groups of the macromolecules. Thus, it will also determine the free energy of transfer of these groups from the hydrophobic
macromolecular interior to the aqueous environment. One can view the
mechanism of the effects of ions (and other perturbants) on T_m as caused
by a "tripartite competition" between the organizing forces, all of which
impose a particular (and different) type of order on the local water. These
three competing, organizing forces are the nonpolar groups of the macromolecule, the ions (or other nonaqueous additives), and the unperturbed
water lattice itself. According to von Hippel and Schleich, structure de-
stabilizers make less water available to Frank–Evans icebergs around the
exposed macromolecular nonpolar groups than is available in the presence
of unperturbed water lattice; whereas structure stabilizers must have the
properties of somehow loosening the unperturbed water lattice, thus making more water available for organization about the nonpolar groups. See
also the recent article by Schleich and von Hippel (1970).

For an excellent survey of the use of infrared spectroscopy with biomolecules, see the article by Susi (1969), dealing particularly with the
study of polypeptides, protein structure, and various nucleotides.

c. *Relation to Bound-Water Problem.* The problem of the concept of
extensive stable hydration hulls has been neatly summed up by Lumry
and Rajender (1971) in their discussion of the subject:

> For some years it was customary in studies of protein hydrodynamical proper-
> ties to incorporate into the calculations a shell of water which was treated as
> moving with the protein. There was considerable experimental justification for
> this course of action provided by the results of drying experiments, calculated
> diffusion coefficients, transport of solvent by protein, bouyancy measurements
> and so on, but the use of partial specific volumes computed on a dry weight
> basis makes it unnecessary to examine this problem in every hydrodynamic or
> equilibrium sedimentation experiment.

Recent history of the bound-water problem seems to stem from specific
suggestions made by Jacobson (1953) about the structure and extent of
the bound-water shell. Since that time there has been much argument
about the matter, but no solutions have been forthcoming. Thus, Lumry
and Rajender (1971) point out that:

> The idea of stable hydration shells has not been very popular in recent years
> despite the skillful championing of the idea by Klotz (289–292). It is probable
> that there are some very unusual features of water at protein surfaces and the
> experiments by Lauffer and co-workers on TMV protein (332–334) are particu-
> larly important in this connection; but the idea of a structured shell with either
> ice-like or clathrate character has not been generally appealing for reasons
> which have been presented by Kauzmann (177). However, if such shells exist
> they are the most obvious places to look for the source of solvent-dependent
> compensation behavior and should have top priority in studies of compensa-

tion. Although the estimates of "hydration" of water differ markedly from method to method, a frequent figure deduced from different types of experiments is 30% of the weight of the protein (see Glasel, 293, 294 for a recent experimental approach to the problem).

Lumry and Biltonen (1969) have also discussed various aspects of bound water. These authors include in the category of bound water those molecules that, at the surface, interact strongly with the charged groups of the proteins and those molecules that are found in deep indentations (caves) and large holes (if such ever exist). They also point out that this water probably is not a stabilizing factor in conformation configuration and also that such surface and "cave" water appears to be the only part of the protein hydration about which everyone seems to be in at least partial agreement.

Meryman (1966) has also discussed the problem of bound water in his introductory article to the volume "Cryobiology." The definitions of bound water reviewed by Meryman range from water that does not freeze (of the order of 5 to 10% of the total amount of water in animal tissue) to water that does not act as a normal aqueous solvent. Meryman also quotes Bull (1943) who compiled a review of no less than fourteen different methods for measuring bound water, which, as Meryman points out, give rise "to almost as many definitions."

d. *Enthalpy–Entropy Compensation Phenomena.* Lumry and co-workers, and most recently Rajender, have carefully reviewed enthalpy–entropy compensation phenomena in aqueous solutions of both small molecules and proteins. In this section is summarized briefly a few of the many important suggestions made by Lumry and Rajender, and speculations are made upon some possible basic, unifying features. In the treatment of membranes as a site of the action of the vicinal water structure, the linear compensation effect will be discussed again in connection with the hemolysis of erythrocytes (the studies by Good and co-workers).

The compensation process implies that there is a linear relationship between the entropy change and the enthalpy change; this is specifically discussed by Lumry and Rajender for a number of processes of small solutes in aqueous solutions. The proportionality constant between the enthalpy and entropy is referred to as the "compensation temperature." In other words, in the general expression

$$\Delta G^0 = \Delta H^0 - T \,\Delta S^0$$

ΔH^0 and ΔS^0 may be related in a compensatory fashion, such that

$$\Delta H^0 = \alpha + T_c \,\Delta S^0$$

This latter equation is an extrathermodynamic statement and no clear-cut explanation can be offered for the relative constancy of observed values for T_c (wherever water is the solvent). The remarkable fact is, however, that the values for T_c all fall in a relatively narrow range from about 250° to 320°K for processes as dissimilar as solvation of ions and nonelectrolytes, hydrolysis, oxidation–reduction, ionization of weak electrolytes, and the quenching of indofluorescence, among others (see Lumry and Rajender, 1971, for details). The interpretation of the existence of this enthalpy–entropy compensation is that it is a manifestation of an intrinsic aspect of the water structure itself. Regardless of the solutes and the solute processes studied (thus, including both thermodynamic equilibrium properties and transport phenomena), it has been discovered that similar enthalpy–entropy compensation regularities are observed among the proteins of functional importance in physiological processes. Hence, it is suggested by Lumry and Rajender that the enthalpy and entropy relationship may be used as a diagnostic test for the participation of and, in fact, the controlling role of the water in the protein processes. The enthalpy–entropy compensation phenomena discussed by Lumry and Rajender certainly emphasize the probable critical and crucial role played by the structure of water in protein and enzyme processes. For a list of authors who have contributed to the subject of linear compensation phenomena, see the article by Lumry and Rajender.

Obviously, temperature plays an enormous role in any consideration of protein and enzyme stability. For an excellent survey of the effects of temperature on the proteins in solution and in biological systems, see, for instance, the recent monograph edited by Poland and Scheraga (1970). There is little doubt that a study of temperature effects on the properties of proteins will suggest a wide and almost continuous range of temperatures of importance in protein chemistry. Thus (apparent) melting temperatures, T_m, may fall anywhere in a large range—usually between 10° and 95°C. However, without a more detailed look at the underlying processes, these temperatures are of little immediate use in an attempt to elucidate the interplay between protein properties (for instance, the helix-coil transformation) and the water. The reason for this is the fact that (apparent) melting temperatures, T_m, will be strongly influenced by other factors, such as the ionic strength, pH, and the concentration and nature of nonelectrolytes present in the medium. Not only does this obscure any specific role that the water might play, but it makes the application to living systems far more difficult because of the delicate and minute differences in pH and ionic activity which must occur in different living cells within the same organism at any time. It is in this connection that the real advantage of the study of the enthalpy–entropy compensation phenomenon becomes obvious. As discussed in this section it is by means of the compensation phenomenon that it is possible to identify a single characteristic temperature. It is this temperature (actually a proportionality constant), the compensation temperature T_c, that reveals far more directly the role of the water structure. Nevertheless, it is possible in some cases to see anomalies in protein behavior, as a function of temperature, superimposed upon the general trend of the data in graphs of properties such as optical circular dichroism versus temperature (to mention but one example), due to the water structure changes. Although a sigmoid shape curve is generally anticipated (as the result of the phase cooperative behavior of the proteins themselves), evidence is sometimes present for smaller, but notable, deviations at or near the temperatures of the thermal anomalies. Again, this is a matter of paying attention to and accepting trends in data even when the total deviations may appear to be within the experimental error on an individual determination.

Values for T_c generally fall within the range of 250° to 320°K, but far more often the values cluster between 270° and 290°K—notably between 285° and 290°K (i.e., 12–17°C). It is interesting to speculate that the compensation temperature T_c may be the "critical temperature" for the stability of one or another type of vicinal water structure, stabilized by the proximity to the interface. In this connection, recall the discussion in Section III,C,1,e of the paradox of the invariance of thermal anomalies with the specific nature of the substrate and also the suggestion that superimposed upon this invariant behavior there must be (perhaps minor) specific influences, for instance, due to a polar versus a nonpolar interface.

As an example of the entropy–enthalpy compensation, Fig. 11 shows the data collected by Arnett (quoted by Lumry and Rajender, 1971). It is worth mentioning that the following passage from the discussion by these authors:

> The aggregate of evidence thus far presented not only emphasizes the dominant role of liquid water but also indicates that neither the type of small solute nor the type of processes is very important. There seems to be a wide variety of perturbations of liquid water which evoke the same type of response in water. Even the alcohol-perturbed examples can be included in these blanket statements since Rajender has found that the characteristic near-infrared spectrum of water in these solutions retains its distinctive isobestic pattern up to the alcohol mole fractions which produce extremum behavior in the examples we have discussed. . . .

Recall here, as dicussed by the present author, that even in relatively strong aqueous solutions there is evidence for structural arrangements characteristic of the bulk structure of water which are unaffected by the presence of the solute, even in rather high concentrations of electrolytes. The same notion has been stressed by Safford (1966).

e. Mutual Effects—Solute–Solvent Interaction. In connection with the mutual interaction of water and the substrate, recall again that not only does temperature change the water structure in general, but also the water structure near an interface and the latter more or less abruptly. Also, the changes in the vicinal water structure must, in turn, influence the nature of the (solid) substrate. Thus, if the ordering in water near an interface changes at one of the transition temperatures, changes may then occur in the properties of the underlying substrate. This may be an overriding

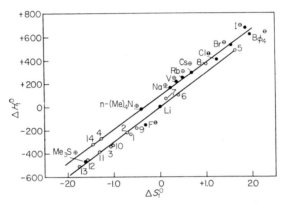

Fig. 11. Enthalpies and entropies of transfer for nonelectrolytes and individual ions from H_2O to D_2O at 25°C. Nonelectrolytes are shown with open circles (○) and ions by filled circles (●). The two lines are drawn arbitrarily through what seem to be the most "typical" members of each series and their relative positions are not significant although their slopes are. The ion correlation line could be shifted merely by using some other ion than Li^+ as reference. (1) Argon, (2) propane, (3) butane, (4) iodine, (5) methyl fluoride (30°C), (6) methyl chloride (30°C), (7) methyl bromide (30°C), (8) methyl iodide (30°C), (9) glycine, (10) DL-alanine, (11) DL-α-aminobutyric acid, (12) DL-norvaline, (13) DL-norleucine, (14) L-phenylalanine. (Arnett and McKelvey, 1969, with permission of Marcel Dekker.)

principle in connection with protein conformation and should perhaps be related to what has been mentioned by Lumry and co-workers as "subtle and not so subtle effects of the water on the protein structure."

With regards to the interaction between water and some of the macromolecules of the biologically interesting systems, particularly the proteins and enzymes, we again quote Lumry and Rajender (1971):

The similarity of T_c values for small solute and protein compensation processes in water has provided the basis for the proposal that all Vaslow–Doherty processes have a common source. Systematic studies of the effects of solvents on the linear compensation processes of proteins now underway should provide definite evidence for or against this generalization, which is only a hypothesis at the present time. In the protein cases there are other alternatives. For example, it is a natural consequence of any cooperative structure that enthalpy and entropy changes must to some extent compensate each other. Protein unfolding processes are weak first-order phase transitions, weak because the cooperative unit is very small so that the transition-temperature range is very much broader that that for the melting of ice. Compensation behavior due to the protein conformation is clearly apparent in the thermodynamic changes in protein unfolding which are discussed at length elsewhere. However, this type of compensation process is not quantitatively similar to Vaslow–Doherty compensation, except for changes involving the exposed nonpolar groups as demonstrated in the ribonuclease and trypsin examples.

Lumry and Rajender note the fact that the compensation process results in observed values of T_c for small solutes which are nearly identical with the values for proteins; they also note that the most likely explanation for this behavior is that water alone is involved as the underlying agent. This, in turn, is astutely correlated with the "fitness of the environment for life," as discussed by Tracey (1968). (See Section VI,A,2 of the present chapter.) Earlier, Lumry and Biltonen (1969) emphasized the probable importance of the enthalpy–entropy compensation phenomenon with the statement, "Indeed, it is probable that the enthalpy and entropy change during conformational changes of proteins is the single most important physical-chemical characteristic of protein function."

f. pH Effects. The apparent melting temperatures for proteins yield an almost continuous distribution of values. Hence, the obvious advantage of a determination of enthalpy–entropy compensation behavior is to simplify complexities otherwise encountered in the "point-by-point" description of the effects of temperature upon different protein properties in various media—strongly dependent upon the ionic strength and other parameters (such as pressure). However, the effects of pH are more difficult to deal with and the compensation phenomenon probably cannot be obtained from conventional rate data alone in most instances where the pH is varied greatly. Lumry and Rajender feel that the degree of complexity in such systems must be considerable and "compensation may be hidden."

g. Protein–Hydrocarbon Interactions. In connection with the mixed clathrate hydrates and the possibility that the nonpolar side chains on the proteins in solution may act as "help-gas" (*Hilfsgase*), it is interesting to consider some of the studies by Wishnia (1962) and by Wetlaufer and Lovrien (1964). Wishnia demonstrated that the solubility of ethane, propane, and butane is greatly enhanced in solutions of bovine serum albumin, hemoglobin, and lysozyme. The increase in solubility was several-fold above that in pure water. It was shown that the solubility enhancement was almost temperature independent over a wide range of temperature (25°), suggesting that the enthalpy of bonding is small, and, hence, that the solubilization process is determined primarily by entropy changes.

Wetlaufer and Lovrien observed reversible changes in viscosity, optical rotation, and other properties of bovine serum albumin (at alkaline pH) and β-lactoglobulin (for neutral pH) in the presence of various hydrocarbons. This suggests then that the hydrocarbons readily become associated with the protein and that the mechanism most likely depends on the incorporation of the hydrocarbon into cavities—probably clathrate or semiclathrate hydrate-like entities—induced in the vicinal water of the macromolecules. As mentioned previously, the structure of the vicinal water will again determine, in part, the conformation of the macromolecules, and, conversely, the macromolecule will influence the structure of the water. Hence, it is not surprising that changes occur in properties such as viscosity or optical rotation of the protein solution upon addition of hydrocarbon moieties into the vicinal water structure. See also the discussion by Schreiner (1968) of the effects of various inert gases on some properties of purified enzymes (such as tyrosinase and acetylcholinesterase).

h. *Notes.* In addition to Lumry and Rajender and Lumry and Biltonen, Brandts (1969) has stressed the need to consider carefully the nature of vicinal water in determining the conformational stability of macromolecules in solution. Brandts notes, "Thus, a complete description of the denaturation process must take cognizance not only of the order–disorder transition of the polypeptide chain itself, but also of the order–disorder transition associated with the solvent in the mode of accommodation of non-polar side chains in the denatured state."

A useful review of the behavior of nonelectrolytes in water is presented somewhat incidentally by Brandts (1969) who discusses the properties of aqueous solutions of nonelectrolytes as a function of the nature and concentration of the dissolved electrolytes. The reader is referred to this article not only for the review of the underlying theory, but particularly for the attempt to apply these ideas to the problem of the stability of biologically interesting macromolecules.

Sidorova and co-workers (see Kayushin, 1969) have studied the state of water in biological tissues by infrared spectroscopy. The article by Sidorova and co-workers should be consulted for some useful suggestions regarding the utility of infrared spectroscopy as a tool in the study of water in biological material. The authors point out that the interpretation of the vibrational spectrum of water and the assignment of certain frequencies is still somewhat ambiguous and involves the use of other methods, such as neutron inelastic scattering, NMR, and dielectric studies for additional structural information. Among the interesting conclusions which were drawn by Sidorova and co-workers is the suggestion that because the data for eggs (egg yolk and egg albumin) differ so little from those obtained for pure water, the water in eggs is identical in structural properties to ordinary water. They further note that the water appears to freeze at the same temperature as pure water and "has the same solvent action." The present author finds this statement somewhat difficult to accept and particularly takes issue with the statement by Sidorova and co-workers that "this conforms to the widely held view that most of the water in protein solutions is in the free state," although this latter statement is possibly correct for globular proteins. Attention is called to the statement in Section V,D,6,b that limitations exist on the use of infrared as a discriminating tool in biological systems because of the relatively poor resolution of the broad bands due to any hydrogen-bonded systems. However, the paper by Sidorova should also be studied for further information because it contains a separate section covering the near-infrared spectra of water, alcohol solutions, and the influence of urea on the solubility of hydrocarbons in water. With regard to urea, Sidorova and co-workers conclude that water–urea clusters probably exist in solution; this should be compared to the opposite conclusion reached by Glasel (1970a) and by Frank and Franks (1968).

For a recent infrared spectroscopy study of the hydration of DNA, see the article by Falk et al. (1970). These authors proposed that even when the surrounding water has frozen into ice (ordinary Ice-Ih) an inner layer of ten water molecules per nucleotide does not freeze. The authors conclude that the biopolymer hydration shells are not "ice-like" in the sense of possessing a crystalline Ice-Ih-like structure; specifically: "The present results demonstrate that, at least for DNA, there is no such ordering. The innermost, least mobile part of the hydration shell is in fact, entirely incapable of crystallizing into the ice structure. Evidently, the preferred configuration of water next to the biopolymer is incompatible with the structure of ice. There is no spectral evidence of 'quasi-crystallization' of any part of the hydration shell at room temperature, and at low temperature it is the water *far* from the biopolymer surface which freezes into ice. The hydration shell of DNA is definitely not 'ice-like' in the structural sense."

7. Enzymes

a. Dixon and Webb's Review. For a general review of enzymes, as well as a brief review of the effects of temperature on enzyme activity, see, for instance, the monographs by Dixon and Webb (1960) or by F. H. Johnson *et al.* 1954). Dixon and Webb note the occurrence of several abrupt changes with temperature in enzyme activities (particularly in quoting data by Massey). Although the authors allow for the possibility that a phase change could occur in the solvent, they have apparently only concerned themselves with a first-order transition (water–ice). Apart from this possibility, Dixon and Webb summarize the following possibilities for the origin of abrupt changes in Arrhenius plots.

1. Anomalies may be present due to the existence of two parallel reactions with different temperature coefficients. The authors note that it is very difficult to get anomalies as sharp as are often observed unless the enthalpies of activation differ by a very large amount.

2. The occurrence of successive enzymic reactions with different enthalpies of activation.

3. Different forms of the enzymes with different activities. These forms would both be active, with different enthalpies of activation. Again, however, this requires that the activation parameters differ greatly.

4. Dixon and Webb quote Kistiakowsky and Lumry (1949) who proposed that the enzyme may exist in two forms, one of which is inactive.

5. The simultaneous occurrence of forward and reverse reactions, but only with the forward reaction exhibiting an anomaly. This, of course, does not explain the occurrence of the anomaly in the forward rate of the reaction.

There is certainly no lack of suggestions to explain the occurrence of thermal anomalies in enzymic rates. However, it seems that a less forced and far more likely explanation for the anomalies may be sought in a change in the structure of the vicinal water, manifesting itself either through the general change in vicinal properties (including entropy, viscosity, and diffusion rates), or resulting from conformational changes of the enzyme, "impressed" upon the macromolecule due to the different states of the vicinal water above and below the thermal transition points.

b. *Rate Expressions.* As mentioned in the preceding section, Dixon and Webb reviewed some of the possible causes for the occurrence of abrupt changes in slopes in Arrhenius plots of enzymic reaction rates. More recently, Brandts has been concerned with this phenomenon (1967; also see F. H. Johnson et al., 1954). One of the most likely mechanisms for achieving notable changes in slope is through a reaction catalyzed by an enzyme which may exist in two forms (a native and a denatured form). Thus, in the expression for the rate of the process, one must allow for a temperature effect on the equilibrium constant between these two forms, characterized by a free energy of transformation. This generally will give a rate expression of the form:

$$\text{rate} = \frac{CT \exp(-\Delta E^{\ddagger}/RT)}{1 + \exp(-\Delta F^0 RT)}$$

where it is customary to assume the free energy is composed of two temperature-independent terms, namely, the enthalpy (ΔH°) and the entropy term $(T\Delta S°)$. The result of such an approach was reviewed by Brandts (1967) and is shown schematically in Fig. 12. The different curves in this hypothetical illustration refer to enzymes of varying thermal stability. The less thermally stable the enzyme catalyzing the reaction, the more truncated does the rate expression become. The present controversy is whether or not this approach (and an extension of this approach) offers an explanation of the observed anomalies in well-known enzyme systems, particularly when extremely abrupt changes in slope are observed. It must be remembered that the equilibrium distribution between the native and

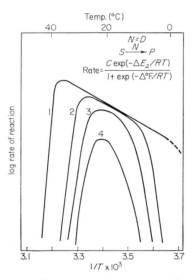

FIG. 12. Schematic representation of rate of enzyme-catalyzed reactions; see text for explanation. (Brandts, 1967, with permission.)

denatured form of enzyme will follow a normal free energy–equilibrium constant (exponential) distribution, thus limiting the abruptness of the transition. The more abrupt the transition, the larger the free energy difference between the two forms of the enzyme must be; but this intuitively seems energetically contrary to the expected behavior of biological systems in general. It is for this reason that the present author proposes the introduction of another factor, namely, vicinal water structure which can act to sharpen the transitions, recalling the mutual interaction between the structure of the vicinal liquid and the molecular conformation of the interface presented to the vicinal water.

  c. *Examples of Marked Anomalies in Enzyme Reactions.* A typical, although certainly not a forceful example, of the occurrence of an anomaly in the properties of an enzyme has been reported recently by Lehrer and

Barker (1970). These authors studied the thermal stability and reactivity of aldolase obtained from rabbit muscle. (Previously, Massey *et al.* (1966) had reported the enzyme to undergo a temperature-dependent transition, and Lehrer *et al.* studied this in some detail.) A possible anomaly was observed in the vicinity of 30°C; a typical result is shown in Fig. 13 as an Arrhenius plot for the cleavage reaction of aldolase. This clearly indicates a nearly linear behavior above and below 28° to 30°C, respectively. The

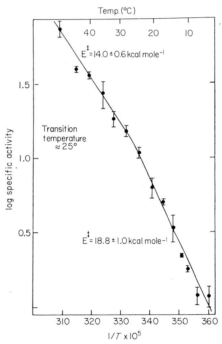

FIG. 13. Arrhenius plot of rate of cleavage reaction of aldolase. (Lehrer and Barker, 1970, with permission.)

authors tentatively ascribed the change to the mechanism proposed by Massey, namely, that the enzyme may exist in two different conformations above and below the transition temperature. However, it is suggested here that the change is a manifestation of water structure; it is possible that there is a conformational change in the enzyme to conform to whatever is the stabilized, vicinal water structure in the different temperature intervals (changing at 30°C).

Another typical example of a remarkably abrupt change in enzyme property with change in temperature was reported by Takahashi and Oshaka (1970). These authors studied the proteinase from the venom of the snake *Trimeresurus flavoviridis*. Again, very notable thermal anomalies were observed. Thus, the enzyme activity of the enzyme on casein was decreased by two orders of magnitude on increase in temperature from 60° to 63°C approximately.

Henn and Ackers (1969) have recently provided another example of an abrupt temperature anomaly of an enzyme property. These authors have observed a very abrupt change between 12° and 14°C in the association constant for the dimerization of the subunit of D-amino acid oxidase. The authors specifically propose that the variation in the enthalpy of the reaction reveals a reversible change in the heat capacity of the protein. This is

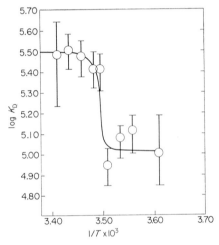

FIG. 14. Van't Hoff plot for dimerization of D-amino acid oxidase apoenzyme. (Henn and Ackers, 1969, with permission.)

interpreted as an isomerization reaction between different conformational states (in addition to an association of the subunits). Figure 14 shows the results of the van't Hoff plot for the dimerization of the apoenzyme. Again, we suggest that the vicinal water associated with the amino acid oxidase apoenzyme may play a crucial role in determining the transition temperature.

A full discussion of the general problem of the structure of water adjacent to proteins in general and enzymes in particular is outside the scope of this chapter. However, a number of other examples of highly anomalous temperature dependencies of enzyme properties are mentioned below; it is proposed that these reflect the discrete and specific influences exerted by structural changes in the water vicinal to these macromolecular solutes. In other words, various types of phase transitions are, indeed, to be ex-

80 W. DROST-HANSEN

pected on purely energetic grounds in the proteins per se, but at least some of the thermal anomalies reported in the literature for protein properties are likely manifestations of changes in the vicinal water structure.

Fischer and co-workers (see Piguet and Fischer, 1952; Meyer *et al.*, 1953) have demonstrated the occurrence of abrupt changes (kinks) in the properties of β-amylase. These authors obtained kinks between 17° and 19°C with differences in apparent energies of activation ranging between 9.3 (0–18°C) and 13 kcal/mole above this temperature range. Anomalies were obtained both for barley and wheat β-amylase.

Subsequent to these initial studies, Markovitz *et al.* (1956) studied the properties of α-amylase isolated from *Pseudomonas saccharophilia*. Again, it was found that the rate of enzyme activity exhibited notable anomalies in the vicinity of 15°C. The difference in the apparent energies of activation for the α-amylase was almost 6000 cal (8500 kcal/mole between 15° and 40°C and 14,400 cal/mole between 0° and 15°C). Similar differences in energies of activation were obtained for α-amylase isolated from another bacterium; but the authors point out that the amylase of malt, swine pancreas, human saliva, and human pancreas show only a single energy of activation over the same temperature interval.

A further example of abrupt temperature changes in another enzyme system has been reported by J. J. Baldwin and Cornatzer (1968). These authors studied the glyceryl phosphorylcholine diesterase. This enzyme shows extremely abrupt inactivation at approximately 60°C.

It is worth mentioning in passing that the enzyme data which in general tend to show the most distinct evidence for abrupt thermal transition are those for which the properties are the most sensitive to changes in pH! Undoubtedly, a very careful study of this effect would throw considerable light on the general phenomenon. The observation of notable changes with pH corresponding to the largest sensitivity to temperature is likely a manifestation of the same underlying cause: as discussed elsewhere, hydrogen ions appear to upset the water structure more than any other single ion and whenever extensive vicinal structuring is present (and thus, able to manifest most clearly changes with temperature), the effects of hydrogen ions are most pronounced. Compare this effect with the results obtained by Glasel discussed in Section V,B,5.

In addition to the few examples mentioned here specifically of abrupt changes in enzyme properties at the temperatures of thermal anomalies, see also, for instance, the article by Kemp and co-workers (1969) who observed a rate change in the succinate oxidation by rat liver mitochondria near 17° to 18°C. See also the article by Staal and Veeger (1969) describing a thermal anomaly near 15°C in the glutathione reductase.

It should be noted that particularly in the literature on enzyme kinetics the existence of thermal anomalies has often been discussed, although no agreement currently exists as to the origin of these phenomena. As a matter of semantics, it is important to note that these anomalies have often been referred to as discontinuities (in Arrhenius plots, for instance). Actually, overt discontinuities are observed only very rarely in log-rate plots. For a discussion of possible types of thermal anomalies in rate processes, see Drost-Hansen (1967a).

d. Cold Inactivation. Graves and co-workers (1965) have studied the cold inactivation of glycogen phosphorylase. The authors found that storage at temperatures less than roughly 15° to 20°C (in a buffer at pH

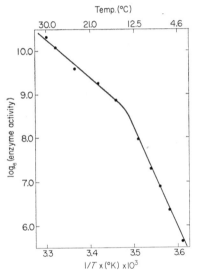

Fig. 15. Arrhenius plot of rate date for phosphorylase *b* activity. (Graves and co-workers, 1965, with permission.)

6.0) led to more rapid inactivation than storage at higher temperatures. The cold inactivation could be reversed upon rewarming. Figure 15 shows the effect of temperature on the enzyme activity of the phosphorylase *b* activity. It is seen that a notable anomaly occurs in the vicinity of 13°C. Again, we propose that the occurrence of the thermal anomaly is a manifestation of the attendant change in water structure at this temperature. Particular attention is called to this example because of the likely importance of the general phenomenon of cold inactivation on the process of vernalization (see Section VI,C,2). For additional discussions of the phenomenon of cold inactivation, see the recent article by Kuczenski and Suelter (1970); the authors observe:

Low temperature instability of proteins indicate that associations between apolar groups, significantly weakened at low temperatures (Kauzmann, 1959; Scheraga *et al.*, 1962), are important in these proteins. The temperature dependence of inactivation both in the presence and absence of FDP suggests a first-step dissociation involving such apolar groups. In the presence and absence of FDP, the initial rates of inactivation, k_2, decrease as the temperature is raised. Although the data are insufficient to quantitate the effect of temperature on the equilibrium concentration of the dimer, examination of the shape of the curves suggests that the concentration of the dimer decreases with increasing temperature but that the rate of dissociation of the dimer, k_3, increases. This suggests a heterologous interaction between sub-units, i.e., hydrophobic forces predominating between the dimers, and electrostatic forces predominating between the subunits of the dimer.

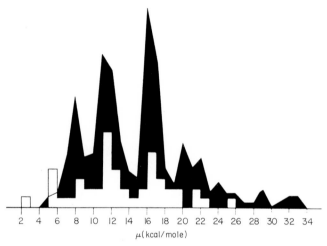

FIG. 16. Distribution of energies of activation of biologically interesting processes. (F. H. Johnson *et al.*, 1954, with permission.)

Finally, we mention the studies by Brandts and Ting (see Brandts, 1967) and Biltonen and Lumry (1969). These studies clearly show maximum stability of chymotrypsin, chymotrypsinogen, and dimethionine sulfoxide chymotrypsin near 10° to 15°C! This observation may be purely accidental (although this is perhaps not entirely likely) in view of the fact that RNase, for instance, has maximum stability below 0°C. Brandts (1967) has reviewed other cases of cold denaturation of proteins. The phenomenon of low-temperature denaturation will no doubt eventually prove of considerable interest in connection with hypothermia, hibernation, and vernalization.

e. Distribution of Activation Energies. It is interesting to compare the relatively narrow range of compensation of temperatures which is normally encountered with the notably discrete distribution of apparent energies of activation for various biological and biochemical rate phenomena. Figure 16 shows the highly nonuniform distribution of energies of activation as reported in the monograph by F. H. Johnson *et al.* (1954). At first sight, this illustration might tentatively be interpreted to indicate that the rate-controlling process in a large variety of biological and biochemically interesting systems is determined simply by a rather small number of discrete specific enzyme reactions with characteristic energetics. However, the possibility should also be considered that this discreteness is a further manifestation of those processes which are determined, or at least notably influenced, by a relatively small number of distinct states of the water of solvation of the macromolecules involved. Unfortunately, it does not seem possible to extract from the presently available data any further information to corroborate this tentative suggestion.

C. LIPIDS AND LIPOPROTEINS

1. *Lipid–Water Interactions*

Lipids, lipid–water, and lipid–polymer interactions in general are exceedingly complex and outside the scope of the present chapter. However, it is necessary to point briefly to a number of phenomena related to lipids (including the phospholipids and lipoproteins) in order to appreciate an origin of thermal anomalies in biological systems which may not be related to the anomalous thermal properties of vicinal water. Higher-order phase transitions—not related to water structure—do certainly occur in proteins and (for other reasons) in pure (anhydrous) lipids. This is not to imply that other phase transitions involving proteins as well as lipids and lipoproteins may not, in addition, reveal phase transitions influenced by or possibly determined by the nature of the vicinal water structure. The possible occurrence of transitions in lipid–water systems caused by the changes in water structure (for which only a very limited amount of evidence will be presented) is not a highly useful piece of information and, in fact, confounds the issue rather than shedding light on the general problem of the temperature response of biological systems. However, one must not despair because of the extreme complexity which is implied in this connection. With the availability of more and more discriminating techniques (such as NMR), it is very likely that the environments and motions of water molecules vicinal to, say, a phospholipid membrane may be delineated with some degree of certainty. For a comprehensive review of lipids, see the article by Shah (1970).

An excellent discussion of an example of a ternary system of lipids has been presented by Mandell *et al.* (1967). This study has resulted in very detailed phase diagrams for some three-component systems (e.g., sodium caprylate–decanol–water). The corresponding visually observed microscopic phase appearances are also shown in this article and a number of suggestions made to explain structuring in such systems. The behavior is truly complex, and it is somewhat distressing that in biological systems the mesomorphic phases are undoubtedly even more complex. In connection with this study, it is also notable, as pointed out by Mandell *et al.*, that water and decanol, which have practically zero mutual solubility, become mutually soluble in all proportions in the presence of a critical concentration of bile acid salts. What is desired to stress here is, of course, the active part which the water may play in this process, namely, through the ubiquitous ability to form bonds in a great variety of ways through the multitude of possible modes of hydrogen bonding (see, for instance, the discussion of the organic hydrates by Jeffrey in Section V,B,2).

For a more recent review of the study of lipids and lipoproteins, particularly in membranes, see the article by D. Chapman and Salsbury (1970). This review is mainly concerned with the use of NMR as a tool for the study of the lipid system. In connection with the phospholipids, see also the discussion by R. A. Chapman (1967). The great complexity observable with anhydrous lipids is amply illustrated in the article by Barrall and Guffy (1967); see also the article by Cyr *et al.* (1967, p. 13) and the article by Flautt and Lawson (1967, p. 26). The complexity of molecular interactions in some mixed lecithins was studied recently by Phillips *et al.* (1970). Finally, see Luzzati and co-workers (1968) regarding water/lipid interactions.

An interesting note on the temperature stability of high-density lipoprotein from egg yolk has been published by Franzen *et al.* (1970) who reported on data for the specific optical rotation of β-lipovitellin as a function of temperature. The data shown have, according to the authors, been "arbitrarily fitted to a linear function. The dashed lines represent one SD from the least squares mean value of the specific rotation which is represented by the solid line." This is a remarkable approach as the data themselves show an impressive trend of temperatures above 30°C; thus, although the authors have suggested an "average" straight line fitted to the data with a positive temperature coefficient, an inspection of the data (above 30°C) clearly indicates a negative temperature coefficient. (Apart from this trend, other trends may possibly be present in the data, such as a minimum in the specific rotation in the vicinity of 15° to 20°C.) If the changes at 15° and 30°C are caused by vicinal water—certainly not an unlikely suggestion—the results are of further interest as specific rotation for other optically active, aqueous systems have also shown anomalies at 15° and 30°C. Compare Section V,D,10 for a discussion of the optical rotation of D-glucose (as studied by Martin-Löf and Söremark, 1969b,c) and the measurements by Kendrew and Moelwyn-Hughes (1940; see Drost-Hansen, 1965a) on some reducing sugars.

2. Lipids in Membranes

A review of the chemistry of membranes, including lipids, is presented by Van Bruggen in Chapter I of Part A. One of the most obvious needs for studies of lipids and lipoproteins is in connection with the very question of the cell interface. This interest stems from the pioneer studies of Gorter and Grendal (1925) and Danielli and Davson (1935) and co-workers (see Davson and Danielli, 1943) who devised the concept of bimolecular lipid leaflets as a possible model of the cell membrane.

Various authors have devised different lipid–water systems amenable in principle to theoretical treatments. However, the treatment of each is often tenuous. Ohki (1970) has considered the following possible forms of phospholipids dispersed in aqueous solution: (1) spherical micelles, (2) cylindrical micelles, (3) bilayers, and (4) lamellar and "bubble" membranes (vesicles). (The latter two are both related to the bilayer structure.) The essential point is that in all cases mentioned the surface-to-volume ratio is large, and it is to be expected that vicinal water structures may play an important role in determining the overall properties of the systems. Differential thermal analysis has proven a discriminating tool by which to study phase transitions in phospholipids—both in the anhydrous state as well as in the lipid–water systems. Again, the complexity is enormous and the reader is referred to articles such as the recent one by Abramson (1970). Other references to the problem of phase transitions in lipid systems, and particularly in connection with membranes, are found in Ladbrooke and Chapman (1969); this article is a lucid survey of lipids as well as proteins and biological membrane material by thermal analysis and contains an excellent review of the extensive writings by Chapman and co-workers on the subject of lipids. See also the monograph on liquid crystals edited by Porter and Johnson (1967).

An interesting discussion of the behavior of lipid systems was presented by Small (1970). Again, the use of tools such as differential scanning calorimetry and petrographic microscopy have assisted greatly in delineating the complexity of cholestrol esters. Notable transitions were observed at a number of different temperatures. However, it is emphasized that while the anhydrous lipids do reveal abrupt, anomalous changes at a number of discrete temperatures, biological systems most likely involve aqueous lipid systems. It is very possible that some, if not all, of the anomalies observed in biological systems originate in lipid–water transitions rather than transitions which occur in the anhydrous lipids [see the discussion of thermal transitions of phospholipids in water presented by Abramson (1970, p. 37)]. Because of the complexity referred to above, it remains to be determined which of the anomalies owe their existence primarily to the water structure changes and which, if any, are primarily determined by the nature of the pure lipid transitions.

Various effects of lipids in membranes have been discussed in a number of articles published in the recent monograph "Biological Membranes," edited by D. Chapman (1968). Particular attention is called to the chapters by Rouser et al. (1968, p. 5), Luzzati (1968, p. 71), D. Chapman and Wallach (1968, p. 125), and Dawson (1968, p. 203).

Recently, Bean and Chan (1969) have studied thermal transitions and electrical conductivity of ultrathin lipid membranes. While recognizing that thermal transitions may occur due to changes in water structure, Bean and Chan seem to prefer an interpretation (in the case of those lipid bilayer membranes which are modified by proteins) in terms of a shift in equilibria between two conductive states in a protein "pore." The discussion by these authors is interesting; they state:

> ...both of the anomalous transitions observed here are in the direction of decreasing conductance with increasing temperature, which would be more compatible with an increase in molecular organization or interaction leading to a decrease in diffusion coefficients for ions in the region of the membrane. Such reorganization might be within the lipid bilayer, at the bilayer interface, or even in the surrounding thickened torus of retracted lipids at the edge of the bilayer. Interfacial reorganization might include changes in water structure (icebergs) associated with the membrane, as suggested by Drost-Hansen and Thorhaug but could also be the result of reorientation of the polar groups of the lipids in the interface, which would also create a shift in water structure.

In connection with the question of lipids as a membrane constituent, it is of interest to note that D. Chapman [1965, quoted in Rose's book (1967, p. 131)] has observed endothermic transitions at temperatures of about 33° and 48°C by differential thermal analysis with human cerebral and central nervous system myelin.

Various aspects of bimolecular films have been reviewed by Tien and James in Chapter VI of this volume (Part A). As mentioned in the following section, other resistance measurements on bimolecular leaflet films have suggested strong temperature-dependent anomalies (T. E. Thompson, 1964). More recently, Ting et al. (1968) have studied black lipid membranes prepared from chlorophyl and chloroplast pigments in sodium chloride solutions. Properties of these chlorophyl lipid membranes are notably different from those of black lipid membranes prepared with naturally occurring phospholipids, synthetic surfactants, and oxidized cholesterol. However, it appears that temperature plays a notable role in the stability of these films. Thus, Ting *et al.* note that, whereas the resistance decreased with increasing temperature (as would be expected) over the range of 16° to 30°C, the membrane was stable only for short periods of time above 34°C while below 16°C a precipitation took place in the membrane.

D. MEMBRANES

1. *Introduction*

The general problem of morphology, structure, and functioning of cellular membranes is outside the scope of the present chapter. We are concerned here only with the role which water may play in membrane stability and functioning. However, in order to appreciate this problem it is necessary to have some acquaintance with the nature of cellular membranes, and the reader is referred to the monographs edited by Järnefelt (1968), Kavanau (1965), D. Chapman (1968), New York Heart Association (1968), several of the symposia published by the Society for Experimental Biology (see S. E. B., 1965), Schlögl (1964), Clark and Nachmansohn (1954), Kleinzeller and Kotyk (1961), Schoffeniels (1967), Snell *et al.* (1970), and particularly Lakshminarayanaiah (1969). The last-mentioned volume (dealing especially with transport phenomena in membranes) contains an extensive bibliography, including an excellent annotated bibliography of recent work; see also the earlier work of Davson and Danielli (1943). Among other reviews of biological membrane structure, see "Recent Progress in Surface Science" edited by Danielli *et al.* (1964, 1970) and also the selection of papers by Branton and Park (1968). Two symposia on membrane properties deserve special mention: "Membrane Phenomena" (see Faraday Society, 1956) and "Biological Membranes: Recent Progress" (New York Academy of Sciences, 1966). See also "Symposium on Cell Membrane Biophysics" (1968).

2. *Vicinal Water in Membranes*

In terms of transport processes the thermal anomalies may be seen as manifestations of water structure changes, resulting in vastly different mobilities of solutes (ions and low molecular weight nonelectrolytes) in the vicinal water in membranes. Again, a change in the vicinal water structure may, in turn, result in a change in the conformation of the substrate. In terms of the holist theory, the thermal anomalies are manifestations of changes in the thermodynamic activities of the solute as the vicinal water structure suddenly changes. It is also possible, of course, that the influence of the changes of the water structure simultaneously affect both thermodynamic equilibrium and transport properties. Thus, by itself, the occurrence of thermal anomalies does not distinguish between active and passive transport, on the one hand, and change of the cell water *in toto*, on the other hand.

A recent study of membranes by dielectric measurements will illustrate the likely (or perhaps maximal) extent of water structuring in membranes. Coster and Simons (1970) have used a Wayne–Kerr conductance bridge to study the capacitance changes in membranes. The main result from this study was the discovery of layers of water near lipid membranes with properties notably different from the properties of bulk water. The apparent thickness of the water layers adjacent to the membrane appeared to approach 40,000 Å! Such a layer (4 μ thick) would obviously have a very pronounced effect on the overall electrical properties of all types of cells if these do, indeed, have characteristics similar to the lipid membranes proposed initially by Danielli and Davson (1935). However, Coster and Simons (1970) call attention to the need to examine carefully some of the underlying assumptions required to extract the value for the thickness of the changed water layer from the experimental data; it is to be hoped that this type of study will continue as it appears to offer a very promising approach to the understanding of vicinal water in biological systems.

More than anyone else, it is probably Stein (1967) who has contributed the most systematic analysis of membrane functioning in terms of the underlying molecular processes, especially transport theories. Stein seems particularly attuned to the possible structural roles of water in the morphology and functioning of membranes.

3. *Examples of Thermal Anomalies in Membranes*

Remarkably sharp anomalies in membrane properties have been observed by Dalton and Snart (1967). It is interesting that in these studies one of the characteristic temperatures—at which an abrupt change in the energy of activation for the conduction through the toad bladder membrane occurs—is very close to 28°C (28.2 ± 0.4°C). It seems reasonable to propose that this change is a manifestation of a change in water structure associated with the vicinal water of the membrane. However, a second (and often very abrupt) anomaly occurs in the vicinity of 37.2°C. The origin of this anomaly is far more difficult to explain within the framework of what has been discussed in the present paper as this temperature does not coincide with any of the known thermal anomalies due to water structure changes. However, as pointed out in Section V,C an anomaly near 37°C could be due to a transition in the lipids of the membrane.

As demonstrated in this chapter, many data from the biochemical and biological literature show evidence of thermal anomalies, but the anomalies have frequently been overlooked or ignored by the authors themselves. A striking example of the recognition that not all functional relationships are straight lines (or, at best, simple smooth curves) is reported in the discussion "Cellular Dynamics" (Peachey, 1968). When Booij (see Peachey, 1968) became aware of the data on the anomalous phospholipid membrane resistances as a function of temperature, as presented by T. E. Thompson (1964; also see Drost-Hansen and Thorhaug, 1967), he reported an observation which he had previously considered too unimportant to publish, namely, that in the range from 17.5° to 27.5°C the permeability of onion scale was relatively temperature independent. However, below this temperature the water permeability dropped, apparently significantly, and above 27.5° the permeability increased rapidly. Booij conjectured (undoubtedly correctly) that these "breaks" might be related to the anomalies reported by Thompson for the electrical resistance of the phospholipid membranes as a function of the temperature.

Siegel (1969) has studied the excretion of β-cyanin by beet roots as a function of temperature in an oxidizing environment. An Arrhenius plot of the β-cyanin rate of discharge versus reciprocal absolute temperature shows two distinct limiting curves—changing from one curve segment to another at 60°C. The difference in the apparent energies of activation is considerable (93 kcal/mole below 60°C and 18 kcal/mole above 60°C).

Among the anomalous, abrupt changes in membrane properties reviewed by Drost-Hansen and Thorhaug (1967) were the rates of diffusion of sodium and potassium chloride across a "butanol membrane" (data from Rosano et al., 1961). Mention was also made of the temperature dependence of the resistance of a phospholipid bimolecular membrane studied by T. E. Thompson (1964) and the highly anomalous conductance of a barium stearate multilayer membrane studied by Nelson, and Blei (1966). Additional data on collodion–potassium oleate membranes were reported by Nelson, and these frequently exhibited maxima near 15°C in bi-ionic potentials. Particularly interesting results were obtained (using 1 M solutions of KCl and NaCl, separated by the membrane) at high temperatures. The results, reversible below 50°C, suggest that significant, anomalous changes occur in these membranes as a function of temperature.

Some interesting results obtained by differential scanning calorimetry were discussed by Steim (1968). The study by Steim involves spectroscopic and calorimetric measurements of biological membrane materials, as well as of some simple aqueous systems, particularly lipid–water systems. The results on the materials from the membrane of *Mycoplasma laidlawii* are especially interesting as they clearly show several transitions near the temperatures discussed in the present article and here ascribed to the changes in vicinal water structure. It is not presently possible to correlate all of the observations made by Steim with the changes discussed in the present chapter; however, the study is important in demonstrating some possible advantages of differential calorimetry over other tools such as high-resolution NMR or optical rotation dispersion studies. Incidentally, Steim concludes from this study that his data can be interpreted most readily in terms of the membrane bilayer hypothesis, but Steim does not emphasize the role of water structure changes in this connection.

Chaudhry and Mishra (1969) measured the diffusion of ^{24}Na across atrial wall segments from the rat heart. The measurements were made with equal sodium ion concentrations on both sides of the wall of the tissue studied. The authors observed transitions in the diffusion versus temperature at 15°, 32°, and 42°C. Because these values are close to those described by the present author, suggesting the occurrence of higher-order phase transitions in water near interfaces, Chaudhry and Mishra observe that this may be taken as evidence for the diffusion of the sodium ion through water-filled pores. While the experiments by Chaudhry and Mishra are remarkable, caution must be exercised in the interpretation. Thus, conceivably the transport through the membrane might still take place via mechanisms other than diffusion in water-filled pores; the notable temperature effects might be due to structural effects in the layers of water adjacent to the membrane rather than *in* the membrane itself. However, the study by Chaudhry and Mishra certainly deserves careful consideration and emphasizes the utility of further detailed studies along these lines.

4. *Permeability Studies*

In an impressive series of studies, Wright and Diamond (1969; also see Diamond and Wright, 1969) have studied the membrane permeability of various nonelectrolytes. The study is not only a profound and extensive, comparative study of various permeability coefficients for a large number of nonelectrolytes, but it is based on a surprisingly simple approach—the nonelectrolyte movement is monitored through the attendant osmotic flow through the membrane, giving rise to a streaming potential.

In connection with osmotic flow, it is important to consider the existence of an unstirred layer; this was discussed in some detail by Wright and Diamond. They correctly point out that the unstirred layer may play a dominant and, in fact, destructive influence on the interpretation of data aimed at calculating activation energies of permeation (from permeability measurements at different temperatures), solute–solvent interactions in membrane permeation, and comparisons of measured and observed reflection coefficient values in connection with different kinetic models of membrane structure and permeation processes. However, in the study of the rabbit gall bladder membrane by Wright and Diamond the effect of the unstirred layer played a considerably less crucial role. According to these authors, the unstirred layer effect is simply to shift the observed reflection coefficients since the method is essentially a comparative method. Nonetheless, it is obvious that the effect can hardly be completely neglected since the unstirred layer will, if structured, obscure differences between the structured elements of water in the permeated membrane material and in the unstirred layer.

Many nonelectrolytes may be diffused through cell membranes, but vast differences are observed in the permeation rates. Thus, Wright and Diamond (1969) note that molecules of approximately the same size and molecular weight and diffusion coefficients in bulk solution may differ by as much as a factor of 10^8 in the rate of diffusion through various membranes.

Wright and Diamond measured the permeability of some sixty nonelectrolytes and correlated the permeabilities with the oil–water partition coefficient (and with the ethyl ether–water partition coefficient). Similar studies have been carried out by Collander (see, for instance, Stein, 1967). Phenomenologically, the data from these studies (as well as many others; see the two foregoing works for references) have provided a large amount of information. However, it appears very difficult to elucidate systematically the relation between the nature of the nonelectrolyte and its permeability. Obviously, size alone will play an important role as will the presence of functional groups, steric aspects, etc. Yet, undoubtedly superimposed on these facets is the probable importance of the hydration of the nonelectrolytes, the hydration of the membrane matrix, and the detailed structure of the vicinal water of the membranes. Unfortunately, it will no doubt be some time before these relationships will begin to be understood quantitatively. Wright and Diamond refer to "anomalous" and "normal" permeability components; there is little doubt that the detailed nature of the matrix will influence relative permeabilities of the solutes. It is interesting that in spite of all the information and its correlation, for instance, with lipid solubility, it is still impossible to make significant statements regarding the likelihood that permeation (especially for water) is primarily via a continuous phase (water-filled pore) or a solvent process in the membrane material.

92   W. DROST-HANSEN

Among the several significant facts noted by Wright and Diamond is
the possible important effects of unstirred layers on tracer permeability
when compared to osmotic permeabilities. The difference was tentatively
identified with momentum transfer between solute and water molecules.
The articles by Wright and Diamond (1969; Diamond and Wright, 1969)
should be consulted for details in connection with this discussion; the
authors point out that:

> All these properties of water in narrow channels are intermediate between
> properties of bulk water and of ice, indicating that water near a charged
> surface assumes a more ordered, ice-like structure. Most of this experimental
> work has been in channels with diameters in the range of 100–1000 Å, whereas
> anomalous non-electrolyte permeation in the gall bladder disappears for
> solutes with more than about three carbon atoms (hydrated diameters larger
> than about 5 or 6 Å).

Further permeability studies have been published by Wright and
Prather (1970; also, see, Prather and Wright, 1970). Finally, it should be
noted that Wright and Diamond (1969) analyzed the available data in
terms of molecular models, taking into account the various types of forces
affecting the nonelectrolytes, such as the permanent dipoles, induced
dipoles, van der Waals forces, including short-range repulsive forces and
inductive effects.

5. Unstirred Layers

For some interesting measurements and, particularly, some interesting
speculations regarding the effects of unstirred layers (in connection with
transport numbers), see the article by Barry and Hope (1969). These
authors are not concerned with the possible structuring of any vicinal
water in or adjacent to the cell wall and cell membranes, but draw attention to the charge accumulation effects which may occur in such regions
due to superimposed electric fields.

The thickness of unstirred layers is usually taken to be more than a few
microns and often as much as 20–50 μ (see, for instance, Curran). For
a discussion of the effect of unstirred layers on the determination of apparent permeability coefficients, see the discussion by Goldup et al. (1970,
p. 244). Recently, another study has dealt with the problem of the effects
of the unstirred layer. Green and Otori (1970) have studied by a direct
optical method the thickness of the unstirred layer of fluid adjacent to
two solid interfaces. The surfaces observed were, respectively, the poste-
rior surface of the rabbit cornea and a glass surface (a contact lens). The
procedure was to study the movement of various types of small,
discrete light-scatterring particles, such as polystyrene (less than 0.25 μ in

diameter) or carmine particles. The thickness of the layer in which there was no notable movement of the suspended particles (other than that which could be ascribed to diffusion) was determined optically. In the unstirred case, the layer thickness appeared to depend on the nature of the solid material, being about 350 μ thick on the cornea and 150 μ thick on the contact lens. The thickness was also measured with vigorous stirring, which reduced the stagnant layers to 65 μ and less than 20 μ, respectively. It is unfortunate that not quantitative estimates were presented for the shear rates. Thus, it is relatively uninstructive to know the motion in the cell was achieved by stirring at 400 rpm with a Teflon-coated magnetic stirring bar. However, even without a quantitative estimate of the degree of agitation at the interface, it does appear as if very notable thicknesses must be allowed for in diffusion studies at membranes.

6. Comparison of Membrane Matrix Effects

In connection with the structural characteristics of the cellular plasma membrane, Schultz and Asunmaa (1970) have studied the possibility of ordered structures of water. The article reviews some of the evidence available for ordering of water near solid surfaces in general and near membranes in particular. The authors appear to adopt the notion of a well-defined thickness of very highly structured water. One of the most interesting observations from the studies of these authors is the suggestion that "It has been demonstrated that the characteristics of ordered water in a porous glass desalination membrane are very similar to those in a cellulose acetate desalination membrane. This result is very surprising and further experimental work is required to see if the same is true in other strongly hydrophilic membranes."

The present author does not necessarily subscribe to the notion of a very sharply defined, structured layer near any solid interface nor to the experimental foundation on which the proposed similarity in water structure between the highly different substrates has been based. At the same time, however, as discussed elsewhere in the present chapter, there is evidence that the detailed chemical nature of the substrate may play only a secondary role. In other words, the proposal by Schultz and Asunmaa does agree qualitatively with the observation that similar, ordered structures appear to be induced by the proximity to different solid interfaces, as evidenced by the occurrence of thermal anomalies at the same temperatures, regardless of the nature of the substrate.

a. Spectroscopic Studies. Only relatively few systematic studies have been made on membranes by spectroscopic means. However, Zundel (1969) has recently published a monograph devoted to the study of hydration and intermolecular interactions with polyelectrolyte membranes by infrared spectroscopy. Although this book deals with physicochemically well-defined types of membrane materials, it is very likely that the general approach will become a model for spectral studies of other types of membrane materials, including biological materials. As could be expected, a large fraction of this study is devoted to the spectroscopic properties of the water of such membranes and membrane materials. It should be stressed, however, that the approach, while sound and extremely funda-mental, does not allow for more sophisticated aspects, such as modified or extensive hydration structures. This is quite natural, of course, consider-ing the relatively insensitive technique used, namely, the spectral changes primarily effected through the hydrogen bond with its attendant large bandwidth. Those aspects that result from the ion hydration or the spe-cific polymeric contributions from the membrane substrate will not read-ily reflect the details of "secondary hydration structure" effects (see the discussion of Sidorova's studies, Section V,B,6,h).

b. *Study by Resing and Neihof.* Resing and Neihof (1970) studied the nature of adsorbed water on bacterial cell walls by an NMR technique. These authors worked with a "representative bacterium" (*Bacillus mega-tarium*) over a wide temperature range with isolated cell walls containing approximately 33% water. The authors found no evidence that the water of these cell walls was "ice-like" nor did they find that the mobility of the water was of the order of magnitude expected for "solid-state like" water (in the temperature range in which the bacterium grows). Specifi-cally, the authors found that "the distribution function necessary to fit the cell wall relaxation time data is so broad that it reaches, with appre-ciable amplitude, from the liquid value to the value in ice. Nevertheless, the median is much closer to the jump time for liquid water than for ice. The conclusion is clearly that the water in cell walls is not 'ice-like' in terms of mobility."

A number of comments are appropriate regarding the study by Resing and Neihof (1970). These authors did not discover any notable reduction in the mobility of the water near the membrane surface. On the other hand, as discussed in the section dealing with NMR studies by Glasel, it has been clearly shown that some macromolecules do not appear to pos-sess extensive hydration structures. Again, notable differences are fre-quently observed between the properties of water in living organisms and in dead organisms (and, incidentally, between dormant and active states). The argument by Resing and Neihof, that removing some of the water (from "at least 90% of the volume") to 33% water (by lyophilizing and rehumidifying) should not necessarily be expected to result in a water structure resembling that in the living organism. However, it might still be argued that since most any solid interface apparently tends to induce vicinal water structures, some kind of structure should still have been observed—even though it might have no relation to the original water structure of the living bacterium. It seems at this time that the study by Resing and Neihof is more an anomaly than a general finding. As such, the study deserves careful consideration but the results certainly cannot readily form the basis of more generalized statements about absence of ordered water structure near biological interfaces!

The membrane samples used by Resing and Neihof (1970) were prepared by storing the specimens in an evacuated desiccator over saturated sodium acetate solutions, corresponding to a relative humidity of 76% (at 20° C). Such drying may significantly influence the nature of the adsorbed macromolecules. Wetzel and co-workers (1969), for example, have shown that pronounced changes in the UV absorption spectrum and the dicroism of oriented films of calf thymus DNA occur as a function of relative humidity. Particularly, these authors found that notable hysteresis effects were observed in the range of relative humidities from 0 to 65% and called attention to the structural changes proposed by Falk and co-workers (1962, 1963) in the structure of DNA in the range from 55 to 75% relative humidity. Other authors have demonstrated changes in configuration in the range from 75 to 92% relative humidity in DNA. Thus, the absence of evidence for structuring at the cell walls studied by Resing and Neihof (in the range of water contents employed by these authors) does not rule out the possibility that structuring may indeed occur in the living cell.

7. Studies by Good

An important contribution to the understanding of the state of water in membranes has come from the extensive studies by Good (see also Coldman) on the hemolysis of mammalian erythrocytes. For over a decade Good has studied the details of hemolysis of erythrocytes as a function of temperature in the presence and absence of various pharmacons. For this study, Good and co-workers have based their approach entirely on the kinetics of the hemolysis of the erythrocyte membranes. The results have been cast in the form of the Eyring rate equation and the results interpreted in terms of the apparent enthalpies, entropies, and free energies of activation (and the activation equilibrium constant K).

The most obvious results obtained by Good and Coldman is the impressive degree of linearity between the apparent entropy of activation as a function of the apparent enthalpy of activation. Stressed by several authors as well as elsewhere in this paper, the detailed information about the apparent enthalpies of activation and the entropies of activation are far more revealing than merely the apparent free energy of activation $(\Delta G^{\ddagger})$. The linear relationships between $\Delta S^{\ddagger}$ and $\Delta H^{\ddagger}$ is often encountered in chemical kinetics. Good points out that the proportionality constant in the expression

$$\Delta H^{\ddagger} = \Delta H_0^{\ddagger} + T_c \, \Delta S^{\ddagger}$$

can usually be related to the influence of the solvent on the kinetic process in question (see the detailed discussion of the enthalpy–entropy compensation phenomenon in Section V,B,6,d).* Coldman and Good (1968b) interpret the compensation phenomenon as implying that the hydration of the membrane plays the dominant, controlling role in the hemolysis of most of the erythrocyte cells (with the exception of cells from cattle and dogs). The authors further go on to note that the value for the apparent entropy of activation ($\Delta S^{\ddagger}$, extrapolated to 0°K) is practically equal to the standard entropy of water and suggest that the process appears to take place in a wholly ordered water structure environment. This is interesting in connection with the recent NMR studies discussed in Section III,F. Coldman and Good (1968a) summarize some of the pertinent conclusions:

> It is concluded from these results that linearity between the Arrhenius activation parameters depends more on cell membrane hydration than on any other single factor.

Furthermore,

> It has recently been postulated that the cell membrane contains an interconnected hydrogen-bonded framework—a hydrate continuum—that permeates the ordered lipoprotein structure (16) and it has also been proposed that changes in the configuration of the lipids may determine the water content of the membrane structure (17). More recently a model has been put forward (18) which assigns to water an important role as an integrated structural component of the membrane protein, and which provides also for extensive cell surface hydration. There are, therefore, grounds for supposing that ordered water of hydration is just as important a constituent of the cell membrane as it is of the intracellular phase, and it may be that cell swelling in a hypotonic medium is not due merely to an increase in volume of the intracellular phase, but also to an increase in volume of the membrane.

In a subsequent series of papers, Good and co-workers studied the effects of various pharmacons on the hemolysis of erythrocytes. Particular interest attached to the study of the effects of the barbituates [also, see, the article by Tracey (1968)]. Again, it was found by Good and co-workers that the kinetics conform with the compensation law; that is,

---

* Note that in Section V,B,6,d we were concerned mostly with equilibrium properties, whereas Good's studies are of rate processes.

that the apparent entropy of activation is linearly related to the apparent enthalpy of activation and thus reveals the likely importance of hydration effects on the mechanism of the process. Furthermore, Coldman and Good (1969) note that the results suggest that the nonpolar side chains of the barbituates play an important role through the attendant hydration phenomenon and relate this observation in terms of the pharmacological activity. Specifically, with regard to hydrophobic hydration (referred to by Good and co-workers as "apolar hydration"), the authors note that:

> ...it is well known (21), however, and its occurrence depends on the structure of the solute and the absence of a direct interaction between solute and water; the solute behaves essentially as an inert support that maintains the first layer of surrounding water molecules in tetrahedral configuration, thus favoring the formation of pentagonal polyhedral hydrogen-bonded structures or clathrate cages of water. The capacity of water to form such cages is almost limitless, and studies with alkyl-substituted ammonium salts show that entities as large as the tetraisobutyl group (22) and the benzene ring (23) can be enclosed by water in this way.

Finally, in a separate study, Coldman and Good (1969) studied the hydrational effects of leptazol and concluded "... that the convulsions induced by leptazol, insulin hypoglycemia and electric-shock treatment may depend on the disruption of cerebral hydration structure."

In conclusion, there is little doubt that the studies by Good and co-workers will become classics in their approach to the understanding of the kinetic behavior of biological systems (through careful studies of temperature effects) and through the information which may be obtained from such studies regarding the role played by the solvent—the water.

8. Other Studies on Erythrocytes

Controversy continues as to whether or not the permeation of water through red cells is through individual pores (by a flow mechanism perhaps resembling Poiseuille flow) or through some "solubility matrix effect" (implying the absence of discrete pores). Recently, Solomon and co-workers (see Vieira et al., 1970) studied the hydraulic permeability of erythrocytes of humans and of dogs as a function of temperature. Based on their measurements, the authors concluded that the product, $L_p \times \eta_w$, was constant over the temperature range studied. This, in turn, was interpreted as supporting the view that over the temperature range of interest (5–39°C), temperature induced no restraints on the equivalent pores. Though it could not be ascertained that Poiseuille flow did, indeed, take place in such small pores, speculations were advanced that at least for the dog erythrocyte membrane, the diffusion of water occurred through a diffusion mechanism similar to that occurring in free solution. Figure 17A shows the normalized hydraulic conductivity coefficient for canine erythrocytes, as shown by Solomon and co-workers. Figure 17B shows the same data redrawn by the present author. It is immediately obvious

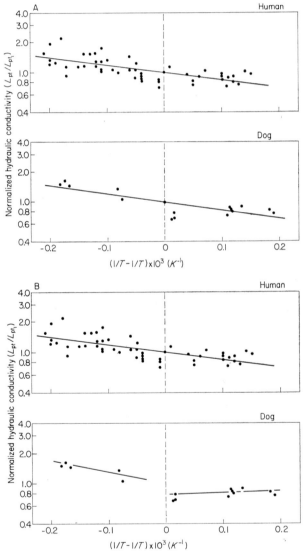

FIG. 17. (A) Normalized hydraulic conductivity coefficient for erythrocytes. (Solomon and co-workers, see Vieira *et al.*, 1970, with permission.) (B) Same data points as in Fig. 17A but curve redrawn by present author.

that the points below approximately 27°C are more-or-less temperature independent, and thus the product of the hydraulic permeability coefficient and the viscosity of the bulk water will not be temperature independent as required for Solomon's interpretation. The difference in the two sets of curves (A and B) comes, of course, from the different ways of assessing experimental errors: Solomon and co-workers merely drew a straight line, fitted by a least-square best fit, through all the data points obtained without any regard for trends. The redrawn illustration (Fig. 17B), however, becomes just as probable when the possibility of a structural change near 30°C has been accepted.

The data obtained by Solomon and co-workers for the hydraulic conductivity coefficient in human erythrocytes may or may not show the same effect, namely, that below 27°C the hydraulic conductivity is essentially temperature independent but increases slightly above this temperature. However, in the case of the experiments with the human erythrocytes, the experimental scatter is too large to draw any conclusions with certainty. The large scatter in the data in the latter case by itself does obviously not prove the converse hypothesis, namely, that the least-square best fit adopted by Solomon and co-workers is the proper analytical functional representation of the temperature dependence of the hydraulic conductivity. For a discussion of the pore concept in membranes, see also Solomon (1968).

The studies by Solomon and co-workers have in the past contributed significantly to our understanding of membrane functioning and particularly to the problem of the mechanism of solvent and solute transport across membranes. The example discussed here is in no way meant to belittle the signal contributions of Solomon and co-workers, but rather to call attention to the dangers inherent in overlooking the importance of trends versus experimental errors in experimental data.

The possible existence of actual water-filled "pores" in cell membranes (of lipidic nature) compared to other possible mechanisms for transport, especially for water, has been discussed by several authors. In addition to the contributions by Solomon et al., see also the article by Ilani and Tzivoni (1968) who suggest that the water, at least in the simple hydrophobic membranes studied by these authors (prepared by impregnating filter matrices with toluene or other organic liquids), does not possess actual "open pores"—hardly a surprising conclusion. The inferences from this study were tentatively considered in connection with the general cell membrane. In this connection, see also the studies by Ting et al. (1966). These authors have considered the possible existence of "soft ice" at the interface between butanol and an aqueous salt solution. The butanol was chosen as an experimentally convenient and conceptually reasonably simple model system of a lipid by Schulman and co-workers (see Rosano *et al.*, 1961). Ting *et al.* note, incidentally, that at least for rubidium, the Arrhenius plot of the rate of transfer of the ion across the interface suggests a break at 15°C. Compare in this connection the note by Drost-Hansen and Thorhaug (1967). Among the many other studies concerned with the problem of the possible existence of water-filled pores in membranes, attention should be drawn to the study by Gutknecht (1968) who concluded that, at least for *Valonia*, it appears unlikely that the protoplast contained specific, water-filled pores.

9. *Bangham's Studies*

S. M. Johnson and Bangham (1969) have studied the effects of anesthetic agents on the phospholipid membranes; Bangham and co-workers (1965a,b, 1966) have contributed greatly to the understanding of membrane properties and processes (as well as anesthesia) over several years. Johnson and Bangham (1969) specifically studied the permeability to potassium ions of 4% phosphatidic acid—96% phosphatidylcholine liposomes. Experiments were performed in the presence and absence of anesthetic agents, including ether, chloroform, and *n*-butanol; the effect of valinomycin was also studied. The rate data obtained were exploited in terms of apparent enthalpies of activation in an Arrhenius equation. Among the results obtained were (1) the observation that a cation permeability barrier was located at the water–lipid interface, (2) "the anesthetics increased the freedom of movement of groups in the lipid molecules near the interface," (3) the increase in permeability of K^+ in the presence of valinomycin was due to an entropy increase in the activated state, approximately 35 cal $\times$ mole^{-1} $\times$ deg^{-1}, and (4) "the increased freedom of movement in the interface when the anesthetic was present allowed the valinomycin to adopt a more favorable orientation in the interface for the exchange of K^+."

The study by Johnson and Bangham is a relatively straightforward application of an Arrhenius equation to the experimental rate data observed. However, the study is a typical example of the dangers which result from neglecting obvious anomalies, such as trends in experimental data. Figure 18A (Johnson and Bangham, 1969, p. 93, Fig. 3) shows the liposome permeability in the presence of chloroform. Valinomycin was present in a mole ratio of $1:10^6$ (lipid). Characteristically, the trends in the data were ignored. Figure 18B shows the same data as reported in the original article, but with the straight lines deleted and more realistic curves drawn in. It is seen that, indeed, rather abrupt anomalies do occur, and these reflect vastly different slopes, i.e., very different activation parameters. Hence,

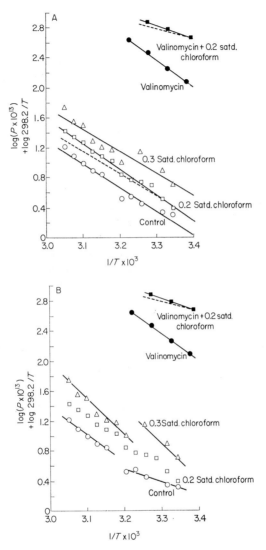

FIG. 18. (A) Liposome permeability; Arrhenius plot. (S. M. Johnson and Bangham, 1969, with permission.) (B) Same data points as shown in Fig. 18A but with curves redrawn by the present author (note—one curve deleted for clarity).

although the present author does not take issue with the general attempt to interpret molecular happenings in terms of a simple kinetic rate expression, he does take issue with the frequently practiced custom of ignoring realistic error limits and the forcing of all experimental data to fit straight lines in order to obtain, by "brute force methods," data for the apparent activation parameters.

In connection with the study by Johnson and Bangham, notice in Fig. 18B that the failure to produce a reasonably smooth curve (i.e., free of thermal anomalies) increases as the concentration of chloroform increases. A tempting suggestion here is that the chloroform stabilizes vicinal water structures through clathrate hydrate formation—the more pronounced this structure is, the more abrupt the attendant thermal transitions will be (compare the studies by Nelson and Blei, 1966). Note also, however, that chloroform is readily soluble in the lipid.

There is little doubt that the curves shown in Fig. 18B fit the experimental data considerably better than the straight lines proposed by Bangham and Johnson. However, it should be noted in Fig. 18B that the anomalies do not appear to occur near 30° and 45°C as would be expected, but rather tend to show an anomaly in the vicinity of 38°C. In this connection, compare the discussion of the phase transitions of the lipids and particularly the studies by Chapman and co-workers, Sections V,C,1 and 2, and the studies by Steim, Section V,D,3.

10. Cellulosic Membranes

The problem of water in and adjacent to cellulose has been a subject of a great deal of research because of the obvious industrial importance, as well as the role cellulose plays in understanding the physiology of plants; many of the temperature-sensitive phenomena observed in plant growth are likely related to the water–cellulose interaction. In this section, however, the stress is primarily on the simplified system of isolated cellulose and water.

Notable contributions in this field have come from studies such as those by Goring and co-workers in Canada, by Forslind in Stockholm and by the researchers at the Swedish Forest Research Laboratory in Sweden. Before discussing these contributions, attention is called to results of Haase and Steinert (1959), discussed by the present author (Drost-Hansen, 1969b). Haase and Steinert observed notable anomalies in the permeability of treated cellulose membranes and in the apparent heat of transport across cellulosic membranes, near 32° and 45°C. The authors noted that no evidence of anomalies had been reported for the cellulose itself at these temperatures. [It should be noted, however, that in subsequent studies by Haase and de Greiff (1965), the effect was not reproduced. In this connection, see the discussion by Kerr (1970), by the present author (Drost-Hansen, 1969b), and in Section VI,J.]

Martin-Löf and Söremark (1969a,b,c) have studied thermal transitions in cellulose utilizing various techniques. These authors have observed abrupt anomalies at a number of discrete temperatures (in some cases, near 0°, 35°, and 60°C); most of the studies were by some type of dilatometry. However, additional experiments were carried out, for instance, with D-glucose, where measurements were made of the optical rotation. This quantity showed a minimum near 28°C and a maximum near 47°C. These results should be compared with the study by Kendrew and Moelwyn-Hughes from 1940 (quoted by Drost-Hansen, 1967a): these authors studied the optical rotation of several different reducing sugars. In all of the available data, it appears that a thermal anomaly occurs in the vicinity of 15°C.

Finally, it is of interest also to note that the studies by Martin-Löf and Söremark also revealed anomalies near 30°C in the NMR linewidth in hemicellulose. The reader is referred for details to the extensive reports by these authors. It should be mentioned that the authors do not necessarily consider the anomalies as manifestations of higher-order (cooperative) phase transitions, but tentatively suggest the possibility (Martin-Löf and Söremark, 1969) that the anomalies may owe their existence to the sudden onset of rotational modes of individual water molecules.

Martin-Löf and Söremark note that Wahba has added both infrared spectroscopy evidence and optical refraction data to suggest the occurrence of transitions in the water–cellulose system; anomalies were also noted by Back who used a sonic pulse technique and by Kubat using a torsion pendulum.

The earlier studies by Ramiah and Goring (1965) were also concerned with water–cellulose interactions. Again, anomalous changes were observed in the expansion of water-swollen materials (cellulose, hemicellulose, lignin). The authors described the changes and perturbations in the water structure caused by the hydrophilic surfaces of the woody macromolecules. For some details and references, the reader is referred to the paper by the present author (Drost-Hansen, 1969b) or the original papers by Goring and co-workers.

Gary-Bobo and Solomon (1971) have studied transport across cellulose acetate membranes of varying porosity. The authors chose these membranes to "gain further insight into the nature of water–membrane interactions," and employed measurements of hydraulic conductivity and diffusion coefficients (using tritiated water) as a function of temperature. One of the aims of the study was to distinguish between influences due to geometry (such as tortuosity, etc.) and the effects of the water–membrane interactions. Previously, Solomon has advocated that viscous flow through membranes with even very small equivalent pore radii is essentially "classic"; Gary-Bobo and Solomon claim also that "all the experimental results can be accounted for in terms of known properties of free water and no anomalous behavior of water needs to be postulated." However, Gary-Bobo and Solomon do point to the importance in the diffusion process of the water–membrane interactions, even across membranes with large equivalent pore radii. The choice of cellulose acetate appears somewhat unfortunate for a study of water–membrane interactions for the very reason the authors stress as being of particular interest. The authors quote Franks (1965) stating that since "hydroxyl groups do not alter water structure much, if at all, the behavior of the water in the membrane might be expected to be similar to that of water in bulk." It would seem that, in order to study water–membrane interactions, it would have been better to choose a hydrophobic membrane or one of a less obvious hydrophilic nature. However, compare in this connection, also, the statement by Tait and Franks (1971) that the hydroxyl groups certainly are "sensed" by the aqueous environment, since water appears able to distinguish between α- and β-methyl pyranosides. In spite of the hydrophilic nature of the membrane material, Gary-Bobo and Solomon demonstrated water–membrane interactions by the notable differences in the observed energies of activation for diffusion in the different membrane materials. Thus, in the membrane with the smallest pore radii, the apparent energy of activation 7.8 kcal/gram mole, compared to 4.8 kcal/gram mole for self-diffusion in water (the value at 20° obtained by Wang, 1965). Gary-Bobo and Solomon also stress the notable difference in the small-pored membranes between the apparent energies of activation for diffusion compared to viscous flow; they suggest that notably different mechanisms are involved and that, in this connection, "viscous flow is a relative motion of portions of a liquid, diffusion is a relative motion of its constituents." It is interesting to speculate that, were it possible to measure viscosity over a very wide range of shear rates, the "limiting value" for the apparent energy of activation for infinitely small shear rates might approach that observed for diffusion! As mentioned briefly in Section VI,J, it is not inconceivable that the viscosity of vicinal water may be shear-rate dependent. Forslind (1968) has previously claimed that water is non-Newtonian. At the same time, the studies by R. J. Miller (1968) failed to demonstrate the existence of a definite critical shear stress; thus, the water at least does not actually "gel" under the conditions of the studies by Miller and co-workers (who worked mostly with clay matrices).

11. *Diffusion Studies*

A vast amount of literature exists on the diffuson of various solutes (of both low and very high molecular weight) in water through various porous materials. However, only a very limited number of studies have been reported where the particle size and the pore diameters do not differ greatly and the dimensions are still large enough to justify *a priori* the use of classic hydrodynamics (Poiseuille flow). Recently, Uzelac and Cussler (1970) have studied the diffusion of monodispersed spherical latex particles (diameter 910 Å) through Millipore filters. These filters had nominal pore diameters of 2200, 3000, 4500, and 12,000 Å. The results of the study are of obvious interest to biophysics as a model for the study, for instance, of red cells passing through vascular capillaries or large macromolecules passing through discrete pores in various types of membranes. Many of the results from this study are important with respect to a theoretical interpretation in terms of the movement of an inert sphere diffusing in a continuum liquid. The authors claim that the temperature dependence is that which would be expected on the basis of the simple expression used by the authors, although an inspection of the data may throw some doubt on the validity of the conclusion. The authors give the expression for the diffusion coefficient (D) as:

$$D = \frac{kT}{6\pi\eta a k_1(a/R)} \tag{1}$$

where $k_1(a/R)$ is a tabulated function. Admittedly, whereas the diffusion coefficients themselves differ by as much as 60%, the product $a\eta/kT$ differs by only 10%. The ratio of this coefficient at 25° and 45°C, respectively, is less than 1 for the larger pore size (0.45 μ). Whether or not significance can be attached to this, the authors point out that the limiting value of the diffusion coefficient for the ratio a/R (particle diameter to pore diameter) $= 0$ is about 10 times larger than the Stokes–Einstein value. The authors claim that this result is not an experimental artifact although they are unable to give an interpretation for the observations. They also call attention to some other studies which have reported enhanced diffusion coefficients—much larger than those predicted by the Stokes–Einstein equation. It is in this connection that it is of interest to consider the possibility that part of the liquid structure in the pores of the Millipore matrix may be disordered, as proposed recently by the present author (Drost-Hansen, 1969b). Some rather tenuous evidence for this has already been noted, based on data for the diffusion coefficient of a number of gases in aqueous suspensions, although there is apparently no general agreement on this point. However, if, indeed, a disordered zone may also exist in reasonably small pores, a greatly enhanced diffusion coefficient might be expected if the viscosity of less structured liquid is lower than that of the ordinary bulk water (and the highly ordered, vicinal water).

12. Active Transport

Electrolyte pumps, required to explain active transport, have been discussed in inordinate detail over the past several decades. The differences in the electrolyte contents between interstitial fluid and cell fluid is remarkable. These different electrolyte solutions are separated by the cell membranes, and the various equilibria (or, rather, steady-state processes) are normally considered to require metabolically derived energy. Apparently, in all active transport processes, the source of energy for separating the different electrolytes is derived from ATP. To "explain" active transport in classic terms, various "carriers" are required; these are usually assumed to be proteins. In this section are discussed (if only briefly) some aspects of an alternative view of active transport. A more detailed discussion of this alternative explanation is presented in Section V,G, while some selected aspects of water structure, ATP, and the ATPase problem will be treated later in a separate paper (Drost-Hansen, 1971; in preparation).

Changes in the structure of the intracellular water is a possible cause of transmembrane movements against a "total stoichiometric" concentration gradient. Changes in "available" ions (i.e., changes in "effective concentrations") have been termed changes in "solubility." This, however, seems a poor terminology, since the total number of ions in solution may remain unchanged; instead, what is changed is the activity of the ions (and other solutes). An inspection of even the simplest forms of the Debye-Hückel expression for the activities of ions reveals that notable changes are effected through a change in the dielectric constant of the solvent.

All cells are considered capable of active transport. In this section, we restrict the discussion to plant cells, based on the extensive writings by Stadelmann with some comments in terms of the water structuring discussed in this chapter. In a recent review, Stadelmann (1970) calls attention to the criteria proposed by Sitte (1969) for active transport. Three criteria must be simultaneously obeyed to consider a movement of ions truly to be of the nature of an active transport: "(1) the process uses energy, (2) there is a stoichiometric relation between the amount of ATP used up and the amount of substance transported, (3) the energy supplied from ATP is used directly for the transport of the substance under consideration."

Difficulties have arisen in the past in attempts to demonstrate the reality of active transport for lack of suitable "trans-membrane carriers"; these "transport proteins" have been termed "permeases," "trans-locases," "transfer locases," etc. Such carriers should have some stoichiometric relationship to the number of ions transported. It is possible that one advantage of the alternate theory—that ions are transported merely in a gradient of ion activity (due to the different aqueous environments)—is the fact that no stoichiometric relationship is necessarily required. The amount (and possibly the rate) of transport may instead be determined exclusively by the changes in activity of the ions and the water molecules (hence, changes in "apparent solubility," in current terminology). Certainly, of the three criteria discussed above, requirement (2) may play the key role in settling the question of active transport versus a movement in an activity gradient. It is likely, in fact, almost certain, that both mechanisms will require energy [requirement (1)], and it will probably be difficult to prove or disprove any one-to-one correspondence between the energy used and the energy directly supplied from ATP for the transport process under consideration. It is also possible that the actual transport may depend on, or be facilitated by, one of the "transport proteins." However, the transport proteins act merely as the vehicle for the transported solute. Effectively, this may reduce the energy barrier to be overcome in the process of moving the ion across the membrane, but it bears no relation to the energetics that determine why the ion is moved against the stoichiometric concentration gradient.

As discussed above, the alternative to active transport is the existence of solute activity gradients, caused by the different aqueous environments within the cells. Relaxation measurements may prove to be the most direct way of obtaining further information regarding the structural characteristics of the intracellular water. If the dielectric properties of the intracellular water were known, and particularly if it were possible to obtain information about the variation in effective dielectric constant as a function of electrolyte and nonelectrolyte contents and as a function of the proximity to the membranes and surfaces of the various organelles, it might be possible to calculate actual ion activities, using suitably modified, classic solution theory. In view of the above, the current work in the author's laboratory is now being directed toward measuring directly the dielectric properties of water near interfaces, especially in and adjacent to membranes, and eventually of intact biological systems. Thus, while it may be a decade or more before a detailed understanding will have been reached regarding the structural characteristics of vicinal water, empirical values for effective dielectric constants (and especially their dependence on distance from the interface—if such information can be obtained) may prove useful in estimating actual activity gradients in cellular systems.

E. Nerves

The functioning of nerves is little more and no less than the operation of membrane processes. We have chosen here to consider separately the nerves as a possible site for the effect of vicinal water, i.e., a critical mode of exerting a notable influence on intact animals (as well as on isolated nerves). This choice is a logical one because of the discrete and readily delineated physiological role that the nervous system plays in animals. The uniqueness of the nervous system effects become most obvious in the discussion of the functional aspects of water structure with regard to the Pauling–Miller clathrate hydrate structure model for anesthesia (see Section VI,E,1). However, the present section is limited to a discussion of a few typical illustrations of the specific effects of temperature on various gross nerve processes without concern for the underlying membrane aspects, lipid transformations, or ionic transport effects. When reading the recent treatise on "Membranes, Ions and Impulses" by Cole (1968), it is interesting that although reference is made to temperature effects and temperature coefficients, most of the data referred to were characteristically obtained at 5°, 10°, and 20°C, or 6°, 12°, and 18°C. Thus, here, as in so many other fields of biochemistry, biophysics, and biology, temperature studies are often restricted to very limited numbers of observations. Temperature has truly been the Cinderella of physical parameters. Furthermore, where additional data are available, the inevitable tendency to draw smooth curves (or even merely straight lines) through all data points—rather than paying attention to individual trends—clearly prevails; as a result, a multitude of information is probably lost. Again, in the study of nerve processes, one should be aware of the unique properties of vicinal water and, specifically, the ability of such water to undergo higher-order phase transitions at discrete temperatures. This phenomenon no doubt influences, and often underlies anomalous, abrupt changes in biological functioning determined through the effects on the nervous system.

The effects of temperature on nerve properties are rather remarkable. Few physiological processes appear to be as dependent upon temperature as some of the processes associated with functioning of the nerve, especially, frequency of discharge of nerve impulses and conduction velocity. This may perhaps be the result of structuring of the water near the nerve interface in agreement with the fact that water structures can be further enhanced or modified by the presence of anesthetic agents of the clathrate-forming types.

A great deal of work has been carried out attempting to understand and describe quantitatively the functioning of nerves; a typical example is the study of the effects of temperature on the conduction velocity of the action potential. The nerve used in this study by R. A. Chapman (1967) was the giant squid axon. The purpose of the study was to test the accuracy of the expression derived by Huxley for the conduction velocity of nerves. It is interesting that the agreement between theory and observed values is quite good in the range from approximately 5° to 25°C. However, above 25°C, a significant decrease in velocity occurs. At 30°C the conduction velocity reaches a maximum followed by a rather steep drop in the range from approximately 32° to 37°C. This drop is not predicted by Huxley's theory; it seems reasonable, therefore, to infer that this notable deviation is related to the structural changes in the vicinal water structure occurring at 30°C.

Even more remarkable anomalies were reported as the results of the studies by Lippold *et al.* (1960), the studies by Hensel *et al.* (1960), and by Hensel (1963). Somewhat similar anomalies were reported earlier by

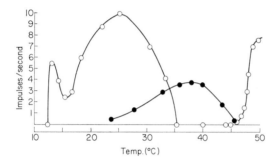

FIG. 19. Rate of nerve impulses as function of temperature in lingual nerve from cat. (Zotterman, 1959, with permission.)

Dodt and Zotterman (1952a,b) and even earlier studies by Laget and Lundberg (1949).

An exceedingly complex response of a nerve to temperature has been illustrated by Zotterman (1959). Figure 19 shows the data discussed by Zotterman for the rate of nerve impulses as a function of temperature in the lingual nerve from the cat. It is seen that the response exhibits several minima and maxima. There is little doubt that the minimum near 15°C is real as are the rapid decrease above 30°C and the rapid increase above 45°C. The latter is the so-called "anomalous cold response." It is certainly not possible at present to explain these extrema quantitatively in terms of changes in the vicinal water structure of the nerve. On the other hand, it would be truly remarkable if the similarity between the temperatures of the anomalous changes in the nerve discharge rate and the temperatures for changes in vicinal water structure was purely a matter of chance. It seems reasonable to seek a rationale for the observed complex behavior in terms of the underlying water structure. It would certainly be hard to find another more traditional mechanism which predict *a priori* that pronounced extrema might occur at these temperatures.

Interesting effects of temperature on the nervous system were described in a review of the effects of temperature on insects (Clarke, 1967). As an example, Fig. 20 shows the drastic changes below 10° and above 35°C in the activity (impulses per unit time) of the tarsal nerve after the insect (*Periplaneta americana*) was acclimatized to 32°C (before being subjected to the various temperatures).

Yamashita and Sato (1965) have studied the effects of temperature on the individual taste units of the rat. In nearly all cases studied the re-

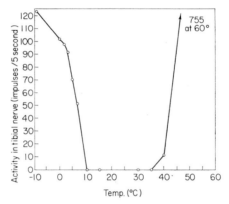

FIG. 20. Rate of nerve impulses in tarsal nerve of insect (cockroach). (Clarke, 1967, with permission.)

sponse to all kinds of stimuli (except 0.01 and 0.03 M sodium chloride) increased with temperature up to 30°C and decreased above this temperature. This effect was observed in the chorda tympani nerve of rats upon stimulation of the tongue with sodium chloride, potassium chloride, calcium chloride, hydrochloric acid, quinine, and sucrose. An inspection of the data reported by these authors clearly shows the notable maxima obtained in general (with the exception of sodium chloride as stated above). In the case of sodium chloride, the "normal" response was obtained also for the more concentrated solution (0.1 and 1.0 M). It is undoubtedly important that such vastly different solutes as various electrolytes and nonelectrolytes all evoke a response with maxima near 30°C, regardless of the nature of the substance tasted. Thus, it must be concluded that the temperature response is a function of the nerve itself rather than of the substance that induces the response. The fact that rather distinct maxima are noted at 30°C may likely be interpreted as reflecting a structural change in the internal aqueous environment of the nerve at the temperature of the thermal transition in vicinal water.

For an impressive survey of membrane behavior and nerve functioning, see the monograph by Cole (1968) referred to above. The author claims that this monograph is not a definitive work, yet, it covers the subject matter admirably. It is equally remarkable that in a book of 550 pages devoted to this topic, the subject index does not once contain the word "water," let alone water structure, or structure of water near interfaces.

F. MUSCLE

Convincing evidence for ordered water in biological systems has come from the NMR studies by Hazelwood *et al.* (1969). These authors obtained the NMR spectra of water in muscle tissue and concluded that at least two different degrees of ordering exist in this system. One of these ordered forms is the result of specific interaction between the water molecules and the native macromolecules in the muscle cell. This water is characterized by restricted motional freedom, since it is more ordered than free water. One of the important observations is the fact that only one broad water peak is observed, which indicates that, on the average, all the water in the muscle experiences a significant restriction in motional freedom. Obviously, a number of uncertainties are present in interpretations of NMR spectra, but it appears that Hazelwood and co-workers have addressed themselves to these difficulties and, indeed, presented a very strong case against each of the possible objections.

Other results by Hazelwood and Nichols (1968, 1969) are the changes in sodium content of muscle as a function of age of the experimental tissue. The amount of ordered water appears to increase with age since gestation and this is accompanied by a reduction in sodium "solubility" in the intracellular fluid. This finding has been further substantiated through correlation with changes in membrane potentials.

Finally, Cope (1967a,b,c) has demonstrated that the sodium content in the interstitial water of actomyosin is markedly reduced compared to solubility in ordinary water, suggesting an organization into a more "rigid" space due to the proximity of the actomyosin molecules. In the more recent studies by Cope (1969), use was made of heavy water in a study of muscle membrane tissue. Again, many of the objections that can be raised to the interpretation of NMR spectra were critically discussed and eliminated, leaving only the possibility that the water in the tissue was greatly enhanced structurally.

The exclusion of intracellular sodium as an organism ages was discussed, for instance, by Hazlewood and co-workers who noted the correlation with an increased degree of ordering (determined by NMR measurements). This may be correlated with the decreased dielectric constant of vicinal water, mentioned by Derjaguin (see Section IV,A,6). The notion that sodium ions are excluded from the cells as the ordering of the water increases is imminently compatible with the idea that the dielectric constant of ordered water is far lower (by an order of magnitude) than bulk water. Of course, the dielectric properties of water in biological systems is still poorly understood and one may question the comparison of such water with that adjacent to a silica gel surface, although (as discussed in Section III,C,1,e, the "paradoxical effect") it appears that the chemical nature of the substrate may not play a dominant role in determining the type of water structure which becomes stabilized vicinal to an interface.

Considering the evidence which is accumulating for notable changes in the structure of water in cells, particularly vicinal to membranes and some macromolecules, it is regrettable that practically all studies in cellular physiology continue to be based on classic physical chemistry of aqueous solutions. To be sure, the use of classic methods was appropriate initially when no evidence was available for changed water structures. However, new approaches must now be sought, even though we presently do not know, for instance, how to cope with problems such as diffusion rates in an aqueous phase, the structure of which is different from bulk water. Very likely the activity of solutes in the cells (of both electrolytes and non-electrolytes) may be vastly different from the activities obtained from bulk measurements. It seems from these considerations that the next step forward in cellular physiology will depend on the availability of far more detailed information about the structure and properties of aqueous solutions in ordered vicinal water.

G. Aspects of Water Structure in Cellular Physiology

We conclude this section with a brief review of more recent studies. Thus, some additional results obtained from NMR studies will be discussed together with some speculations regarding active transport across membranes. These topics, in turn, lead to a brief review of cooperative effects in biological systems. See also the discussion in Section V,D.

The idea that thermal anomalies of the type discussed in this chapter are manifestations of higher-order phase transitions was apparently first stressed by Schmidt and the present author (1961). Higher-order phase transitions are generally ascribed to cooperative processes. Independently, Ling (1962) introduced his "association–induction hypothesis," building on a suggestion by Troschin and involving cooperative effects. Ling coined the expression "adsorbed polarized multilayers" for the ordered water structures, but this phraseology is perhaps somewhat unfortunate. However, the ideas from Ling's extensive studies have proved highly useful as conceptual models and recently as inspirational guidance to many NMR spectroscopists. Nevertheless, it should be recalled that ion–dipole and dipole–dipole interactions are not likely to be the cause of extensive structuring of water near interfaces and are anyway restricted to ionic or highly dipolar surfaces. Other possibilities for structuring do exist, however, including stabilization of clathrate cage-like entities. Furthermore, the "paradoxical effect" discussed in this chapter suggests that the detailed nature of the substrate may play only a minor role. The structures ultimately stabilized at a water/solid interface are most likely structures that are "almost stable" in the bulk liquid phase, were it not for the continuous disruption due to thermal fluctuations. Incidentally, one possible case of actual "polarized multilayer adsorption" may exist at the interface between an advancing ice surface and an unfrozen aqueous solution (or pure water); for a discussion of this aspect, see the paper by Drost-Hansen (1967b).

Recently, Cope (1971) and Hazlewood and co-workers (1971b) have dicussed structuring of water, complex formation of sodium and potassium ions in biological systems, and the nature of cellular water and water/macromolecule interactions. The two papers review the general field of NMR studies of biological systems and contain excellent collections of references. It is interesting that it has been possible to make such notable advances based essentially upon the results of only this one type of measurement (although the more recent papers by Cope and by Hazlewood do contain references to other types of evidence, demonstrating structuring and cooperative phenomena in biological systems).

The paper by Hazlewood and co-workers (1971b) reviews the NMR measurements on water in the gastrocnemius muscle from rats. The interpretation of the results follows the ideas advocated by Ling, based on the "association–induction hypothesis" (discussed in Section IV,A,5). Among the suggestions emphasized by Hazlewood is the possible intracellular structuring of all the protoplasm, providing the driving force for a sodium transfer due to the changed solvent properties and thus eliminating the need for a specific sodium pump. The paper by Cope (1971) reiterates, and very lucidly discusses, the evidence for structuring of water in cells as based on the available NMR studies. Cope also (as discussed above) reiterates the evidence for the complexing of the sodium in the structured water of the cell. In addition to the experiments with gastrocnemius muscle of the rat, Hazlewood and co-workers (1971a) have also studied the interaction of water molecules with the cells in cardiac muscle. It was again found that "dramatic shortening of relaxation times suggests that the water molecules within the myocardial tissue significantly interact with cellular macromolecules."

In summary, the recent NMR studies from many laboratories strongly suggest the existence of ordered water structures in biological systems. Thus, as emphasized, the studies by Hazlewood and co-workers have shown that a large fraction of intracellular water appears highly ordered and, furthermore, that in the immediate postnatal period, the intracellular sodium concentration decreases as the degree of ordering of the cell water increases. On this basis, it has been suggested that membrane transport may not require the operation of various active transport mechanisms, and issue has particularly been taken with the need for a sodium pump. At the present time, the evidence for ordering of water in cells appears conclusive—both as the result of the highly specific NMR results as well as the result of the type of general evidence reviewed in this chapter.

Considering the theorized need for specific solute transmembrane pumps, Tait and Franks (1971) observed that "the state of intracellular water seems crucial to these arguments. It has been found (reference to Kushmerick and Podolsky, 1969) that the diffusion coefficients of various solutes in an intracellular environment are a factor of two lower than in aqueous solutions, but the fact that the diffusivities of some cations, anions, and nonelectrolytes are reduced to the same extent is in direct conflict with the concept (reference to Cope, 1966) of selective ion binding by the macromolecular cell components." I feel that the point made by Tait and Franks is well taken. Furthermore, if structuring of intracellular water is caused merely by proximity to an interface, as advocated by the present author, the difficulties encountered due to the "specific adsorption on polarized multilayers" is avoided. Thus, instead of "complexing" of the ions (for instance, the sodium ions) by the macromolecular solutes, nothing more may be needed than the formation of clathrate hydrate cages around the ions. See also the paper by Vaslow (1963), who discussed the possibility of small ions being found in clathrate cages, even in pure bulk solutions.

In connection with the problem of the nature of water in biological systems, and the standard approaches to solution chemistry in cell physiology, attention is called to the paper by DeHaven and Shapiro (1968). While these authors did not specifically discuss the nature of structural changes in vicinal water, they emphasized the probable changes in solute activities that must result from the effects of other solutes on the dielectric properties of the intracellular fluids. The authors outline how "classic" solution theory may be employed in the study of transport processes across

VI. STRUCTURE AND PROPERTIES OF WATER AT BIOLOGICAL INTERFACES   115

membranes if one takes into account the changes in dielectric properties of the intracellular water. I find this approach both valid and useful, but must emphasize that, in addition to the induced changes in "effective dielectric constant" of the intracellular water (as the result of the presence of other solutes), changes must also be expected due merely to the proximity of the water to the various macromolecular interfaces, resulting in different structures (and hence different dielectric properties) in the vicinal water.

The "polarized multilayer adsorption" envisioned by Ling, corresponds essentially to the structuring tentatively suggested by the present author for water adjacent to an ionic or highly polar interface. This is the type of interaction depicted in Fig. 8, page 32. However, as discussed in this chapter, and particularly in an earlier paper (Drost-Hansen, 1969b), a different and possibly more frequently occurring type of structure may be present. This type of vicinal structuring (see Fig. 9, page 33) is likely, for instance, to be most prevalent adjacent to nonpolar hydrocarbon side chains in proteins in solution or adsorbed on membrane surfaces, etc. However, in this connection, recall also the "paradoxical effect": the occurrence of thermal anomalies in the properties of water adjacent to both ionic surfaces, polar surfaces, and near such nonpolar surfaces as the air/oil interface and possibly even the air/water surface. This occurrence of the thermal anomalies—i.e., cooperative processes—near nonpolar surfaces argues against the suggestion by Ling that the cooperative effects are due to "polarized multilayers," organized by interactions with surface charges.

VI. Functional Role of Water in Biological Systems

A. INTRODUCTION

1. *Temperature Ranges of Life*

In a universe where the temperature ranges from near $0°K$ to millions of degrees, life as we know it is restricted to a fantastically narrow interval—at most, about $100°C$. That this is the range in which water (at 1 atm) remains liquid is hardly a coincidence. For plants, the range over which survival may occur may be as large as $90°C$. However, the interval of temperature tolerance for most commercially valuable plants is usually limited to about $30°C$ or less (see Rose, 1967, p. 245).

It is likely that the study of organisms capable of living at temperatures higher than present ambient temperatures* will become more and more important if continued energy production (with its attendant thermal inefficiency) continues to increase the "average, ambient temperature" of the earth (presently about 12°) and/or if the combustion of carboniferous fuels continue to increase the carbon dioxide in the atmosphere, enhancing the "greenhouse effect" with its inevitable world-wide increase in temperature.

For an extensive review of temperature effects on biological systems (and particularly growth), see the monumental monograph "Temperature und Leben" by Precht et al. (1955) or the excellent and more accessible reviews in the treatise "Thermobiology" edited by Rose (1967).

In reviewing the literature on temperature effects on biological systems, particularly microorganisms, the reader is encouraged to look carefully for evidence of discrete effects near the temperatures discussed in the present chapter, namely, near 13° to 16°, 29° to 32°, 44° to 46°, and 60° to 62°C. These are the temperatures at which some vicinal water structures undergo abrupt changes. Let it be reemphasized that the phenomena that are not determined primarily by details of water structure may not reveal any unusual dependencies on temperature (compare the discussion of Glasel's studies, Section V,B,5). As a corollary to the first statement is the implied admonition that future work concerned with temperature effects in biological systems ought to be carried out at closely spaced temperature intervals, lest these studies miss what may possibly be important hints as to underlying mechanisms, that is, mechanisms which may reveal structural influences due to the unique properties of vicinal water.

It is worth while to recognize that it is impossible to separate temperature completely as an isolated independent variable. Other simultaneous dependent and independent variables will influence most of the systems of

---

* A few organisms may live at temperatures higher than 100°C if the pressure is sufficient to prevent boiling, but such organisms (as well as others capable of living at what are ordinarily unphysiologically high temperatures, 60°–75°C, for instance) are of relatively little importance in this context. However, there is likely great virtue in focusing attention on the forms of life such as the thermophilic algae capable of living near 70°C (surviving to somewhere below the boiling point of water): the life processes in this temperature range, say, above 60°C, suggest that living systems may be able to choose selectively those constituent proteins, enzymes, membrane materials, and lipids that are compatible with far more disorganized water structures. However, an alternative, even if somewhat farfetched, is the possibility that these organisms have been able, through evolution, to select, enhance, and depend on those structured macromolecules which have the greatest ability to order vicinal water. In other words, the organisms have selectively produced those macromolecules that have the conformational stability required for physiological functioning, as well as the ability to influence strongly the vicinal water structure. For a review of "life at high temperatures," see the article by Brock (1967).

interest to biologists. Since few of the measurements made have any "equilibrium thermodynamic" meaning, time alone will be a parameter of significance. In addition to time (in an absolute sense) there are unexpected but apparently real natural fluctuations of cosmic origin [as studied by Piccardi (1962) and co-workers] as well as more obvious variations such as diurnal changes related to the circadian rhythm. An interesting example of the latter phenomenon is described by Damaschke and Becker (1964). The results from this study show that the oxygen consumption rates vary in discrete and abrupt manners with temperature, especially near 30°C, but that superimposed upon this intrinsically complex dependency, an additional effect of diurnal rhythm influences the results significantly.

2. Role of Plant Alkaloids

An inspired view of the possible role of water in plants has been presented by Tracey (1968) in an article, "Fitting the Environment by Modifying Water Structure." It is Tracey's contention that many of the low molecular weight solutes in the plants primarily serve the function of suitably modifying the cellular (and likely vicinal) water structure to enhance the possibility of survival of the plant. Tracey sees the controlling effect essentially as a distribution of availability of water between the components in the complex water–pore system, where, on the one hand, the water is required for the hydration of membrane constituents (primarily cellulosic, but including the lipoproteins) and, on the other hand, for hydration of the various solutes—the lower molecular weight solutes as well as nucleotides, proteins, etc.

Although a study of solute effects on the properties of dough may seem a very indirect way of probing for structural effects on water, it is interesting that in the hands of Tracey this approach has led to some intriguing results. Tracey has studied the effects on the rheological properties of dough of known central nervous system depressants compared to stimulants. Tracey notes that some 2000 plant alkaloids are known, which have pronounced effects on the central nervous systems of animals while they are apparently of little obvious, functional use to the plant. Since these alkaloids seem to serve no protective function so far as plant–animal interaction is concerned, the presence of the alkaloids in the plants is perhaps a device employed by the plants to optimize the use of the available water. It is interesting to note that apparently plants do not often suffer from a shortage of energy, but instead are subject to shortage of raw materials—primarily water, carbon dioxide, and combined nitrogen. Hence, Tracey speculates that "if some alkaloids enable a plant to 'immobilize' water, then a response to water deficit might be expected to be increased alkaloid production." Alkaloid yields are, indeed, higher if plants are kept short of water. Tracey speculates further that the presence of the alkaloids may be the vestiges of a device which was effective in water control on the molecular level at a time when the availability of (fresh) water was the major hardship for the plant material, for instance, at the time when the plants began to colonize dry land.

B. METABOLISM AND GROWTH

1. *Distribution of Optimum and Lethal Temperatures*

Metabolism and growth are complex physiological processes. Complexity is taken here to mean the simultaneous and consecutive involvement of a large array of individual chemical processes (reactions in the classic sense, diffusion, active transport, etc.). In 1956 the present author suggested that for such complex systems, temperature optima and minima might be predicted from a simple consideration of the structural changes in water.* One may assume that many reactions which might potentially be rate determining in a complex biological system may undergo notable changes at the temperatures of thermal anomalies. Based on this assumption, it was proposed that during evolution, biological systems have tended to avoid the temperature regions near the sudden changes in the (vicinal) water structure and, hence (at least in the case of the mammals), have optimized body temperature as far away as possible from a lower and a higher thermal anomaly (at 30° and 45°C, respectively). Were the thermal anomalies to occur exactly at 30° and 45°C, the body temperature would then be expected to fall near 37.5°C. Figure 21 shows a histogram of frequency of occurrence of body temperatures for approximately 160 mammals. It is seen that this distribution does, indeed, center very closely around 38°C with a remarkably narrow distribution. (Note 37°C equals 98.6°F.) A small number of exceptions are indicated by the cross-hatched area under the curve, around 27° to 34°C. This group includes the anteater, the sloth, the echidna, the armadillo, and a few other species such as the duckbill platypus. It seems legitimate in a first approximation to neglect these exceptions because of the somewhat unusual nature of the animals compared to most other mammals.

The distribution of body temperatures of birds appears to be centered

* At that time it was felt that evidence was available for the existence of thermal anomalies in all properties of water and aqueous systems. As pointed out elsewhere in this chapter (Section III,C,1,d), it now appears that the anomalies are most pronounced and possibly occur only in vicinal water rather than in bulk water.

around 41.5°C. It was proposed by Drost-Hansen (1965a) that the displacement (by 3° or 4°) toward higher temperatures might be a concession to the highest possible rate of energy production required for flight (approximate Arrhenius type of activation). It is interesting to note that both for all mammals and all birds studied, 45°C (±1°) appears to be an absolute, upper thermal limit (lethal temperature). It is also interesting to note that those birds that do not fly, such as the ostrich, the kiwi, and the penguin, appear to have normal body temperatures around 38° to 39°C. This would tend to substantiate the proposed explanation for the higher body temperatures of birds. It is also well known that 30°C is a tempera-

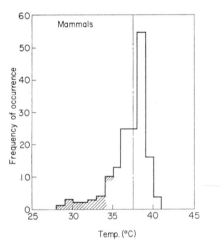

FIG. 21. Distribution of body temperatures of mammal (frequency distribution). (Drost-Hansen, 1965a, with permission from the New York Academy of Sciences.)

ture of considerable physiological importance in all mammals *and* birds. This will be further discussed in the section on hypothermia.

By analogy with the reasoning presented above, it has been suggested that an optimum might exist somewhere near the middle of the temperatures between 45° and 60°C. Indeed, a majority of thermophilic bacteria and thermophilic fungi are known to possess optima around 53° to 55°C. It is also well known that pasteurization temperatures usually tend to be 60° to 62°C; this suggests that the pasteurization temperature is a direct manifestation of the structural changes in the vicinal water near this temperature.

Finally, by the same type of argument it is proposed that optimum activity may be encountered for a group of organisms (plants and animals) between 15° and 30°C. A large number of different types of organisms appear to have optimum activity between 22° and 26°C, including many insects (though not all), many fishes, and soil bacteria. Thirty degrees Centigrade is known to be an important temperature physiologically for both fishes and insects. For specific examples, see Drost-Hansen (1965a).

Allowable ranges of temperatures for poikilotherms are often 15° to 16°C—truly a vanishingly small range out of the total span of temperatures in the universe. As will be discussed in another section, thermal adaptation may occur, but more frequently than not the adaptation is merely a slight change in temperature of, say, optimum activity (for instance, for growth or reproduction) rather than a notable change in the low-temperature tolerance limit or the upper lethal temperature. However, for unicellular organisms, a special form of adaptation may take place, namely, through the development of multiple optima for growth. Mitchell and Houlahan (1946) reported distinct binodal distributions for growth of a mutant of *Neurospora crassa*. Somewhat similar results were subsequently obtained by Oppenheimer and Drost-Hansen (1960) studying a sulfate-reducing bacterium. Later experiments (Schmidt and Drost-Hansen, 1961) tended to suggest similar behavior for *Escherichia coli*. More recently, Davey and Miller (1964) have also obtained very distinct multiple optima for growth of a number of microorganisms. Four different types of bacteria were used to cover the temperature range from 5° to 70°C; in all four cases, multiple growth optima were obtained. Oppenheimer and Drost-Hansen (1960) suggested that such organisms might be able to grow optimally in two different temperature intervals by utilizing different metabolic pathways. Some preliminary evidence was obtained for this proposition through a study of changes of pH of the medium on which *E. coli* was grown and from a qualitative study of the pigmentation in *Serratia marcescens* (Schmidt and Drost-Hansen, 1961).

2. *Examples of Thermal Anomalies in Growth Processes*

An interesting example of unusual temperature effects is described for the rate of mycelial growth of a fungus (*Waitea circinata*), studied by Agnihotri and Vaartaja (1969). These authors found that the mycelial growth of the fungus was strongly temperature dependent and, moreover, that exudate from pines (*Pinus cembroides*) further stimulated this mycelial growth. For both (without exudate as well as in the stimulated mode of growth), mycelial growth showed indications of an inflection point near 15°C and a relatively notable drop in rate of growth above 30°C. The same effect—a maximum in radial growth of the mycelium—was obtained in the presence of various nutrients such as aspartic acid, malonic acid, glutamic acid, and arabinose, with notable drops in growth above 30°C.

Recently, Walker (1969) has studied the effects of temperature in 1° increments on the behavior of maize seedlings. The temperature range covered was from 12° to 36°C, and several anomalies were observed. Thus, significant irregularities in the concentration of many of the nutrients in the shoots of the plants occurred near 15°C. Minor but persistent anomalies were also noted in total leaf length of the seedlings, and irregularities occurred in growth rate at 29° and 30°C. The author was not convinced that all of the anomalies observed were real or whether some were caused by experimental artifacts. However, an inspection of the data of, for instance, dry weight of 23-day-old maize seedlings strongly suggests an anomaly near 29° to 30°C. Walker correlates his observations with the similar anomalous results by Davey and Miller (1964) at 15°C for the uptake of potassium by wheat. Walker, although aware of the claim for the existence of thermal anomalies in (vicinal) water, did not make any conclusive association between the water structure changes and the observed anomalies.

A good example of the abrupt change in growth at 30°C is shown in the data by Buetow (1962), referred to by Farrell and Rose (1967, p. 162, Fig. B). These results clearly show the dramatic change in the specific growth rate of *Euglena gracilis:* a sharp maximum occurs in the vicinity of 28° to 30°C.

Attention is called here to the monograph by Andrewartha and Birch (1954) (in particular Part III, Chapter 6). A large number of examples are discussed which clearly show that critical temperatures for many organisms frequently coincide with the thermal anomalies stressed in the present chapter. It is impossible to go over all the examples discussed by Andrewartha and Birch, but in particular the logistics curves are important (discussed in the section on "Weather: Temperature," Chapter 6, pp. 129–205). In this connection, the frequency with which excellent agreement can be achieved between some empirical or semitheoretical logistics curves over the interval from about 15° to 30°C is noteworthy, as is the frequent failure below 15°C and almost invariable failure above 30°C.*

Levinson and Hyatt (1970) have studied the effects of temperature on the activation, germination, and outgrowth of spores of *Bacillus megaterium.* The study is particularly interesting as it was designed specifically to determine if there was evidence of thermal anomalies in these stages of bacterial spore growth. Measurements were made at closely spaced temperatures. The authors concluded that they "found no evidence of thermal

---

* It is unfortunate that many authors have apparently chosen to study various organisms "exactly" between 15° and 30°C.

discontinuities, or 'kinks' in these biological processes, but we felt nevertheless that our data on the response of spores to small temperature increments had sufficient intrinsic value to warrant publication." The negative conclusions drawn by Levinson and Hyatt is quite astounding in view
of the data reported. An inspection of their illustrations might equally
well have suggested that anomalies do occur. Thus, Fig. 4 in the paper by
Levinson and Hyatt suggests a distinct change in slope near 16° to 18°C
for the germination temperature with a relatively abrupt peak or change
in slope near 28° to 32°C. The authors felt these changes were not significant but offer little additional information to substantiate this conclusion. The authors further note "there was some suggestion of a sharp
increase in germinability after heating at 56°C. However, as seen in the
semi-log plot (Figure 1) activation appeared to be an exponential function of activation temperatures from 52 to 60°C." The authors do not
point out, however, that above this temperature, the change in optical
density is practically constant over the range from approximately 62° to
78°C; again revealing a rather notable change in the vicinity of 60° to
62°C. The point intended here is not that the study by Levinson and
Hyatt provides strong evidence for the existence of thermal anomalies in
biological systems. Instead, it is merely emphasized that the findings by
these authors are not inconsistent with the notion of the occurrence of
thermal anomalies and that no other current theory for germination and
growth of spores is likely to predict the shape of the observed curves.

It is interesting that Levinson and Hyatt (1970) quote Thorley and
Wolf (1961) as having observed three temperature optima (near 3°, 25°,
and 41°C) for the germination of *Bacillus cereus* strain T. spores.
Levinson and Hyatt go on to "explain" that the multiple optima were attributed by O'Conner and Halvorson (1961) to the use of a suboptimal
concentration of L-alanine. In other words, in the presence of sufficient
L-alanine there is no evidence of thermal anomalies in the response of the
spores to temperatures of germination. It appears that Levinson and
Hyatt have, indeed, missed the point: as stressed by Oppenheimer and
Drost-Hansen (1960) and by Schmidt and Drost-Hansen (1961), it is on
minimal substrates that the multiple optima are to be expected. The fact
that the anomalies can be "swamped" by excess nutrient supply does not
explain away the nature of the growth on minimal media. In the latter
cases, limitations are imposed upon the organisms with respect to the
available metabolic pathways and the choice is limited, therefore, with the
result that the metabolites and/or appropriate enzymes are only those that
are most compatible with the structure of the vicinal water in the respective temperature ranges.

Observations of interesting anomalies around 15° (to 20°C) have been reported by Nishiyama. Because many of the papers by Nishiyama and co-workers as well as other Japanese authors are not available in English translation, we mention a number of these studies in some detail, based on a recent personal communication to the present author from Nishiyama.

Nishiyama has been concerned with the effects of relatively low temperatures on a number of plant phenomena (Nishiyama, 1969, 1970). In the most recent article, "What is Between 15° and 20°C?" Nishiyama suggests that general physiological (and pathological) changes occur in the temperature range between 15° and 20°C. An inspection of the illustrations in this article suggests to the present author that the rate of these changes is frequently the greatest around 15° to 17°C. Nishiyama (1969) specifically proposes that the changes may be due to "the phase transition point of water (crystal) in protoplasm." Further, Nishiyama (1970) has suggested "various crops are injured by low temperatures—below 15 to 20°C. One such example is a sterile type injury in rice plants (Figure 1-A in our report, Nishiyama, 1969)." Although Nishiyama clearly recognizes the importance of the role of water and the possibility that it may undergo some type of phase transition, he also draws attention to the fact that "it is to be noticed that the critical temperature varies with varieties and conditions of cultivation. We must consider the participation of protoplasmic substances, such as proteins and lipids, other than the water itself." Nishiyama goes on to mention cold injury to plants discussed by other Japanese researchers. Thus, injury to soybeans and red beans, and a type of delayed injury in rice plants, is observed for low temperatures, that is, temperatures below 15° (to 20°C).

Nishiyama has added several other examples in support of thermal anomalies in plant physiology and pathology. Thus, he states:

> Dr. Yamashita et al. claim that there is a changeover temperature for day length requirement [in "Control of Plant Flowering" (Y. Goto, ed.), pp. 54–57. Yokendo, Tokyo, 1968 (in Japanese with an English summary)]. The temperature was estimated about 17.5°C in several plant species. Various fruits and vegetables after harvest are susceptible to cooling below and about 15°C. These include bananas [T. Murata, Plant Physiol. **22**, 401 (1969)], oranges and lemons [I. L. Eaks, Plant Physiol. **35**, 632 (1960)], apples [A. C. Hulme et al., J. Sci. Food Agr. **15**, 303 (1964)], cucumbers [I. L. Eaks and L. L. Morris, Plant Physiol. **31**, 308 (1956)], cucumbers and pimentos [L. L. Morris and Platenius, Proc. Amer. Soc. Hort. Sci. **36**, 609 (1938)] and sweet potatoes [T. Minamikawa et al., Plant Physiol. **2**, 301 (1961)].

3. Thermal Classifications of Microorganisms

The traditional classification of bacteria into cryophiles, mesophiles, and thermophiles may possibly be seen as a tendency for these groups of organisms to exhibit maximum activity (usually optimal growth) between various consecutive thermal anomalies in the vicinal water. It should be mentioned in this connection that one of the difficulties in making a clear-cut distinction results from the fact that multiple temperature optima are often encountered. Thus, growth curves over an extended temperature interval may show merely a broad and, at times, rather flat peak around 30°C! The studies by Schmidt and Drost-Hansen (1961) have suggested that this may result from considerable overlap of two growth peaks (each with optima near 23° to 25° and 37° to 39°C, respectively). Experimentally, we have noted that growth on "minimal media" tends to separate the overlapping peaks. Likewise, distinctly binodal growth curves are sometimes seen in very old cultures—long after the cessation of the logarithmic growth phase.

4. Thermal Conduction in Biological Systems

In connection with the problem of metabolic processes, the question arises as to how the cell dissipates the heat produced in the cellular processes. Naturally, the component of the cell water which is more or less bulk-like will have limited thermal conductivity (but rather high heat capacity). However, once a steady state has been reached, the heat evolved must be dissipated to maintain isothermal conditions in the homeothermic organisms and in the poikilotherm organisms in "equilibrium" with the surroundings. In this connection, recall that the heat conductivity of ice is almost an order of magnitude greater than the heat conductivity of bulk water. It seems eminently reasonable to suggest that the ordered water of the cell interface facilitates the conduction of heat from the interior of the cell to the surroundings. Heat conductivity studies of water between closely spaced mica plates have been carried out in Russia by Metsik and Aidanova (1966; also see Derjaguin, 1965). These studies demonstrated notably enhanced heat conductivity of vicinal water—as much as an increase by 50 or more (for thicknesses less than 0.1 μ) (see, also, Section VI,F,2 on hyperthermia).

It is interesting to speculate that shivering may reduce the amount of ordered, structured water, somewhat similar to the breakdown, on agitation, of "set gels." In this fashion, the heat conductivity of the cellular water might be reduced and thus minimize heat flow to the environment upon cold exposure.

5. Notes

As suggested by Oppenheimer and Drost-Hansen (1960), temperature adaptation may, indeed, take place. Thus, some bacteria have definite binodal distributions of growth rates as a function of temperature, with a minimum near 30°C. As mentioned, similar results were obtained on a mutant of *Neurospora crassa*. The tentative proposal, by Oppenheimer and the present author, is that in different temperature intervals those different metabolic pathways are chosen which are best "suited" for the organism at that temperature interval. Adaptation then may, in part, be the proper choice of the substrate on which the organism is grown. Characteristically, for bacterial studies, one gets the qualitative feeling that thermal anomalies are enhanced when the organisms are maintained on a "minimal medium." This would imply that the organism does not have the normal "availability" of metabolic possibilities. It is also possible that genuine adaptation—resulting from modification of, say, the controlling protein structure—may be achieved through adaptation to a different water structure in a different temperature interval. However, this is not a likely possibility and is certainly not easily achieved. Hence, as will be discussed in the section on paleozoogeography, it is undoubtedly correct to say that the thermal anomalies at 15° and 30° (and perhaps 45°C) have, in the past, imposed a significant "barrier" leading to geographical zonation of multicellular organisms, dependent on the local average (or maximum) temperature. Thus, in a sense, throughout evolution the water structure changes have imposed inviolable, "invariant" constraints. Stehli, among others, has invoked this possibility in connection with paleozoogeographic studies (see Section VI,D,3).

As discussed in the present section, it appears that such phenomena as body temperature of mammals, optimal temperatures for many organisms, as well as maximum and lethal temperatures are determined by structural changes in water (Drost-Hansen, 1956). Later (Drost-Hansen, 1965a), it was more specifically noted that the interaction between the vicinal water structure and the nature and conformation of the underlying substrate is the result of mutual interactions:

> The cooperative action between many water molecules in the water clusters of the solvent water may well be expected to influence drastically the rather large amount of water associated with the proteins or membrane material. In other words, the structural transitions in water may exert a direct and profound influence on the immediate environment of the macromolecules of the biologic systems; the effects of the transitions are not merely "solvent effects" manifested by minute changes in the solvent viscosity, dielectric properties or activity!

C. Germination

A vast amount of literature exists on the subject of germination (and vernalization). It is interesting that these studies have often considered the effects of temperature in some detail. However, as in a number of other fields in biology, such as thermal adaptation, a vast number of complications occur due to other concomitant changes, such as changes in relative humidity (water activity), light, and pressure. Hence, with the exception of a relatively small number of studies, it is difficult or impossible to make significant systematic comparisons between the structure (and thermal anomalies) of vicinal water and the specific effects on the processes of germination and vernalization. Obviously, the study of the influence of water structure on these processes is further complicated by the fact that frequently the systems have not been studied as a function of temperature at closely spaced intervals, and the systems are, in addition, sensitive to various electrolytes and nonelectrolytes, which undoubtedly exert specific influences through direct chemical interaction, for instance, with singular functional groups in some controlling enzyme or at some membrane site.

A few examples of abrupt changes in germination rates with temperature were discussed by Langridge and McWilliam in Rose's monograph (1967, p. 244). The authors state ". . . the optimum temperatures for the germination of most seeds fall between 15° and 30°C, although higher optima (35° to 40°C) have been reported in tropical species, such as *Paspalum* and *Saccharum*." Also (Rose, 1967, p. 26), ". . . similarly, it has been shown that potato tubers immediately after harvest are able to sprout only within a narrow temperature range above 30°, which presumably protects them from premature sprouting in the autumn."

P. A. Thompson (1969, 1970a,b) has studied the germination of seeds in considerable detail by a variety of techniques. He has made use of a thermogradient bar—a polythermostat somewhat similar to the one used in the studies by Oppenheimer and Drost-Hansen (1960). In some cases, the results are in excellent qualitative agreement with similar results obtained for the rate of growth of a number of bacteria showing abrupt changes near 15° and 30°C, for instance, for *Silene tartarica* and *Silene coeli-rosea*. In other cases, vastly different temperature responses were obtained in the sense that critical temperatures occurred, for instance, near 25° or 36°C depending on the geographical origin (and, hence, climate) of the plants studied. There is little doubt that the study by Thompson will prove a most important contribution. He observed that a temperature difference as small as 0.5° may play a discriminating role in the germination of seeds. Thompson also introduced alternating temperatures in order to study an environment more nearly identical to that encountered in nature. Furthermore, multiple growth optima were also encountered on occasion. Thus, for *Fragaria vesca* Linn and, particularly, *Ajuga reptans* Linn, multiple growth optima were observed with minima near 30°C. At this point, attention is called in particular to Fig. 22 showing the percentage of germination curves as a function of temperature of two different species of *Fragaria vesca* Linn. The germination curves as a function of temperature clearly exhibit binodal character with relative minima between 26° and 29°C. This behavior strongly resembles the multiple temperature optima obtained for the growth of a number of bacteria studied by Drost-Hansen and co-workers (Oppenheimer and Drost-Hansen, 1960; Schmidt and Drost-Hansen, 1961).

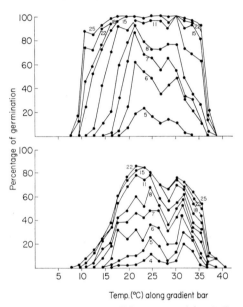

FIG. 22. Percent germination for two different species of *Fragaria vesca* Linn. (P. A. Thompson, 1970b, with permission.)

Finally, attention is called to some studies of the effects of gibberellins on the germination of some seeds. Thompson (1969) concludes

There would appear to be no requirement for a close relationship between responding species in taxonomic terms, nor is it easy to find any similarities from one to another, suggesting a common bond in terms of the conditions required for germination in normal circumstances. Gibberellins will substitute for light in dark-grown seed, for chilling treatments, and for fluctuating temperatures: they will procure germination at temperatures normally too high and also temperatures too low; and they will replace complex conditions for germination such as the combination of leaching and chilling required by the seed of *Meconopsi cambrica*, and the combination of light and fluctuating temperatures required by *Lycopus europaeus*.

The fact that the gibberellins may act in such a variety of ways and mimic such vastly different functions suggests to the present author that their effect is not based on a specific chemical reaction, such as interaction with one particular functional group in a controlling enzyme or substrate. It is suggested, therefore, as an alternative hypothesis, that the effect is due to some general influence and this most likely is through the action of the gibberellins on the structure of the water vicinal to the site of control of dormancy in the seed.

1. *Vernalization*

Vernalization is the induction of seeds to germinate after (often prolonged) exposure to low temperature. The subject is obviously of enormous practical importance. The need for prior cooling, before germination can take place, is the principal means whereby freshly discharged seeds from a plant are prevented from germinating upon release in the autumn which would expose the young plant to the cold of winter.

It is proposed here that vernalization is the relatively slow restructuring (and probably the increased ordering) of water adjacent to some critical component in the seed, probably a membrane or a protein. Only after the vicinal water structure has changed to conform to the lower-temperature range is the seed latently capable of germinating. Recall in this connection that, whereas the substrate undoubtedly influences the nature of the vicinal water, the converse must also hold true, namely, that the structure of the vicinal water must influence the nature and conformation of the underlying macromolecular substrate. That the process is slow is perhaps related to the thermal memory effect discussed in Section VI,J, probably reflecting the difficulty in inducing order by merely removing available thermal energy (thermal energy [kT] tends to disorder structures and, conversely, a lowering in temperature increases the ordering).

Recently, Levitt (1969) has surveyed the growth and survival of plants at extreme temperatures and presented a unified concept. Levitt proposes that temperature exerts a controlling role in the response of plants through the state of denaturation of the proteins. Specifically, he proposes the simple scheme shown below for interrelations between the native (N) and the denatured (D) forms of many enzymes:

$$N \underset{\text{below 40°C}}{\overset{\text{above 40°C}}{\rightleftarrows}} D$$

and

$$N \underset{\text{above 10°C}}{\overset{\text{below 10°C}}{\rightleftarrows}} D$$

In other words, Levitt proposes that only in a range, essentially 10–40°C, do the necessary enzymes occur in a native form, and that above and below these temperatures, denaturation may occur. Denaturation by low temperature is the cold inactivation discussed in Section V,B,7,d. Levitt considers the ratio of hydrophobic to hydrophilic groups of the proteins. He specifically suggests that the properties will not only depend on conformation but also on the proportion of amino acids to hydrophobic groups of the nonpolar side chains in the proteins.

Levitt leans heavily on the previous studies referred to in this chapter by Brandts regarding the relationship between water structure and the nature of the protein. Levitt directly and explicitly invokes the role of the water structure in conformational stability of the proteins which, in turn, are seen as controlling factors in both cold and heat resistance of plants. In discussing chymotrypsinogen and chymotrypsin, Levitt notes that these proteins differ primarily only in the conformation in the vicinity of SS groups. He mentions that in one conformation the SS groups are "protected" (from the aqueous environment), whereas in the other conformation such protection does not occur. Levitt further discusses the role of water structure:

> In other words, during the hardening period there must be a nearly complete turnover of proteins. This means that each newly formed protein chain must fold at a temperature which prevents the formation of hydrophobic bonds. They must all therefore fold with the hydrophilic, reactive groups within the folds and the hydrophobic, weakly reactive groups on the outside. All the newly formed proteins must therefore be resistant to aggregation and the plant can therefore undergo freezing dehydration without injury.

The present author is greatly impressed with the qualitative ideas embodied in the proposals by Levitt for the survival of plants at both low and high temperatures. However, as mentioned elsewhere in this chapter, he takes issue with the classic explanation for the existence of abruptness in various rate processes (those involving proteins). Indeed, Levitt illustrates the Arrhenius nonlinearity by quoting a study by Talma (1918) for the rate of growth of roots of *Lepidium sativum*. This process shows extremely sharp changes at exactly $15°$ and $30°C$—changes in slope far more pronounced and far sharper than those which can be allowed for through the Arrhenius plot for, say, enzymically controlled (simultaneous and/or consecutive) reactions, as discussed by Brandts and others (see Section V,B,7,b).

D. GENETICS AND EVOLUTION

1. *Adaptation to Thermal Changes*

The problem of thermal adaptation in organisms is of enormous importance: it plays a prominent role in ecology and paleozoogeography, in thermal pollution, in the effects of fever on pyrogenic organisms, and in many other areas of immediate, biological interest. In the present chapter only one facet of thermal adaptation is discussed. Although thermal adaptation may, for instance, result in the ability to change the optimum temperature for various types of activity (growth, mobility, etc.) by several degrees, we stress here that frequently lethal temperatures are invariants. Thus, those lethal boundaries which are determined by changes in the associated water structure are probably time invariants, boundaries not likely to be transgressed easily—by any mechanism of thermal adaptation —without some very major changes in the biochemistry (see below).

Of interest to the biologist are the intact organisms; vastly different temperature effects may be expected at various points in the evolutionary states of the organism. Ushakov (1968) has pointed to the need to separate the temperature effects on the total organism from the temperature effects on the individual cellular level and, indeed, the role of temperature at the molecular biochemistry level. Considering the role of cells and proteins in the process of adaptation, one must take into account that any adaptation is a result of natural selection. The object of selection is the intact organism and, hence, adaptations are accomplished only at organismal and epiorganismal levels of organization. From this point of view, the term "cellular adaptation" may be applied only to unicellular or multicellular organisms at early stages of ontogenesis when the organisms pass through oligocellular stages of development.

It is the contention of the present writer that thermal adaptation can, indeed, occur, but that adaptation which significantly extends the range of biological functioning is very limited, if it exists at all. On the unicellular level, multiple growth optima are well known (discussed in Section VI,B,1). In higher organisms, however, the limiting crucial temperature-sensitive element (processes) may not be related to metabolism but rather to other phenomena, such as membrane transport and/or nerve activity. Hence, it is unlikely that intrinsic thermal boundaries may be significantly altered through the adaptation process, as this would probably require the emergence of organisms with completely different sets of proteins, enzymes, and lipids—with the ability to be compatible with a completely different vicinal water structure in higher (or lower) temperature intervals (that is, on either side of the boundary of the thermal transition ranges). However, one must allow for the possibility that a slightly higher thermal temperature limit may be achieved through suitable modification of the substrate, resulting in increased stability of the water structure "lattice." The stabilization may derive from slight modifications in the substrate conformation or change in (nonfunctional) nonpolar side groups or through the introduction of low molecular solutes which may become available to stabilize the vicinal water structure.

Important contributions to the problem of adaptation of organisms to different environmental temperatures have been made by Hochachka and Somero (1968). These authors have studied the adaptation of enzymes to temperature; the enzymes were obtained from a number of different fish the natural habitat of which ranged from the Arctic (*Trematomus borchgrevinki*) to the Tropics [the South American lungfish (*Lepidosiren paradoxa*) adapted to living in the tropical swamp waters of Brazil]. The intermediate temperature ranges of fishes were represented by various brook and lake trouts (salmonids) and tuna.

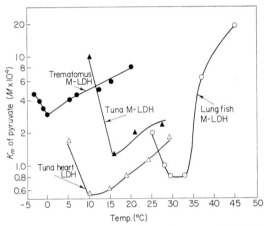

Fig. 23. Apparent Michaelis constant, K_m, of pyruvate lactate dehydrogenase (LDH) homopolymers from various fishes. (Hochachka and Somero, 1968, with permission.)

Figure 23 shows the effects of temperature on the apparent Michaelis constant, K_m, of pyruvate lactate dehydrogenase (LDH) homopolymers from the lungfish muscle, tuna heart and muscle, and muscle from *Trematomus*. In all cases notable minima in the Michaelis constant are exhibited near the temperature for optimum activity for the three fishes examined (corresponding to a maximum value for LDH activity). Hochachka and Somero (1968) discussed the possible mechanism for both long- and short-term adaptation to temperature. They conclude:

> Our data on LDH from species facing a variety of thermal environments demonstrate that the above parameters (activation energies, thermal optima for maximum velocities, and heat denaturation) need not correlate with environmental temperature. Rather, in the adaptation of enzymes to temperature, the parameter most sensitive to selective pressure appears to be enzyme affinity

for ligands. Thus, the sensitivity of a given reaction to temperature can be minimized by compensatory changes in enzyme affinity for substrate, as in our study, or from modulators, as suggested by other work (Ingraham and Maaløe, 1967). In this connection, it is interesting that this mechanism apparently also operates in short-term acclimatization. At least two kinds of control processes are involved in acclimation (Hochachka, 1967; Smith, 1967): (1) control of the level of enzyme types already present in the cell, and (2) control of the level of enzyme variants (LDH isozymes, in our study) uniquely suited to function at particular environmental temperatures. Such enzyme variants, "induced" during short-term acclimation appear to be selected during evolutionary adaptation to temperature.

For another interesting example of enzyme kinetics and adaptation, see the recent article by Somero (1969). The Michaelis constant for two forms of pyruvate kinase was measured for, respectively, a cold- and a warm-adapted type of enzyme. The cold pyruvate kinase showed a sharp increase in the Michaelis constant at about 10° to 12°C, whereas the warm form of the enzyme showed a notable minimum around 12°C. As stressed elsewhere in the present chapter, a correlation is often found between the degree of abruptness in response to pH changes of the enzyme (or protein) and the degree to which these macromolecules exhibit thermal anomalies. In the examples studied by Somero, a remarkably sharp peak is observed in the activity of the enzyme between pH values of approximately 5.5 and 5.8—the resultant curve looking almost like a Dirac delta function.

The thermal properties suggest that the two types of the enzyme are created by an interconversion dependent on the temperature. This interconversion has adaptive significance: as the temperature is lowered, the warm enzyme is converted into the cold enzyme; the opposite situation obtains when the temperature is raised. Interestingly, Somero notes that "the two variants of the pyruvate kinase do not appear to be isoenzymes in the conventional sense. Electrophoretic and electrofocus analysis revealed only single peaks of activity." These findings may support the suggestion made earlier by Oppenheimer and the present author that adaptation (at least in the case of unicellular organisms) may consist of the selection of the proper metabolic routes, each consistent with the predominant water structure in the different temperature intervals. In the case discussed for the Alaskan king crab (Somero) and the trout (J. Baldwin and Hochachka (1970); see below), the adaptation may consist of the selection of the proper enzyme in the metabolic processes. Somero's observation that the isoenzymes are not "isoenzymes in the conventional sense" may mean only that their identical behavior in electrophoresis and electrofocus analysis reveal that differences exist only in the hydration structure rather than in the composition (or, in other words, rather than differences in specific functional groups).

Another interesting study of the temperature acclimatization of iso-enzymes was reported recently by J. Baldwin and Hochachka (1970). Two distinct forms of acetylcholinesterase was found in the brain of trout. One of these forms (the warm variant) was obtained in trout acclimatized at 17°C, whereas the cold variant occurred after acclimatization at 2°C. Acclimatization at intermediate temperatures resulted in the occurrence of both types of enzymes.

Baldwin and Hochachka found that the acetylcholinesterase from the trout acclimatized at 17°C showed a relatively sharp minimum in the Michaelis constant near 17°C, whereas the cold-adapted trout showed a broad minimum around 2°C—increasing rather rapidly above 10° to 12°C. The authors concluded, "In evolutionary terms, it appears that there is a strong selection for enzymes permitting large changes in activity in response to physiological changes in the substrate concentrations. This is reflected in the pattern of enzyme variants produced during acclimatization and those selected during evolutionary adaptation." The results of the study by Baldwin and Hochachka and the results by Somero are interesting when put in context. Thus, Somero (1969) has concluded that "the results suggests that the 'warm' pyruvate kinase and the 'cold' pyruvate kinase are formed by a temperature-dependent interconversion of one protein species."

In the introduction to this section on adaptation to thermal changes, it was contended that adaptation, which significantly extends the permissible range of the biological functioning, is very limited if it exists at all. The important and, indeed, valuable studies by Hochachka and Somero and co-workers suggest that acclimatization does occur, but the present author has not become convinced that this type of example of adaptation ultimately leads to the ability of the organism to live optimally over wide ranges of temperature. Although the trout or the king crab may develop slightly modified enzymes to cope with environmental temperatures from 2° to 17°C, it remains to be proved that this will allow these species to live and reproduce optimally over that wide a temperature range. The zoological stratification described, for instance, by Jones and quoted by the present author (Drost-Hansen, 1965a) suggests, as an example, the almost complete absence of commercially valuable fishes in the South Pacific below the 15° isotherm; see, also, the discussion of thermal pollution, Section VI,K, and particularly the review by DeSylva (1969). In conclusion, therefore, it appears that thermal boundaries set by the changes in the structure of vicinal water are not readily transgressed.

In connection with the study by Hochachka and Somero (1968) of the enzyme adaptations to various climatic environments, compare also the article by Cowey et al. (1969). These authors studied the LDH from cardiac and skeletal muscles of plaice (*Pleuronectes platessa*). They point out that the behavior of the isoenzymes of cardiac and skeletal muscles is indistinguishable as determined by the effects of urea. However, it is possible to distinguish between avian and mammalian heart and muscle LDH on the basis of temperature stability as the heart enzyme is more thermally stable than the muscle enzyme. Compare this with the finding by Hochachka and Somero that the enzymes of the tuna heart appear to have a slightly lower temperature of maximum activity than the muscle enzyme.

It is proposed that the isozyme best suited to function at a particular environmental temperature is selected from those isozymes for which the changes in vicinal water structure is most compatible with the average

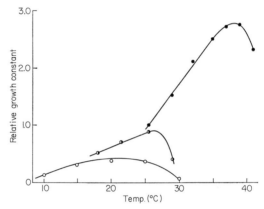

Fig. 24. Growth of three different strains of *Chlorella* as a function of temperature. (Fogg, 1969, with permission.)

environmental temperature. Very likely, only very minute changes in the specific nature of the amino acid sequences may determine the nature and degree of the vicinal water structure of the enzymes (as discussed in connection with the study by Glasel, Section V,B,5).

With respect to adaptation, Fogg (1969) has presented an interesting discussion (see Fig. 24). This illustration shows the growth of three different strains of *Chlorella* as a function of temperature. The two curves shown with maxima below and above 30°C are, respectively, two species of cold water strains—*Chlorella pyrenoidose* from Sweden and the Emmerson strain of same organism. The organism with optimum for relative growth constant above 30°C is a high temperature strain, *Chlorella sorokiniana*. This example clearly shows that organisms of the same genus (*Chlorella*) may adapt to vastly different temperature optima. However, note that for the cold water strains the maximum (lethal) temperature occurs near 30°C, whereas the high-temperature strain appears to decrease in activity above 37°C with a notable drop at 40°C.

For a far more general discussion of adaptation to temperature changes, see the monograph "Molecular Mechanisms of Temperature Adaptation," edited by Prosser (1967).

Finally, mention is made of an interesting study of thermal adaptation of enzymes, reported by Licht (1967). The study of Licht is impressive in that it provides a very careful study of enzyme activities observed at very closely spaced temperature intervals (the enzymes obtained from lizards). The temperature responses in the article, however, are not all readily interpreted in terms of the notion of water structure changes discussed in the present chapter.

An interesting discussion on the limits of microbial existence was presented by Skinner (1968). In addition to delineating in a descriptive fashion the extremes for limits of microbial existence, Skinner also discusses a few aspects of adaptation and mechanisms of tolerance. With regard to life at high temperatures, Skinner mentions two possible mechanisms, namely, that thermophilic organisms operate on biomolecules which are intrinsically thermally more stable than those of the mesophiles or that the protein of the thermophiles is somehow protected against denaturation. Finally, proteins of thermophiles may be denatured as readily as the proteins of the mesophiles, but the thermophilic organisms may be able to "repair the damage" more readily. At the other end of the spectrum, Skinner has described bacteria that may live in strongly saline pools in Antarctica at approximately −23°C, while a flagellate alga (*Dunaliella salina*) has been observed in brine pools at −15°C.

Skinner has also dealt briefly with barophiles (ZoBell's strain of a thermophilic, reducing bacteria, which did not grow at 85°C at atmospheric pressure, but was able to grow at 104°C under a pressure of 1000 atm), halophiles, osmophiles, and microorganisms with the ability to live in acids with pH values of around 0.5 (although the internal pH, even in these cells, is about 6 or 7). Skinner also mentions a blue-green alga able to grow at pH 13. Both the halophiles, osmophiles, and proto- and hydroxy-bacteria are interesting in the requirements for structural stability imposed on the cell membrane to retain a "reasonable" internal water environment.

2. Rate of Chromosome Aberration

A remarkable example of abrupt effects of temperature on a biological system was described by Wersuhn (1967) who studied the rate of chromosome aberration as a function of temperature during microsporogenesis.

The plant used for this study was the broadleaf bean, *Vicia faba*. Eleven different temperatures were employed between 19° and 35°C, and the results are shown in Fig. 25. It is seen that near 30°C the rate of chromosome disordering reaches a sharp peak. Wersuhn points out that each datum point represents 200 counts on each of five plants and that the notable anomaly at 30°C is statistically significant! Wersuhn also notes that this result agrees with the findings of Chira who apparently observed an increase in the disturbance of sporogenesis near 30°C for *Taxus*.

In connection with the spontaneous chromosome aberration discussed here the tentative analysis of the data in Section III,C regarding the properties of vicinal water near 30°C should also be noted. It was proposed specifically that below this temperature range, one form of vicinal water

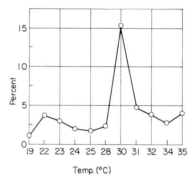

Fig. 25. Rate of chromosome aberrations as function of temperature during microsporogenesis. (Wersuhn, 1967, with permission.)

was stable, whereas above, say, 30° to 32°C, another form was preferred. In the transition, elements of both structures might coexist, together with a number of smaller, kinetic entities, possibly monomeric water molecules (associated with the large entropy of surface formation noted around 29° to 32°C). Comparing these results with the results by Wersuhn, it is proposed that near 30°C the vicinal water structure is highly disturbed and that in the process of cell division, the stabilizing influence of the vicinal water on the RNA–DNA genetic transfer system is disturbed. Thus, the genetic message in this temperature range becomes "scrambled." This problem will be referred to again briefly in Section VI,K dealing with thermal pollution effects.

Regarding the rate of mutation, see the example quoted by Farrell and Rose (1967, p. 197, Fig. 17). The results (quoted from Ogur *et al.*, 1960) clearly show a very dramatic increase of rate of mutation above approximately 30°C. Indeed, the increase is approximately fifty-fold over the average (relatively constant) value attained between 15° and 30°C.

3. Paleozoogeography

No doubt the characteristic properties of vicinal water are truly invariant properties with respect to time. Thus, in the past, as now, changes must have occurred in the structure of vicinal water at the temperatures where thermal anomalies now are observed. Hence, it is probably correct to conclude that in the processs of evolution, the temperatures of thermal anomalies have presented "invariant boundaries" which in general it has not been possible for multicellular organisms to transgress. It is known, of course, that thermal adaptations can occur. As discussed in Section VI,D,1, however, these changes are often matters of only a few degrees or even less; whereas the occurrence of a multiple temperature optima suggests that, at least in lower forms of life—especially for unicellular organisms—alternative ways have been developed to cope with temperature variations over more extensive ranges.

Stehli (1957) has used the time invariant property of the thermal anomalies in vicinal water in a study of paleozoogeographic stratification, arguing that present-day species with well-delineated upper and lower lethal temperatures are probably not significantly different from earlier forms of the same organisms. When certain fossil species from the same geological period occur in various geographical localities, it is likely that this reveals that the same average thermal environment must have prevailed there as that which is presently conducive to these particular organisms. This has been further elaborated upon in Ager's monograph "Principles of Paleoecology" (1963).

Ager discusses the distribution of various faunas in the past geological periods. Thus, as Stehli has suggested, Ager notes that the 15° winter isotherm in Permian times corresponds roughly with the 55° northern latitude and goes on to point out that this agrees qualitatively very well with other faunal distributions. Incidentally, both Stehli and Ager seem to agree that the distributions suggest a climatic zonation parallel to the present equator and thus do not support hypotheses of polar wandering. Ager also discusses the 30°C anomaly. An interesting point in this connection is raised by Ager, namely, the fact that the temperature of the open sea presently rarely exceeds 30° to 35°C. In the past a "hypertropical belt" with a distinctive fauna might have existed. This warm belt may conceivably have disappeared as the climate toward the beginning af the Tertiary period began to turn cold and this, in turn, account for the extinction of a number of groups of organisms. Ager quotes Cailleux who calls attention to the fact that the absence of Paleozoic reef corals in the tropics may be due to their inability to tolerate unusually high temperatures.

E. NARCOSIS

1. *Pauling and Miller's Theories of Anesthesia*

The theories of anesthesia by Pauling (1961) and by S. L. Miller (1961) imply that hydrates form in the water of the neurons (and around the neuron network)—both in the cells of the neuron and in the synaptic regions. These structures represent obstacles in some of the electrically charged chains. Thus, they interfere "as sand grains in small gears" with the movement of ions in the synapsis and in the neurons. This, in turn, enhances the impedance. It is probably worth pointing out that the picture is somewhat crude. However, as we already discussed, one need not invoke the existence of stable solid crystalline forms of material (of the kind most to the liking of the solid crystallographer), but rather merely partially induced and stabilized "fluctuating" structures. On the average, it is probably necessary only to stabilize the half-life of such structured units by an order of magnitude (or perhaps a few orders of magnitude) rather than attempt to invoke any microcrystalline entities. However, this is probably mostly a matter of semantics, although it is far from a trivial point.* For instance, chloroform hydrates may be formed readily from an emulsion of chloroform in water, but this hydrate is stable only below 1.5°C. Thus, "hydrate crystals" are very likely not present in the brain at 37°C.

S. L. Miller (1968) observed that upon lowering temperature to, say, 27°C, "... there might be sufficient icelike water present owing to the protein side chains to make the hydrate lattice stable in the absence of the anesthetic gas. This could be responsible for the anesthesia observed on lowering the temperature of warm-blooded animals (hypothermia)."

The idea that the gaseous, low molecular weight, clathrate hydrate formers may be simulated in function by the side chains of various amino acids in proteins was discussed by Klotz (1965). Thus, Klotz compares the functioning of methane to alanine, propane to valine, isobutane to leucine, and mercaptan to cystine, etc. Compare mixed gas hydrates and the facilitated formation of such hydrates in the presence of a "help gas" (Section IV,C,4).

For a lively discussion of molecular aspects of the pharmacology of anesthesia, see Federation Proceedings, "Symposium on Molecular Pharmacology of Anesthesia" (1968). The likely role of water in anesthesia was particularly well covered in this vigorous discussion.

It is interesting to note that, if, indeed, the theories of Pauling and Miller are correct, they may provide a simple means for explaining the in-

---

* For a discussion of the idea of structure, see Section II,A,2 and Eisenberg and Kauzmann (1969).

creased intellectual acuity experienced by persons subject to a helium-rich atmosphere, such as a 20:80 oxygen–helium mixture. Presumably, intellectual activity can be sustained in such an atmosphere for as long as 20 hours per day. Ascribing some ability on the part of the nitrogen molecule to form hydrates (dissociation pressure of the nitrogen hydrate at 0°C is 160 atm) and remembering that the helium atom is too small to form a clathrate hydrate, we see that the increase in nervous activity experienced in a helium-rich atmosphere may be interpreted as the removal from the brain of the clathrate-forming nitrogen by the helium. Thus, perhaps we all suffer from a slight case of "nitrogen stupor" in the normal atmospheric environment.

In this section some aspects of anesthesia have been discussed very briefly. These aspects were concerned with molecular facets of the functioning of the anesthetic agents on the neural level. Obviously, an important consideration in the administration of gaseous anesthetics is the transfer of the agents via the lungs into the bloodstream. It seems quite natural to expect that water structure plays an important role here also; the mass transport involved (on a detailed level of description) must still be concerned with transport properties across a water–air interface and the diffusion through a water-filled (semirigid) matrix. It is not the purpose of the present chapter to discuss the mechanistic aspects of gaseous anesthetics, but attention is called to the unusual interfacial tension phenomena which must exist in the lung, and these certainly are influenced by the presence of surface-active agents, the functioning of which must, in turn, depend on their interaction with the water at the air–water interface.

Schreiner (1968) has discussed a number of biological effects of the inert gases. One of the main points made by Schreiner is that notable biological effects are observed due to the inert gases (particularly those larger than neon) at all levels of biological activity ranging from whole body (man) to individual molecular phenomena (such as enzyme processes). Among the interesting results quoted by Schreiner is the correlation between the mycelial growth response of *Neurospora crassa* and the inert gases; the study shows a well-defined linear decrease in growth rates with the square root of the molecular weight of the gas. Schreiner notes, however, that such rather inert gases as SF_6 and N_2O also affect the rate of mycelial growth in the mold but that the effect cannot be correlated with the square root of the molecular weight of these two gases. As a result, Schreiner considers the possibility that the inert gas behavior, including nitrogen, may be fortuitous. Schreiner has also discussed previous results (with Buchheit and Doebbler) suggesting that the polarizability of the inert gases correlates with the inert gas pressure required for 50% inhibition of the growth of *Neurospora crassa*. Again, it is probably true that such studies by themselves do not present a choice between the Pauling and the Miller theories of anesthesia on the one hand, and the lipid solubility hypothesis for anesthesia (Meyer-Overton) on the other hand. However, the results do clearly show not only the notable physiological effects of inert gases, particularly argon, krypton, and xenon, but also that these effects appear to a first approximation to correlate well with the clathrate hydrate-forming capabilities of the gases.

Unfortunately, most studies of the effects of inert gases on biological systems which have been reported in the past have all been done at one, or at most, a few different temperatures (see, however, Catchpool, 1966). It would be interesting, for instance, to study the rate of growth of plants

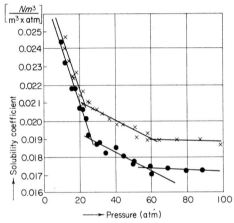

Fig. 26. Solubility of hydrogen in water as a function of pressure. (Schröder, 1969a, with permission.)

or fungi as a function of temperature at closely spaced intervals in the presence of fixed, relative pressures of oxygen and carbon dioxide for various clathrate-forming gases, including the inert gases. Thus, it is interesting to speculate whether an organism which exhibits multiple temperature optima (for a culture grown on a minimal medium) when grown under nitrogen (anaerobic) or under nitrogen–air mixtures (aerobic) might show a different temperature response in the presence of helium only (or, for the aerobic case, in a helium–oxygen atmosphere). Schreiner points out that "While it is therefore quite likely that helium group gases are not essential to life (1), we still cannot exclude with certainty the possibility that rigorous deprivation of an animal organism of a suitable non-metabolized inert gas, such as nitrogen, may not result in the eventual development of physiological abnormalities."

2. Schröder's Pressure Anomalies

Recently, Schröder (1968, 1969a) has described some unexpected anomalies in the solubility of gases in liquids. It appears that these results are destined to play a great role in our understanding of the effects of low pressures on biological systems. The results will probably be particularly interesting in connection with hyperbaric surgery, diving physiology, and

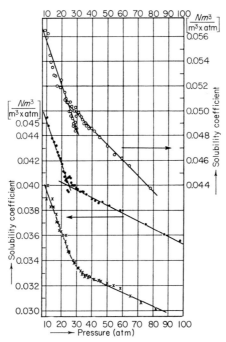

FIG. 27. Solubility of argon in water at three different temperatures (0°, 10°, and 20°C) as a function of pressure. (Schröder, 1969a, with permission.)

the physiology of those marine organisms possessing a swim bladder (or other internal, gaseous phases).

Schröder determined very precisely the solubility of a number of gases (hydrogen, methane, nitrogen, oxygen, and argon) in various liquids, particularly water and aqueous solutions. Schröder (1969b) used these results for some intriguing speculations on the void volume in liquids. It suffices here to stress the empirical results and postpone a discussion of the possible mechanism to a subsequent paper. The main finding by Schröder is that the solubility coefficient [Bunsen's solubility coefficient, $Nm^3/(m^3 \times atm)$], when plotted as a function of pressure, displays abrupt changes in slope ("kinks" or "knees"). For hydrogen in water at 20° and 40°C, the changes in slope occur at close to 22 to 24 atm and near 60 ± 2 atm. Figures 26, 27, and 28 show typical examples of this effect. Attention is called particularly to Fig. 27, which shows the solubility of argon in water, determined at 0°, 10°, and 20°C. It is seen here that the solubility coefficient possesses a sharp change in slope between two rectilinear asymptotic curve segments in the vicinity of 24 atm (at the lower two

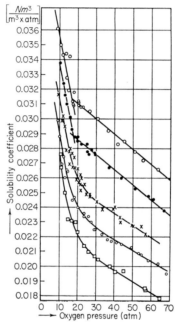

Fig. 28. Solubility of oxygen (at 25°, 30°, 40°, 50°, and 70°C) as a function of pressure. (Schröder, 1969a, with permission.)

temperatures), whereas at 20°C the curve has assumed a smooth, continuous shape. Somewhat similar results are shown in Fig. 28 for the solubility of oxygen at 25°, 30°, 40°, 50°, and 70°C, respectively. Again, a sharp knee is observed near 18 atm at lower temperatures, whereas no such anomaly is present at 40°C (or above).

Table XI shows the temperatures above which the sharp change in solubility has disappeared. This temperature is dependent upon the chemical nature of the gas. However, the pressure at which the anomaly is observed is independent of the nature of the gas. This is also evident in Table XII which shows the critical pressures for a number of gases in various solvents.

The fact that anomalies are observed with liquids as different as water and benzene tends to suggest that the phenomenon is independent of the structure of water. This is, indeed, possible as the effect may owe its origin merely to a general attribute of any liquid structure, namely, the possible existence of voids such as discrete, physical, actual "fluidized vacancies" or free volume.

It is seen in Fig. 27 that metastable conditions may exist, giving rise to two separate curve segments in the range from approximately 24 to 30

TABLE XI

CHARACTERISTIC TEMPERATURES FOR GAS SOLUBILITY ANOMALIES

Gas	Temperature (°C)
CH_4	75
N_2	45
O_2	35
Ar	10

TABLE XII

SOLUBILITY ANOMALIES IN VARIOUS SOLVENT SYSTEMS

Solvent	Boiling point at 760 torr (°C)	Gas	Temperature of experiment (°C)	π (atm)
Methanol	64.7	H_2	50	19
Benzene	80	N_2	10	23
Water	100	H_2 a.o.	20 a.o.	ca. 24 or 65
n-Butanol	117.5	H_2	50	20
Propylenecarbonate	242	H_2	50	45

atm. A metastable liquid structure is synonymous with a "memory effect" in structural properties*—a fact which undoubtedly will prove over the years to be of great significance to biological systems. Other studies have recently been carried out along the lines thermal memory effect, particularly by Bach (1971; see Section VI,J). The effect observed in the case of the gas solubility may possibly not be a thermodynamic equilibrium effect, but rather a kinetic effect.

* It should be observed that the total amount of gas dissolved in any of the systems studied may be relatively small compared to the amount of water which may have occurred as a residual impurity in the organic solvents used and, thus the anomalies may still be related to the structural characteristics of water. However, unfortunately, Schröder has not discussed in sufficient detail his experimental procedures to allow an estimate of the degree to which he was, indeed, dealing with water-free solvents.

A tempting explanation of the results obtained on aqueous systems is that the results imply the existence of discrete sites or voids for solute molecules. This would be imminently compatible with a theory of water structure based on the clathrate hydrate model. It would be particularly interesting in this connection to carry out solubility measurements as a function of pressure and temperature on solutions and suspensions of biologically interesting macromolecules and colloids. However, even in the absence of such information it is obvious that the phenomena must be of biological interest since pressures of the order of 18 to 25 atm are likely to be encountered by marine organisms such as fishes which easily reach depths far below the corresponding 180 to 250 meters. Furthermore, dives are presently being carried out to depths greater than 200 meters. In this connection, it is to be expected that both the kinetics and the equilibrium aspects of gas solubilities in blood and other body fluids may play a crucial role in decompression aspects of diving physiology. Should the hysteresis phenomenon shown in Fig. 27 turn out to be of general importance, and if it occurs with either nitrogen or oxygen, the kinetics of decompression may need to be seriously reconsidered.

Phenomena of the type discussed here show clearly the possibility of elucidating complex molecular mechanisms through simple physicochemical measurements. It is not possible to understand such changes without considering in detail the molecular structure of the liquid. Certainly, it would be difficult to understand observations of this type in terms of a continuum model for liquids; we are thus again impressed with the possibility that a clathrate hydrate model for water is essentially correct as described in this chapter to explain at least part of the observed structuring of water near biologically interesting interfaces. The fact that the critical pressures, at which the solubility regime changes from one domain to another, are independent of the chemistry of the dissolved gas lends further credence to this. At the same time, the types of lattice structure being induced or supported by the presence of the guest molecules no doubt depends on the "chemistry" of the dissolved gas, possibly through no more complicated parameter than size.

3. Deep Diving Pressure Effects

In connection with deep diving narcosis, it is of interest to speculate that clathrate hydrate formation may occur due to the presence of argon as well as nitrogen. At high total air pressures, the partial pressure of argon is also notable. Since this gas certainly is known to form clathrate compounds with a fair degree of ease, it might be conjectured that the diving narcosis phenomenon is due, in part, to argon. Hence, it would be of interest to determine if the narcosis phenomenon is as pronounced in dives where pure oxygen–nitrogen mixtures are used as in dives employing compressed air at the same total pressure.

Although deep diving physiology is currently of great interest, the amount of systematic work available that can throw light on some of the problems involved here is relatively limited. A large amount of work has been done; however, much of it has been rather sporadic. Apparently, among those concerned with deep diving narcosis, several authors appear to prefer the Meyer–Overton theory of anesthetic action, implicating lipid solubility rather than vicinal clathrate formation. With respect to the possible existence of an anomaly at approximately 20 atm, as implied in the study by Schröder, E. B. Smith (1969) mentions some effects of high pressures on mice in which tremors occur at about 50 atm. He also observed, "these tremors may be of similar origin to those observed with men in deep diving routine below 600 feet of seawater." The present author has not located any references regarding the possibility that the partial pressure of argon in compressed air (under sufficiently high total pressure) may reach levels where the anesthetic effect becomes notable. The reader is referred to the chapter by Bennett (see E. B. Smith, 1969) who, incidentally, strongly favors the Meyer–Overton hypothesis of the anesthetic action of inert gases.

F. HYPOTHERMIA

1. *Physiological Responses to Low Temperature*

Hypothermia as used in the present context is the state of reduced body temperature in mammals.

Hypothermia, as a result of exposure to cold, has obviously been seen clinically since the beginnings of medical science; only fairly recently, however, has the artificial reduction of body temperature attracted some clinical interest. Thus, major cardiovascular surgery is occasionally performed on the hypothermic patient. In addition, hypothermia has been used earlier (with apparently relatively poor success) in the treatment of cancer: the rate of proliferation of malignant cells is reduced at low temperatures (as are all other growth processes), but apparently cancer cells are only slightly more susceptible to low temperature than normal tissue. Unfortunately, the difference in susceptibility is not sufficient to allow clinical use of this sensitivity. Mild hypothermia has apparently been used in the treatment of severe eclampsia, postoperative tetany, and some mental disorders as well as toxemia, septicemia, peritonitis, and hemorrhage. It is notable that mild hypothermia usually does not involve reduction of body temperature to lower than 33° (to 35°C). Pronounced effects are observed as soon as the temperature of a mammal is reduced below 30°C (see Drost-Hansen, 1965a). Among the effects particularly seen around 30° to 32°C are loss of consciousness, rapid and drastic reduction in metabolism of various tissues, and loss of ability to restore body temperature to normal levels (when placed in a cold environment). Above approximately 33°C, physiological changes attending hypothermia are not pronounced, and control over the clinical state is good. It is interesting that the muscular activity in most animals in the form of shivering is at a maximum at temperatures of 28° to 30°C.

The effect of cold on the nervous system was particularly well delineated in a symposium "The Problem of Acute Hypothermia," edited by Starkov (1960). Table XIII, from this reference (Starkov, 1960, p. 38) shows the effects of reduced temperatures on conditioned reflexes.

The marked, abrupt changes at 30°C were further emphasized by a number of authors in the symposium on acute hypothermia. Thus, near 30°C, drastic changes in excitability of nerves, muscles, and motor centers in rabbits were described by Karpovich (1960, p. 69ff). Similar changes at 30°C were described for vasomotor centers by Klykov (1960).

J. A. Miller (1957) observed the critical nature of temperatures around 28° to 32°C and emphasized the low-temperature effects on various organs. For instance, heart action appears to be particularly sensitive to temperatures below 29°C. As mentioned in connection with the studies by Hazelwood and co-workers (see Section V,F), Miller also notes the difference between neonatal and mature organisms.

Elsewhere the present author (Drost-Hansen, 1965a) discussed the studies by Kuznetsova on gas exchange in and metabolism of hypothermic rabbits. In "deep hypothermia," more pronounced, often dramatic, and sometimes irreversible changes occur around 14° to 16°C (also see the article by Starkov, 1960). That 15°C is a critical temperature in hypothermia is also clearly emphasized by Andjus (1969) who reviewed some possible mechanisms for mammalian tolerance of low body temperature. Andjus notes (p. 361) that, in the rat, breathing and heartbeat are arrested shortly after cooling to a body temperature of 15°C. Deeper hypothermia is, indeed, possible; however, 15°C appears to be the limit for unassisted survival. Furthermore, cooling below 15°C induces changes characteristic of anaerobic conditions, whereas cooling to temperatures above 15°C is not associated with hypoxic changes, as shown by metabolic parameter measurements. Andjus shows a number of interesting graphs of various physiological responses as a function of temperature, some of which also suggest the criticality of 15°C.

The effects of cold upon the central nervous system and muscle activity are probably the direct and proximate cause of other significant changes seen near the temperatures of thermal anomalies (particularly at 30°C and, to a lesser extent, at 15°C, i.e., deep hypothermia). Prokop'eva (1960) notes the effects on the rate of blood flow. In initial stages of hypo-

TABLE XIII

EFFECT OF VARIOUS BODY TEMPERATURES ON CONDITIONED REFLEXES[a]

Experiment	Temperature at which the conditioned reflex disappeared during overcooling (°C)	Temperature at which the conditioned reflex appeared during warming (°C)
Bobik		
No. 1	27	30
No. 2	29	32
No. 3	30	(Only after repeated combinations with short intervals)
Kashtanka		
No. 1	29	32
No. 2	28	29
No. 3	30	—
Pestrukha		
No. 1	30	Overcooling to 25°C (conditioned reflex reaction
No. 2	30	not restored when temperature increased)
No. 3	30	
No. 4	30.5	(35°C only after repeated combinations with short intervals)
		31.5
		35
No. 5	29	Conditioned reflex could not be produced when
No. 6	30	temperature increased, even by combinations
No. 7	29	with short intervals
		35
Ryzhik		
No. 1	33	(Only after repeated combinations with short intervals)

[a] From Karpovich (1960).

thermia, the rate of blood flow increases when the body temperature is between 32° and 33°C, but the reduction of temperature by about 1° below 32°C results in a notable decrease in blood flow. Another example of the effects of cold is discussed by Kuznetsova (1960, p. 257ff); he notes a sharp maximum near 30°C in, for instance, the amount of carbon dioxide excreted or the rate of metabolism in the hypothermic rabbit.

Cardiac and cardiovascular surgery has been attempted frequently on hypothermic patients. This procedure is associated with considerable danger of ventricular fibrillation. Intravenous glucose transfusions, administered to the hypothermic patient, have been found to lower the incidence of ventricular fibrillation considerably (Virtue and Burnett, 1958, personal communication). The glucose concentrations used have often been remarkably high. It is of interest to consider some possible mechanisms by which the glucose may prevent ventricular fibrillation. It has been shown (Angelakos et al., 1957) that dogs rendered hypocalcemic by administration of sodium ethylenediaminetetraacetic acid (Na-EDTA) (which forms a complex with calcium ions) were able to tolerate cooling to 14 to 18°C without the occurrence of ventricular fibrillation. Conversely, increased calcium ion activity (effected by intravenous administration of calcium) resulted in fatal ventricular fibrillation, even at temperatures as high as 22 to 27°C. It also is of interest to note that Binet (1957) reported that an increase of calcium in the plasma of the rabbit from 11 mg% to 30–40 mg% results in hyperglycemia (for ambient temperatures of 6 or 28°C). This would suggest that the increased calcium ion concentration evokes an increase in glucose concentration; the effect may possibly be a specific protective mechanism.

The interplay between calcium ions and glucose in physiological functioning appears to be complex; it seems possible that the glucose may directly complex with the calcium ion (similar to the effect of EDTA). This possibility seems reasonable and it is of interest to note that solubility data for calcium hydroxide in aqueous glucose solutions show a notable increase in dissolved $Ca(OH)_2$ with increasing glucose concentration (data by Balezin; see Seidell and Linke, 1952). To investigate the possibility of calcium ions complexing with glucose, the present author initiated in 1966 some measurements of calcium ion activities in glucose solutions (Thorhaug, 1967) using the then available specific calcium ion electrodes [manufactured by Orion and by Corning]. Unfortunately, the functioning of these electrodes appeared to be very sensitive to the presence of glucose in solution and no definite results were obtained except to note that no direct evidence was obtained for strong complexing effects of the glucose. However, experiments of this type ought to be continued with the improved ion selective electrodes now available. Finally, it is possible that glucose may specifically influence only vicinal water structures, and that this effect, in turn, alters the ion activities, rather than directly affecting the calcium ion activities through complexing. In this connection, see the discussion of the paper by Apffel and Peters (1969); see also the "classic approach" in the paper by DeHaven and Shapiro (1968).

In summary, although very complex physiological activities and processes are involved in the overall physiology of hypothermia, the abrupt changes occurring near 30° to 32°C indicate that a narrow temperature interval is extremely significant. It is proposed that the dramatic effects associated with this narrow temperature interval are manifestations of the sudden changes in the structure of the vicinal water of the biological systems, most likely through the effects on nerve conduction, membrane permeabilities, and enzymes kinetics.

2. Hibernation

There is no sharp distinction between the concept of hypothermia and hibernation (see, for instance, the discussion in Lyman and Dawe, 1960). Characteristic examples of the critical role of temperature on hibernation are also discussed in an article by Bullard et al. (1960); thus, in both the ground squirrel and the hamster the heart rate drops as the temperature is lowered from approximately 35° (or 37°) to about 33°C. Between 30° and 33°C, however, the heart rate increases, while the metabolism of both animals drops simultaneously until a rectal temperature of about 30°C is reached; the metabolic rate then levels off at a more-or-less constant value over a wide temperature range. (These experiments were carried out in an atmosphere of reduced oxygen tension: 5–6% oxygen in nitrogen.)

Hibernation is far more difficult to explain than is clinical hypothermia from the point of molecular physiology. The dependence upon ambient temperature in competition with homeostatic temperature control leads to exceedingly complex behavior. The processes are slow, yet at no time do the systems approach thermodynamic equilibria. The heat conduction problem, particularly in furry animals, makes a direct comparison between ambient temperature and body temperature highly time-dependent and of little analytical usefulness. Yet, in spite of all these difficulties, an inspection of the available literature on mammalian hibernation (see, in particular, Lyman and Dawe, 1960) again reveal the critical nature of the temperature range from approximately 30° to 32°C. Somewhat similar anomalies occur in deep hibernation for temperatures around the 15°C anomaly (see, for instance, the article by Eisenstraut, 1960, p. 31).

Entrance into the hibernating state occurs in the squirrel around 30°C. In an extensive review by Strumwasser (1960) of the physiological processes regulating the hibernation in squirrels, it is noted that "The induction into the hibernating state occurs . . . when the squirrel's brain temperature had reached 32°, since, as a general rule, once this temperature is reached, there never occurs a spontaneous turning back." Strumwasser also observed changes in the heart rate during entrance into and arousal from hibernation as a function of temperature. The heart rate, according to

Strumwasser, declines as the temperature is lowered from ordinary body temperature to near 34.2°C. Over the following interval of only 0.6° (from 34.2 to 33.6°C), the heart rate drops from 153 beats to 68 beats per minute.

Here, as in most of the examples discussed in this section, factors other than temperature obviously play an important role. However, the point of the present discussion is that, superimposed on the various other influences, is the omnipresent effect of the structural changes in the vicinal water of the systems under consideration. Whenever the controlling mechanism involves some aspect of the structure of vicinal water, thermal anomalies will be manifest in the behavior of the system under consideration.

Subsequent to the two monographs mentioned on the problem of hypothermia and hibernation, a symposium was held on dormancy and survival (see, in particular, the article by Lyman and O'Brien, 1969).

G. CRYOBIOLOGY

The structure of water as well as the structure of ice play an important role in cryobiology. Some more or less standard studies of the freezing of biological systems will be discussed very briefly in this section. The freezing of biological systems upon exposure to cold is obviously an example of a nonequilibrium and, in fact, a nonsteady state phenomenon. Hence, it is not surprising that factors such as the heat conductivity of both ice and water as well as of the protoplasm and membrane materials, specific heats, and the heat of fusion of ice enter into the problem.

In this section is reviewed briefly the process of freezing in biological systems, as discussed most recently by Mazur (1966, 1970). The treatment by Mazur is both detailed and informative, but as in practically all other current studies of cryobiology, it is based primarily on the implicit assumption that "fine details" of water structure changes near interfaces can be neglected. (One credible, notable exception mentioned by Mazur is the possibility that some macromolecules may protect cells from freezing damage through hydrogen bonding or clathrate stabilization.)

Mazur (1970) begins his discussion of cryobiology by pointing out that intracellular water generally is readily supercooled to −10° to −15°C, even when crystalline ice is present in the external medium (the "equilibrium" freezing point of the cytoplasm is about −1°C, corresponding to an osmolar concentration of about 0.5 M). Thus, the cell membrane apparently can prevent the initiation of growth of ice in the cytoplasm in the cell in spite of the "crystalline" environment. Mazur also points out that, by the same token, the interior of the cell contains no elements which make good nucleators of supercooled water.

Mazur (1970) next invokes (a) Raoult law to predict the difference in vapor pressure between the internal water of the cytoplasm and the external water and (b) an equation for the rate of flow of water out of the cell in response to the (vapor) pressure gradient established. [This approach assumes the cell is permeable only (or primarily) to water.] A study of the two equations is interesting, but since they are based on equilibrium thermodynamics they apply only to "equilibrium," i.e., they apply only to exceedingly slow cooling rates, say, less than 1°/minute (in studies of yeast and erythrocytes). For higher rates of cooling, the super-cooling of the cytoplasm increases significantly. It should be noted that experimentally it is reasonably easy to obtain freezing rates varying between 1° and 10,000°/minute, although undoubtedly with rather poor control or significance other than for an "instantaneous value." When freezing does occur intracellularly, it is likely that the nucleation is the result of a "puncturing" of the cell membrane, leading to the nucleation of the interior solution by the external solid phase.

Two types of freezing damage occur to the cell: either the cell becomes dehydrated by the outflow of water (because of osmosis) or the cell may undergo intracellular freezing. A loss of water to the external environment may greatly increase the solute concentration and, thus, produce drastic changes within the cytoplasm. This type of damage is referred to as a "solution effect." As would be expected, rapid cooling produces small ice crystals. However, the smaller the crystal, the greater the tendency for the crystallites to grow to larger crystalline units as the temperature is slowly raised during rewarming; this gives rise to an optimum rate of freezing which will do the least damage, in the sense of ensuring the greatest rate of survival of the cell. Figure 29 shows some effects of freezing on the survival of vastly different types of cells. The difference between the different types of cells is impressive.

An inspection of Fig. 29 shows that both erythrocytes and yeast cells survive poorly at temperatures of −190°C for a cooling rate of 100°/minute. However, the reason for this low survival is entirely different: in the case of the erythrocytes the cytoplasm does not freeze but loses water to the intercellular ice, whereas in yeast the damage is due to intracellular freezing.

Mazur also discussed the better known reports of protective agents for cryobiological purposes such as glycerol, dimethylsulfoxide, sugars, and polyvinylpyrrolidone (PDP). However, in spite of considerable study, it does not seem possible at this time to present a "molecular picture" of the protective action of these materials in general. Compare the cryoprotective agents glycerol, dimethylsulfoxide, and some polyhydroxy compounds with their "typical nonaqueous" behavior discussed by Franks and mentioned in Section II,C,3.

In connection with the readily obtained supercooling of about 10° to 15°C, there is little doubt that it is the membrane surface which prevents the ice from permeating into and nucleating the water of the cytoplasm. This bears on the problem of the possible existence of water-filled pores (see, for instance, the studies by Solomon). Actually, very small pores (say, 5 Å, the order of magnitude discussed by Solomon, 1968) would most likely represent radii of curvature so small as to make the crystalline ice phase unstable with respect to the liquid phase. However, even if larger

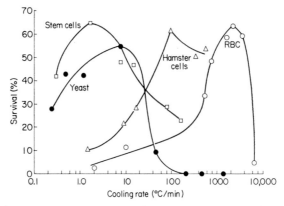

FIG. 29. Survival of various types of cells as a function of cooling rate. RBC—erythrocytes. (Mazur, 1970, with permission.)

pores were present, it is possible that freezing might not propagate through these pores because of the ordered nature of the water in the membrane.

Mazur (1970), finally, mentions the possibility that the cell water may not necessarily be treated as water or normal aqueous solutions. In fact, he calls attention to the possible existence of "bound water" and refers to a possible "ice-likeness." Observing that the water in cells and around macromolecules shows little ability to act as a nucleating agent for super-cooled water, Mazur proposes that the ordered water, at least to this extent, fails to be "ice-like." Of course, as pointed out elsewhere in this chapter, the terminology "ice-like" is probably very unfortunate, as it is unlikely that water, either in bulk or as vicinal water, contains elements with actual Ice–Ih-like characteristics.

There is little doubt that the classic approach discussed so far has brought a certain amount of understanding to the problem of cryobiology, although Mazur points to a great number of unresolved problems. The present author, however, takes issue with the statement that the equations used are consistent and "essentially picture a cell as a membraneous bag of dilute solution, the solvent being water of normal properties." Mazur claims the model satisfactorily predicts which physical events will occur in cells during freezing. He also claims that the amount of heat absorbed by frozen cell suspensions during warming (as measured by differential thermal analysis) is in semiquantitative agreement with the view that the bulk frozen cytoplasm behaves like an ordinary frozen dilute solution. In view of the many questions which Mazur raises at the conclusion of his article of poorly understood or completely obscure phenomena associated with freezing, it is somewhat extravagant to claim that the model—based on the equations for the freezing process—accounts "satisfactorily" for many physical events which occur during freezing. It should also be noted that many different ice structures (ice polymorphs) possess lattice energies differing only relatively little from ordinary ice and, thus, may go undetected in a standard differential thermal analysis.

Meryman (1966a) has reviewed some of the evidence for ordering of water in biological systems and particularly calls attention to some studies by Hori (1960). Meryman has redrawn the original data obtained by Hori and concludes that for thin films of water:

> It is interesting that the vapor pressure approaches zero at a greater film thickness between glass than quartz. Perhaps the lattice parameters of the quartz structure are too foreign to that of water to potentiate structure but, instead, propagate disorder. In any event, Hori's data show dramatically that, between glass at least, water films with a thickness of 0.1 μ, extremely thick by biological standards, neither freeze nor have a measurable vapor pressure. Although the circumstances in biological materials may differ, the point is well made that adjacent structure can dramatically affect water activity.

(Compare the more common notion that quartz is supposed to be responsible for extensive ordering of water by epitaxy because of the similarity between the ice and the quartz lattice.)

For general information about the field of low-temperature biology, the reader is again referred to the impressive monograph "Cryobiology," edited by Meryman (1966b). This volume probably contains the most extensive reviews of the state of the art of cryobiology at its time of publication; also see the recent article by Mazur (1970) referred to above.

For an excellent review of the effects of freezing on aqueous enzyme systems, see the article by Tappel in "Cryobiology" (1966).

H. Hyperthermia

1. Upper Lethal Temperatures

As discussed in Sections VI,B,1 and 2, rather abrupt, critical upper thermal limits are frequently encountered for various organisms. A direct application of this notion is discussed, for example, in the section on thermal pollution (Section VI,K). For mammals and birds, 44°–46°C appears to be an absolute, upper thermal limit for survival. It should be noted that Belding (1967) has suggested that a body temperature of 45°C may be sustained by birds over relatively long periods of time without injury, whereas a rise to about 47°C is lethal. In humans, body temperatures above 43°C (for instance, due to infectious diseases) is considered highly dangerous and the prognosis is generally poor. However, when the elevated body temperature is produced by external means, such as diathermy or hot baths, the probability of survival is significantly increased. This will be elaborated upon in the next section.

A rather detailed discussion of temperature effects on poikilotherms was presented by Fry (1967). The reader is referred to the article by Fry for many interesting details, especially regarding definitions and methods of determining thermal limits. Measurements are made by maintaining organisms for various lengths of times at constant temperatures and determining the time of death due to exposure to a particular temperature. Alternatively, the temperature may be recorded at which the animal dies (or suffers loss of motor ability) in an environment of constantly increasing temperature. The rate of temperature change now plays a crucial (and ill-defined) role. Each approach has its advantages and disadvantages. The problem is not amenable to a description in terms of physicochemical parameters because of its nonreversible, nonsteady state aspects.

In connection with structural changes at 45°C as the causative factor in death, it can be questioned if evidence exists for the operation of only one causative mechanism, or, if more than one direct cause is involved, what are the different possibilities. Specifically, What is the mechanism of thermal injury? Is it possible to determine the site or sites where water structure changes may be the direct cause of death of the organism? Unfortunately, there appears to be very little information available on which to base any judgment in this matter. Fry reports a few sets of data which tend to suggest the occurrence of two different, distinct mechanisms of death. Among possible mechanisms are failure of osmoregulation, increased production of lactic acid (an unlikely effect), or asphyxia and/or damage to the central nervous system. Fry mentions also that much of the work by investigators such as Precht, Prosser, and Ushakov tends to suggest that the animal dies from the cessation of some regulatory activity, rather than from "collapse of its cells."

2. Hyperthermia Therapy

Recently, von Ardenne and co-workers (1965, 1966a,b) and Kirsch and Schmidt (1966) have actively pursued the treatment of cancer by hyperthermia. Initially, the treatment consisted merely in heating the patient, while more recently a "multi-step therapy" is employed, combining extreme hyperthermia (heating of the patient's body to near 43° to 44°C) with the prior, simultaneous, or subsequent administration of certain drugs (pharmacons). With the development of more effective pharmacons, the need for extreme hyperthermia has been reduced somewhat and present therapy uses temperatures as low as 42°C.

The treatment of cancer, both experimentally and clinically, by high temperature is by no means new. Cavaliere and co-workers (1967) have discussed the heat sensitivity of cancer cells and reported both some biochemical and some clinical studies. These authors mention that as early as 1866, Busch described the complete disappearance of a histologically proven sarcoma after the patient suffered two attacks of erysipelas (an acute infection of the skin by a Group A hemolytic streptococci, characterized by sharply delineated, red, swollen local areas with general fever and malaise; temperatures as high as 42°C are often observed). The article by Cavaliere and co-workers should be consulted for a number of examples and a rather careful review of previous studies. It is interesting and, in fact, impressive, that the previous studies as well as the work of Cavaliere and co-workers and von Ardenne and co-workers clearly demonstrate the increased heat sensitivity of malignant cells to temperatures ranging from 43° to 44°C. Although Cavaliere and co-workers were relatively successful in the clinical treatment of 22 cases of cancer of the limbs, they present their findings with considerable reservation—because, among other reasons, of the attendant clinical risk in any high-temperature treatment as well as other complications.

No molecular mechanism has yet been postulated for the role that temperature plays in that part of the original therapy which relied primarily on the effects of temperature alone. It was suggested (Drost-Hansen, 1966) that the effectiveness of the increased body temperature depends primarily on the structural transition near 45°C in water associated with the cells (rather than, for instance, merely increasing the activity of any administered pharmacon). Burk and Woods (1967), Olmstead (1966), and Szent-Györgyi (1965) have pointed out that cancer cells generally have a much higher water content than normal cells. As an example, ordinary liver cells possess about 67% water compared to Rous sarcomas of chicken containing 93% water; in fact, there appears to be a good correlation between malignancy and the water content of cancer cells. It is now suggested that the molecular mechanism underlying the effectiveness of the hyperthermia therapy may, in fact, be due to one or a combination of several processes. The first of these possibilities involves the greater ratio of bulk-like water to structured water in the cancer cells as compared to normal cells. Since vicinal water (and solutions) appear to be stabilized near an interface (where it undergoes a transition at or near 44° to 46°C), the cancer cells may be more susceptible to temperatures in this vicinity than normal cells, merely because of the larger amount of bulk-like water.* If the pharmacons administered tend to accumulate in the malignant cells, the effectiveness of these pharmacons may be due to their ability to alter the water structure in such a fashion that a lower transition temperature is obtained. Of course, the pharmacons (such as alkylating compounds) administered at the time of the treatment will exert a direct influence on the biochemical processes involved in the metabolism of the cancer cells. It is of interest to note that Burk and Woods have shown that at 43°C and above, 9-α-fluoroprednisolone accelerates loss of the Pasteur effect and metabolic death of the cells. In addition, the Pasteur effect in cells of mouse melanoma S91 remains essentially constant (for 1 to 2 hours) at any given temperature below 40°C. However, above 43°C the aerobic acid production (glucolysis) increased markedly more than the anerobic glucolysis. [See in this connection the study by Haskins (1965) who investigated the effects of sterols on the temperature tolerance in a fungus of the genus *Pythium*.]

An alternative hypothesis is based on the assertion that cancer cells probably present a far more disordered interface to the cell fluid than do normal cells: since cancer cells generally are much less differentiated than normal cells, they do not present the intracellular water with the same degree of stabilization near the interface that is provided by normal cells (Szent-Györgyi, 1965). This effect would be particularly important near the point where an abrupt transition in the water structure occurs.

Both of the mechanisms suggested above may play a role simultaneously. In general, the effect of elevated temperatures is to produce greater instability in the water structures associated with the cells, and this effect may be further enhanced near the critical transition point due to the pharmacons administered at the time of treatment. The structural changes in water are suggested here as an important factor in the molecular process

* (See, however, discussion of the paradox of relatively invariant thermal transition temperatures, Section III,C,1,2.)

underlying the phenomenon of enhanced lethal effects of high body temperatures, but this aspect is obviously only one factor in an exceedingly complex molecular system and many other factors may play equally important roles.

In connection with hyperthermia treatment, it is of interest to note that some of the phenylenediamines have been alleged to have a synergistic effect in the hyperthermia treatment of cancer. Apparently so does dopa. The phenylenediamines have the interesting property that their solubilities increase extremely rapidly over very narrow temperature ranges. Thus, conceivably, these compounds are highly "sensitive" to the structural details of the aqueous environment. This would again suggest that a more detailed understanding of the effects of various solutes on the structure of water and, particularly, the structure of vicinal water may aid in predicting the types of pharmacons which may be most useful in the treatment of cancer by multistep hyperthermia therapy.

It is interesting to speculate on a rather simple mechanism for the increased heat sensitivity of malignant cells over normal cells in terms of intracellular heat conductivity (in the vicinity of $44°$ to $45°C$). If it is assumed that in hyperthermia treatment, malignant cells and normal cells are heated to the same temperature and if it is assumed that the metabolic rates in these cells are roughly equal (but obviously not identical in the two types of cells), it is seen that because of the greater amount of ordinary water in the malignant cells, the local, internal temperature and likely the "internal structural temperature" of the malignant cells may be notably higher than that of the normal cell. Recall that the malignant cell is generally characterized by a considerable increase in the amount of total water and this water is undoubtedly less structured (more bulklike) than the water in normal cells. If, indeed, the ordered water possesses higher thermal conductivities as suggested by Metsik and Aidanova (1966), the normal cell will then be able to dissipate (by thermal conduction) the energy produced, faster than the malignant cell. Thus, the malignant cell will be subject internally to a somewhat higher, local temperature. It may be by this mechanism that the heat sensitivity of the malignant cells is enhanced.

Finally, assuming again that the rate of energy production is approximately equal in normal and malignant cells, assume also the heat conductivities may differ by a factor of 70 between ordered water (in normal cells) and bulk water (in malignant cells), based on Metsik and Aidanova's data, rather large differences in internal temperatures may then be expected for the two types of cells. Spanner (1954) has considered the relation between the "heat of transfer" and the equivalent "osmotic" pressure of the cells. The present author does not necessarily accept the development on which Spanner's estimates are made; however, if Spanner's calculations are correct to an order of magnitude, it is interesting that a temperature difference of 1°C produces an osmotic (thermomolecular pressure effect) pressure difference of about 130 atm, where $\Delta P/\Delta T \approx -132$ atm/°C (note the minus sign!). Thus, a difference in temperature as slight as 0.01°C may then be expected to cause changes in the rate of water permeation (driven by hydrostatic forces) equivalent to a pressure well over 1 atm. Considering that the differences in heat conductivity may be as large as one and even two orders of magnitude, temperature differentials of the order of 0.01°C are not at all unlikely within any given system of the cells presumed to be in isothermal (and isoosmolal) equilibrium.

In connection with the role of water in malignancy, attention is called to an article by Apffel and Peters (1969). These authors discussed the role of hydration of various macromolecules, in particular the glycoproteins, first noting that the glycoproteins may have distinctly varying capacities to bind water. The degree of hydration appears to depend directly on the specific nature of the monosaccharides in the saccharides of the glycoproteins and of the polysaccharides. They discuss the general phenomenon of tissue hydration in malignancy, noting the increase in amounts of water, for instance in liver, during carcinogenesis and tumor growth. The authors also consider, qualitatively, the conformational changes that may result from differences in hydration, leading to significant differences of interaction at the surface of a cell or a macromolecule. Thus, "Hydration shells are proper to microorganisms and cells coated with sialoglycoproteins, and to a much lesser degree, to solvated or dispersed macromolecules of that nature. Adsorption of sialoglycoproteins to the surface of cells produces a unique situation where all the oligosaccharide chains protrude into the medium in a single, committed direction. Because of the close association thus brought about, there is a tendency toward gelification. The resulting semi-rigid shell of hydration water is tantamount to a volume forbidden to many solutes, depending on their size, charge and shape." Finally, Apffel and Peters call attention to the note by Good (1967) who also stressed the importance of hydration phenomena superimposed on charge–charge interactions between cells or between cells and macromolecular solutes. It is particularly interesting, as noted by Apffel and Peters (1969) that Good suggests that "at interfaces, as between cells and medium, the hydration of charged ions is more stable because there is less thermal gyration and less mechanical disturbance [see Good, 1967]." It should be observed in this context, however, that the type of hydration discussed by Apffel and Peters (and by Good, 1967) is primarily the very direct (and energetically strong) interactions between ionic sites and the water molecules (or strong dipole–dipole interactions). In this chapter, stress is placed instead on the (very likely) much weaker, but possibly far more extensive hydration phenomena involving energetically only slightly different states of water. However, the basic idea regarding the stabilization of water structure near an interface suggested by Good (and quoted by Apffel and Peters) is the same as the one advocated in this chapter, namely, the reduced "thermal gyration and less mechanical disturbance" which results from the "momentum sink effect." No doubt, the study of the hydration of cell surfaces and macromolecular solutes will prove an important requirement for further advance in a detailed molecular understanding of the role of serum proteins in cancer.

I. CELL ADHESION

Adhesion in general and cell adhesion in particular are extremely complex phenomena. Attention is called here only to the types of interactions that do not depend on attractive forces deriving from functional groups of the membrane materials. Pethica (1961) has reviewed the type of forces which may exist between cell surfaces; these forces include attractive as well as repulsive forces. In addition to such forces as chemical bonds between the opposing surfaces, ion pairing, image forces, and van der Waals forces, Pethica mentioned a "hindrance" to attraction due to steric barriers such as "inert capsules and solvated layers." Pethica points out that the latter do not actually represent a force "except that the entropy effect due to the mutual disordering of adsorbed layers, as the surface is approached, might be regarded as a force. The effect of adsorbed inert layers may more usually be to increase in range between otherwise active groups, and to attenuate the attractions between the surfaces to the point where reversible collisions can take place."

As discussed in Section III,C,1,a, Peschel and Adlfinger (1967) have determined the disjoining pressure between surfaces (specifically, quartz surfaces), and the major feature of their results may probably be generalized to cell surfaces, at least to the extent that anomalous temperature dependencies may be expected in any quantitative data on cell adhesion.

Pethica has considered the role of image forces in the attraction between cells in solution. For this purpose, the cells were modeled as two thick planes of material with much lower dielectric constants. Using the expression (applicable to a structureless dielectric material of dielectric constant ϵ), Pethica calculates the osmotic pressure (π_l) from the free energy (ΔG) expression:

$$\Delta G = \frac{3e^2}{2\epsilon_1} \tag{2}$$

Hence

$$\pi_l = \pi \exp\left\{-\frac{3e^2}{2\epsilon kTl}\right\}. \tag{3}$$

where e, k, T have their usual meanings, and l is the separation between the cells.

From this one obtains the net attractive force (per unit area):

$$\pi - \pi_l = \pi\left[1 - \exp\left\{-\frac{3e^2}{2\epsilon kTl}\right\}\right] \tag{4}$$

and the work (per unit area) required to bring the two opposing surfaces together (from ∞ to $y = 2$)

$$W = \pi \int_y^\infty \left[1 - \exp\left\{-\frac{3e^2}{\epsilon kTy}\right\}\right] dy \tag{5}$$

Although the present author does not take issue with the use of a very low value for the dielectric constant for the wall material, assuming it, for instance, to be a lipid (with $\epsilon_w = 2$ or 3), it should be pointed out that ϵ in Eqs. (2) through (5) is the dielectric constant for vicinal water, and this value is likely to be different from that of bulk water. Vastly different results will be obtained from those arrived at by Pethica [by graphic integration of Eq. (5)], since the "true" value for ϵ may possibly be an order of magnitude lower than the value for bulk water (see the discussion in Section IV,A,6 where Derjaguin quotes values for ϵ of ≈ 8 to 10 near interfaces). Again returning to Eqs. (4) and (5) and recalling the abrupt changes which have been observed in dielectric constants for vicinal water, it is not surprising that cell adhesion may show notable anomalies as a function of temperature. Pethica has suggested that only for separations of about 100 Å will the majority of forces considered, including the van der Waals forces, play a notable role. However, it is the contention of the present author that because the properties of water often show anomalies which appear to extend over as much as 1000 (to 10,000) Å, the structural changes in water may well play the dominant role in quantitative theories of cell adhesion. See also the article by Pethica and co-workers on possible ranges of structurally modified water near certain polymer surfaces (G. A. Johnson *et al.*, 1966). Finally see also the studies by Weiss (1967).

Abdullah (1967) has studied the aggregation of platelets *in vitro*. He notes that a distinction is usually made between platelet aggregation (interparticle association) and platelet adhesion (some standardized measure of adsorption of platelets onto a standard glass surface). For both processes, Abdullah suggests that "some platelet aggregating substances act by increasing ice-likeness (ordered structure) of water around platelets" and that the active, initial process is followed by a "chain reaction" which results in the accretion of many layers of platelets onto the first-formed layer. Abdullah measured the effects of various nonelectrolytes on platelet suspensions, following the degree of aggregation optically. Very interesting results were obtained with a number of normal aliphatic alcohols (pentanol, hexanol, octanol, and decanol), two tetraalkylammonium salts, and argon and xenon (in oxygen-rich mixtures). From the data obtained, Abdullah suggested that water becomes ordered in the vicinity of an interface and this ordering acts as an "entropic trigger" which carries the system (i.e., the platelets or the platelets–glass interface) over a small potential energy barrier; in other words, the entropy change decreases the internal energy, allowing the net entropy of the system to increase.

Garvin (1968) has reported some very unusual temperature dependencies for cell adhesion, specifically the "recovery" from adsorption onto solid surfaces of polymorphonuclear neutrophiles in human blood. Garvin noted that above 45°C the percent recovery increases almost linearly from 0 to 100% over less than 4°C. In other words, above 45°C there is a very rapid decrease in the tendency for the neutrophiles to adhere to the solid substrate. This suggests that the phenomenon of cell adhesion (and cell–cell interaction) may be influenced drastically by the structural change of the vicinal water around the cell (in this case, the neutrophile) surface at 45°C.

J. THERMAL HYSTERESIS EFFECTS

The role of time is one of the essential differences between the study of the thermodynamics of purely physicochemical systems (however complex they may be on the molecular level) and the study of biological systems. With biological systems, measurements as functions of temperature and pressure as independent variables are also invariably measurements of the same parameters as a function of time. At best, steady-state may attain; more often, growth or "decay" occur simultaneously. In principle, we can allow for the effects of time, but effects for which corrections cannot be made may occur when, for instance, both time and temperature change.

Memory effects in physicochemical systems have rarely attracted much attention as the presence of hysteresis invariably suggests lack of rapid approach to equilibrium and thus prevents true equilibrium thermodynamic parameters to be measured. Yet, in kinetic studies, memory effects, or at least time-dependent behavior, is noted in some instances. Characteristically, practically all liquids, water in particular, may be significantly supercooled, whereas the ice lattice (or crystalline hydrates) apparently never superheats. Clay suspensions, once agitated, may "reset" at rest or at low shearing rates, whereas an initial disturbance can lead to the immediate disruption of the prevalent structure (which is causing the gel rigidity).

Recently, the present author and his co-workers have had ample opportunity to note strong time-dependent variations. Among these have been memory effects (or at least, time-dependent effects) in the properties of membranes. Thus, thermal anomalies have, from time to time, been observed in membrane properties when studied during slow heating—either through discrete increments of temperature or using continuously variable temperatures. Likewise, Kerr (1970), working in the author's laboratory, has observed thermal anomalies in the viscous damping of a water-filled vibrating quartz capillary. The anomalies are particularly pronounced during heating. The anomalies are sometimes completely absent (or notably displaced) upon repeating the measurements with decreasing temperatures. The most significant contribution, however, to the study of the possible existence of thermal hysteresis has come from the work by Bach (1971). Bach observed a memory effect in the structural properties of water deposited on a silver (or silver oxide) surface, using a differential thermal analyzer (DTA), and on glass, using a vapor phase osmometer in a differential manner.

It is not difficult to propose an explanation for time-dependent effects. As temperature is increased and thermal energy thus enhanced, any ordered matrix or array is readily disturbed into a more disordered state (i.e., a state of higher entropy). However, upon cooling, removal of a "like amount" of thermal energy does not readily cause a reordering of the system into the original order of the crystalline lattice. It is obviously far easier to disrupt and disorder a lattice than to perform the converse: to induce a specific order by merely lowering the available thermal energy fluctuation. Characteristically, thermal hysteresis phenomena have been observed especially with aqueous systems near interfaces (although supercooling does illustrate a bulk phenomenon of similar type). Near 15° and 30°C, for instance, water in biological systems (or at or near almost any aqueous interface) will undergo a phase transition, as discussed in previous sections. However, since as increase in temperature (for instance, from 28° to 33°C) will have resulted in a disruption of a structured matrix, it is possible, and sometimes likely, that a similar decrease in temperature may not reversibly lead to the "same" change in biological functions—at least, not the same change at the same rate of change. Thus, thermal hysteresis must be expected in living systems also.

In summary, near a biological interface, water structures are stabilized which differ from the bulk structures. Disruption of these structures by increasing the temperature is readily achieved. However, structures that are stable at low temperature may not readily be reformed upon lowering the temperature; thus, the possibility exists for a significant lack of "symmetry" in the behavior of biological systems under temperature cycling. The attention of the experimental biologist to this possibility may likely prove rewarding. It is also of interest to note that the clathrate formed by hydroquinone and argon is "stable" (or, rather, may be kept in a bottle nearly indefinitely) once formed, although at room temperature its (equilibrium) vapor pressure is several atmospheres. The reason for this (meta) stability is the high energy of activation required to break a number of H-bonds in this hydroquinone lattice in order to release the trapped argon (see van der Waals and Platteeuw, 1959).

It is interesting that in many studies, particularly on biological system, the experimentally observed errors often tend to be larger in the vicinity of the temperatures of the thermal anomalies. Based on the studies by Bach (1969), Kerr (1970), and Thorhaug (1971) (all formerly working in the author's laboratory), it is suggested that this may be related to thermal hysteresis effects. Bach, in particular, has proposed that whether or not the system has been cooled or heated immediately prior to an experiment may influence the physical states attained. Thus, the possibility exists that a particular experiment near the temperature of one of the thermal anomalies may find the system in question in one of two states, corresponding, respectively, to either the structure which is stable above the transition temperature or the structure stable in the lower temperature range. Since these states will have different properties (for instance, reaction rates), it is not to be wondered at that the scatter in these cases occasionally are larger than toward the middle of each temperature interval.

K. THERMAL POLLUTION

The general question of thermal pollution is obviously not immediately related to the specific discussion of the structural and functional role of water near the cell surface (and in biological systems in general). Furthermore, the question of thermal pollution is an exceedingly complicated one, but not merely in the ordinary sense of complications as they are encountered in the study of any biological phenomenon, say, metabolism. In the case of thermal pollution, factors enter which are extraneous to a conceptually homogeneous approach to the problem. Thus, as an example, the effects of elevated temperatures will influence the entire life cycle of any of the multitude of organisms making up the ecological network and include as well additional "external" factors such as environmental temperature fluctuations (frequency and amplitude of variations) and attendant changes due to additional stresses such as salinity fluctuations, chemical pollutants, and politicians. However, one specific aspect of thermal effects on the structure and properties of water near interfaces may play a singularly important role in determining the overall response of the entire ecosystem. What is implied here obviously is the abrupt and likely relatively invariant constraints imposed on any biological system due to the sudden changes in vicinal water structure at the temperatures of the thermal anomalies; hence, from this point (and this point only) is discussed the more obvious aspects of the possible existence of guidelines for allowable thermal pollution limits.

Fundamental to the problem of delineating permissible temperature intervals for biological organisms—a problem of crucial significance in any thermal pollution study—is the simple statement (Drost-Hansen, 1965a) that "If we are correct in assessing the importance of the structural changes in water for the behavior of biologic systems, it may be possible to delineate ranges of environmental temperatures that are conducive to life." It is, indeed, this idea which was elaborated upon subsequently in the paper by the present author (Drost-Hansen, 1969c) on thermal pollution limits.

In connection with the effects of temperature on biological systems in nature as distinct from laboratory studies, we must take into account the effects of varying temperature. This undoubtedly plays a crucial role, as is already known from both marine biological studies as well as physiological studies, for instance, on land plants. The overall dynamics are further complicated by variations in light intensity, availability of inorganic nutrients, etc. The question here is whether or not it is more appropriate to be concerned with the extremes of temperature rather than the average temperatures. Certainly, a "steady" temperature of 28°C could conceivably be compatible with growth and reproduction of an organism, but even relatively short-time excursions of ±4° from this average (to temperatures between 24° and 32°C) might lead to catastrophic results in the narrow temperature range from, say, 30° to 32°C.

For marine fishes, the probably critical nature of temperatures around 30°C was carefully reviewed by DeSylva (1969).

Elsewhere the present author (Drost-Hansen, 1969c) has emphasized that in thermal pollution studies it is necessary to be concerned with the effects of temperature on each of the different stages of life development.

A simple example of the different requirements for optimal development is illustrated in the study by Calabrese (1969). In this study the effects of the salinity and temperature on some marine bivalves were studied in considerable detail. Specifically, Calabrese studied the effects of temperature and salinity on the development of embryos and larvae of *Mulinia lateralis* over wide ranges of the two variables. The percent of embryos developing normally shows a notable peak as a function of salinity at approximately 25 ppt. This sensitivity to electrolytes is paralleled with a notable maximum in the number of embryos which developed normally as a function of temperature. The survival of larvae and the number of eggs developing normally both showed maxima in the range between 15° and 30°C with precipitous decreases in normal development above 30°C. It is interesting also that the percent survival of the larvae as a function of the combined effects of salinity and temperature shows a rather wide range of survival: between temperatures of 7.5° and 27.5°C, and for salinities ranging as high as 35 ppt and as low as (10 to) 15 ppt. However, the percent increase in the mean length of these larvae as a function of the change in the same parameters showed a notably more restricted domain of optimum development, namely, between temperatures of 7.5° and 22.5°C and salinities between 20 and 35 ppt.

The abrupt changes in biological functioning, which have been mentioned in this chapter, near the temperatures of the thermal anomalies likely play a crucial role in thermal pollution. The literature provides a vast number of such examples of abrupt changes near 15° and 30°C. See, for instance, the cases discussed in the paper on thermal pollution by the present author (Drost-Hansen, 1969c). Figure 30 shows the percentage of normal development, compared to the development of major anomalies, in the frog (*Rana cyanophlyctis*). It is seen that, above approximately 33° and below 15°C, none of the progeny develops normally, whereas 100% normal development occurs between approximately 21° and 31°C. Undoubtedly this is not a unique example. Compare, for instance, the discussion in Section VI,D,2 on rates of mutation (chromosome aberrations). Furthermore, it should be stressed again that the effects of temperature in an ecological system must be fully concerned with the effects of temperature on *all* stages of the life cycle, ranging from the egg and sperm stages through the development of mature individuals (Drost-Hansen, 1969c): "obviously, even if only *one* life stage is sensitive to the temperature changes around the thermal anomalies, the ecological significance may be great."

In summary, then, it is suggested that significant and, in fact, possibly disastrous results may occur to the ecology of a particular area should the temperature for any length of time exceed the temperature of one of the thermal anomalies (the anomaly which occurs above the range of optimum temperature for the majority of organisms in that locale). This suggests that it may be possible, on the basis of purely physicochemical observations, to propose rather clearly delineated limits for thermal pollution. This is particularly true in cases such as in subtropical and tropical climates where the temperature may already be close to 30°C.

VII. Review and Conclusions

The status of our current understanding of the structure of water and aqueous solutions was reviewed in Section II of the present chapter. It is

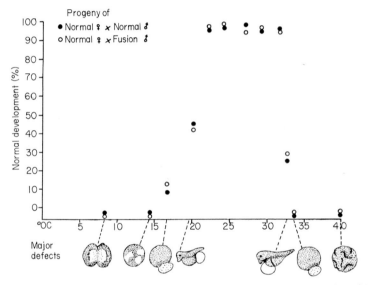

FIG. 30. Development of major anomalies in the frog (*Rana cyanophlyctis*) as a function of temperature. (Data by DasGupta and Grewal, 1968.)

still not possible to decide unequivocally if the structure of water can best be represented by a "continuum model" or a model invoking structural elements. The evidence in Section III suggests that vicinal water likely consists of or includes structured entities. For aqueous solutions, much of the available evidence, especially for solutions of nonaqueous electrolytes, also suggests that the water in such solutions may be structured. However, the brief review of the properties of aqueous alcohol solutions amply demonstrates that even such relatively "simple" systems are exceedingly difficult to interpret in terms of structure. This, together with the general lack of a definitive theory for the structure of water itself, makes it difficult to speculate on the structure of water in biological systems. The present author contends—but certainly cannot prove—that Ice–Ih-like elements are not likely to exist in bulk water and aqueous solutions (nor in water near interfaces). On the other hand, some evidence does exist for clathrate formation in solution, especially near interfaces and likely in the water of the biological systems. Stress has been placed on the fact that "ordering" and "structuring" in aqueous systems implies the existence of some specified geometric arrangement of the water molecules, but it also (and perhaps in particular) implies enhanced temporal stability. Thus, structuring may simply mean an increase in average lifetime of a particular type of "flickering cluster" or other structured entity, by a few (or more) orders of magnitude (say, from 10^{-11} to 10^{-9} or 10^{-8} second).

A number of examples of thermal anomalies in the properties of vicinal water were discussed. These anomalies likely reflect the existence of cooperative phenomena such as higher-order phase transitions. Since order–disorder require the existence of large structured entities in order to exhibit cooperative properties, the thermal anomalies thus reveal (notably) enhanced ordering in water adjacent to an interface.

The more-or-less abrupt thermal anomalies in the properties of vicinal water indicate that for large surface-to-volume ratio systems, great caution must be exercised in attempting interpretations of data in terms of, for instance, Arrhenius- or Eyring-type equations. This was discussed particularly in connection with the careful and closely spaced data reported for the diffusion coefficient of thiourea (Section III,C) and further illustrated in the discussion of Bangham's permeability studies (Section V,C,9).

Very tentative speculations were put forward regarding the nature of possible structured elements near interfaces. An apparent paradox was also stressed; namely, that it appears that thermal anomalies occur in the properties of vicinal water at almost identical temperature ranges regardless of the specific nature of the material in contact with the water. Specifically, in the case of water near various "solid interfaces" (ranging from some types of silica surfaces to certain macromolecules), the detailed chemical nature of the substrate appears to play, at most, a minor (but not negligible) role. This was interpreted as the result of the substrate acting as a momentum sink for thermal fluctuations which, in the case of bulk water, would have disrupted the particular structural arrangement present. The vicinal water thus gains stability merely by the proximity to the solid. Various aspects of the structure of water in biological systems were reviewed in Section IV. It was stressed that even in those cases where there appears considerable evidence for order and structure, it must be kept in mind that a dynamic situation prevails.

A review was given in Section IV of Hechter's proposed "dualistic theory," invoking the existence of both transmembrane pumps (driven by metabolic energy) and the more-or-less complete ordering of the intracellular water. Possible structuring of intracellular water has also been proposed, for instance, by Szent-Györgyi, by Jacobson, and by Ling, and, more recently, by Hazlewood and co-workers, by Cope, and by Fritz and Swift (based on NMR studies). One particular facet is stressed by the present author, namely, the likely important role of mutual interactions between solvent and substrate. The point is made (and referred to in several places in this chapter) that within a certain temperature region primarily only one structure of vicinal water is stable. This structure may, in turn, impose restrictions on the substrate to which it is adjacent and, indeed, in some cases, possibly play the dominant role in determining the conformation (particularly of macromolecules in solution). Conversely, some specific structural characteristics of the substrate may affect the detailed, structural characteristics of the vicinal water. A brief review of clathrates was presented and the likely occurrence of these as structured elements of vicinal water was discussed. It was mentioned that in ordered systems of water (especially ice), some reaction rates may be enhanced, although it appears too early to make any definite conclusions as to the importance of this phenomeonon in biological systems.

The last section of the chapter illustrates that vicinal water notably affects a great many functions of actual biological systems and the question, therefore, arises as to the possible "sites of action" of these water structure effects. A number of possibilities were considered, although far from exhausting the possible number of such "sensitive sites." Among the more likely sites of water structure effects, various solutes were mentioned [ranging from low molecular, organic nonelectrolytes, gel-forming materials, such as polysaccharides, to peptides and nucleotides (including RNA and DNA), proteins, and, specifically, enzymes]. Other possible sites discussed were the lipids and particularly the lipoproteins, especially in membranes. The evidence for structured elements of water in membranes was reviewed briefly, including some examples of thermal anomalies in membrane properties. The difficulties in distinguishing between ordered water near membranes and in the pores of membranes (if/or where such exist) was discussed, as were the effects of unstirred layers. Among other possible sites of water structure effects, the functioning of nerves and finally, muscles was mentioned. In connection with the discussion of water structure near various macromolecules—mimicking biomolecules—the study by Glasel was considered in some detail. His results are particularly important as they suggest that some macromolecules may be able to order water structure vicinal to their surface, whereas other macromolecular solutes appear to lack this ability. Some of the materials that do possess relatively extensive water structures may exhibit this property only over a limited pH range. These findings are especially pertinent in connection with the discussion of anomalies in biological systems—although there is strong evidence that thermal anomalies in vicinal water occur and are reflected in the behavior of many biological systems, it is equally certain that other biological systems do not show evidence of thermal anomalies. If the "controlling factor" in the latter systems is determined by the type of biomolecules with little or no vicinal water structure, the absence of thermal anomalies in these systems is easily understood.

Both in the discussion of proteins and in the discussion of the hemolysis of the erythrocyte membrane, the enthalpy–entropy compensation phenomenon—most recently discussed in detail by Lumry and Rajender— was considered in some detail. It appears that the compensation phenomenon reveals the dominant role of the solvent structure and, presumably, the vicinal water structure in particular. This role was further emphasized and illustrated through examples of marked anomalies in the temperature effects on enzymic rates. It is contended by the present author, although not proven, that the anomalies observed are often far too abrupt to be accounted for by any reasonable extension of the idea of competing simultaneous and/or consecutive reactions. Cold inactivation (the tendency for cold denaturation of some proteins at low temperature) was also discussed and seen as a possible result of the effects of the stabilized vicinal water structure (for instance, below 15°C) imposing conformational restraints on the macromolecules in solution.

In the final part of this chapter, the functional role of water in actual biological systems was examined. A discussion was presented of optimal and lethal temperatures in terms of domains of stability of structured vicinal water and further examples of abrupt anomalies in biological systems were described. Here, and particularly in connection with the hyperthermia treatment of cancer, the role of the increased heat conductivity of ordered water (compared to bulk water) was stressed—based on the measurements of heat conductivity of vicinal water between mica plates (showing notable effects over distances of the order of 1000 Å). Control of germination and vernalization was also discussed in terms of nonspecific effects related to the general changes in vicinal water structure and the constraints thus imposed upon, for instance, some controlling enzyme or membrane system. Thermal adaptation and, particularly, the rate of genetic damage as a function of temperature were also discussed in terms of changes in vicinal water structure at the critical temperature. This information, in turn, was applied to paleozoogeographic studies, based on the assumption that the thermal anomalies likely represent truly "time invariant" boundaries.

The problem of narcosis, particularly the theories of Pauling and Miller, was reviewed briefly. A "pressure anomaly" in the solubility of gases was reported by Schröder, who noted the existence of abrupt changes in gas solubility at around 20 atm. This pressure corresponds to depth in the sea within easy reach of present-day deep diving technology. Hypothermia was discussed in general terms and attention called to the frequent occurrence of very drastic physiological changes around 29° to 32°C. These effects were observed also in the physiological processes of hibernating animals. Further cooling, eventually resulting in the freezing of cells and tissue, was discussed briefly in the section on cryobiology. Finally, hyperthermia was discussed in connection with the lethal effects of temperatures near 45°C (for mammals and birds) and the use of hyperthermia in the treatment of malignancies. The enhanced sensitivity to high temperatures of malignant cells does correlate extremely well with the structural change in water near 44° to 45°C, although only tentative mechanisms can be suggested for the specific selectivity. The problem of cell adhesion was mentioned briefly and again seen as a matter of interaction between neighboring vicinal water structures. Thermal hysteresis effects were also discussed and attention called to the likely biological importance of this generally ignored subject. The effects of structural changes on biological systems were finally related to thermal pollution problems, and it was proposed that it may be possible, on the basis of data gathered to date, to delineate allowable thermal pollution limits.

VIII. Summary

The purpose of this chapter is (a) to call attention to the fact that water near interfaces frequently appears to undergo notable changes in properties and structure at a number of discrete temperature ranges (namely, near 13–16°, 29–32°, 44–46°, and 60–62°C) and (b) to attempt to correlate these changes with abrupt unexpected changes in the functioning of biological systems near those temperatures. The anomalies in the properties of vicinal water are most likely caused by higher-order phase transitions, due to cooperative processes. It appears that vicinal water structures, giving rise to thermal anomalies, may occur adjacent to highly dissimilar types of surfaces, including some (but not all) ionic, strongly dipolar, and completely nonpolar interfaces. The existence of at least four thermal anomalies suggests the occurrence of a minimum of five different vicinal water structures. The cooperative processes probably reflect relatively long-range ordering of the water molecules vicinal to the interface. It appears that very small energy differences between different structural arrangements may play an important role. The effect of the vicinal structuring appears to influence the activities of both the ions and the water in cells; the implications of this phenomenon for the problem of active transport are briefly discussed. The likely "sites of action" of the structural changes of the vicinal water are discussed (for instance, in terms of water interactions with proteins, enzymes, membranes, etc.). Finally, a number of examples are reviewed where it appears that the vicinal water plays a dominant role in the functional behavior of the biological systems. Some difficulties in interpretations have been considered, such as the complexities caused by lipid transformations (known to occur in completely anhydrous lipid systems); however, in spite of obvious exceptions (and some possible alternate mechanisms in individual, specific cases) it appears inescapable that a cause–effect relationship exists between thermal anomalies in the properties of vicinal water (due to structural changes) and anomalies in a large number of vastly different biological systems.

Acknowledgments

The author wishes to acknowledge the continued financial support from the Office of Saline Water for his research on the structure and properties of water and water near interfaces. The author also gratefully acknowledges the support by the Federal Water Pollution Control Administration (Environmental Protection Agency, Grant No. 18050 DET).

The author wishes, moreover, to thank Professor Rufus Lumry for making available a number of manuscripts prior to publication and also for reviewing the present paper. The conscientious literature searches and the assistance in editing by Miss Sharee Pepper have been most helpful and the assistance of Miss Lynda Weller in preparing the manuscript has been invaluable. Dr. George Safford is thanked profoundly for allowing the author to use his extensive tables of current water structure and hydration models. Finally, the author wishes to thank a number of individuals who have contributed through continued encouragement, helpful discussions, and correspondence, particularly Drs. S. A. Bach, Dean Burk, P. D. Cratin, F. Franks, C. F. Hazlewood, W. Luck, and A. Szent-Györgyi.

REFERENCES

Abdulla, Y. H. (1967). J. Atheroscler. Res. 7, 415–423.
Abramson, M. B. (1970). In "Surface Chemistry of Biological Systems" (M. Blank, ed.), pp. 37–54. Plenum Press, New York.

Ager, D. V. (1963). "Principles of Paleoecology." McGraw-Hill, New York.
Agnihotri, V. P., and Vaartaja, O. (1969). *Can. J. Microbiol.* **15,** 1319–1323.
Andjus, R. K. (1969). *Symp. Soc. Exp. Biol.* **23,** 351–394.
Andrewartha, H. G., and Birch, L. C. (1954), "The Distribution and Abundance of Animals." Univ. of Chicago Press, Chicago, Illinois.
Angelakos, E. T., Deutsch, I. S., and Williams, L. (1957). *Circulation Research* **5,** 196–201.
Apffel, C. A., and Peters, J. H. (1969). *Prog. Exp. Tumor Res.* **12,** 1–54.
Arnett, E. M. (1967). *In* "Physico-Chemical Processes in Mixed Aqueous Solvents" (F. Franks, ed.), pp. 105–128. Elsevier, Amsterdam.
Arnett, E. M., and McKelvey, D. R. (1965). *J. Amer. Chem. Soc.* **87,** 1393–1394.
Arnett, E. M., and McKelvey, D. R. (1966). *J. Amer. Chem. Soc.* **88,** 5031–5033.
Arnett, E. M., Kover, W. B., and Carter, J. V. (1969a). *J. Amer. Chem. Soc.* **91,** 4028–4034.
Arnett, E. M., and McKelvey, D. R. (1966b). *In* "Solute-Solvent Interaction" (C. O. Ritchie and J. F. Coetzie, eds.). Marcel Dekker, New York.
Bach, S. A. (1968). Unpublished studies.
Baldwin, J., and Hochachka, P. W. (1970). *Biochem. J.* **116,** 883–887.
Baldwin, J. J., and Cornatzer, W. E. (1968). *Biochim. Biophys. Acta* **164,** 195–204.
Bangham, A. D. (1969). "The Liposome as a Membrane Model." A. R. C. Institute of Animal Physiology, Babraham, Cambridge.
Bangham, A. D., Standish, M. M., and Watkins, J. C. (1965b). *J. Mol. Biol.* **13,** 238–252.
Barrall, E. M., and Guffy, J. C. (1967). *Advan. Chem. Ser.* **63,** 1–12.
Bangham, A. D., Standish, M. M., and Weissmann, G. (1965a). *J. Mol. Biol.* **13,** 253.
Bangham, A. D., and Papahadjopoulos, D. (1966). *Biochim. Biophys. Acta* **126,** 181.
Barry, P. H., and Hope, A. B. (1969). *Biophys. J.* **9,** 700.
Bean, R., and Chan, H. (1969). *In* "The Molecular Basis of Membrane Function" (D. C. Testeson, ed.), pp. 133–146. Prentice-Hall, Englewood Cliffs, New Jersey.
Belding, H. S. (1967). *In* "Thermobiology" (A. H. Rose, ed.), p. 479. Academic Press, New York.
Ben-Naim, A. (1965). *J. Phys. Chem.* **69,** 1922–1927.
Ben-Naim, A. (1967). *J. Phys. Chem.* **71,** 4002–4007.
Ben-Naim, A. (1968). *J. Phys. Chem.* **72,** 2998–3001.
Ben-Naim, A. (1969a). *J. Phys. Chem.* **72,** 2998–3001.
Ben-Naim, A. (1969b). "Thermodynamics of Dilute Aqueous Solutions of Non-Polar Solutes." Preprint. Department of Inorganic and Analytical Chemistry, Hebrew University of Jerusalem.
Ben-Naim, A. (1971a). *J. Chem. Phys.* **54,** 1387–1404.
Ben-Naim, A. (1971b). *J. Chem. Phys.* **54,** 3682.
Ben-Naim, A. (1971c). *J. Chem. Phys.* **54,** 3696.
Berendsen, H. J. C. (1966). *Fed. Proc., Fed. Amer. Soc. Exp. Biol.* **25,** 971–976.
Berendsen, H. J. C. (1967). *In* "Theoretical and Experimental Biophysics" (A. Cole, ed.), pp. 1–76. Marcel Dekker, New York.
Berendsen, H. J. C., and Migchelsen, C. (1966). *Fed. Proc., Fed. Amer. Soc. Exp. Biol.* **25,** 998–1002.
Bernal, J. D. (1960). *Nature (London)* **185,** 68.
Binet (1957). *P. Compt. Rend.* **244,** 1094–1096.
Blandamer, M. J., and Fox, M. F. (1970). *Chem. Rev.* **70,** 59–93.

Blandamer, M. J., Hidden, N. J., Morcom, K. W., Smith, R. W., Treloar, N. C., and Wotten, M. J. (1969a). *Trans. Faraday Soc.* **65,** 2633–2638.

Blandamer, M. J., Hidden, N. J., Symons, M. C. R., and Treloar, N. C. (1969b). *Trans. Faraday Soc.* **65,** 1806–1809.

Blandamer, M. J., Hidden, N. J., Symons, M. C. R., and Treloar, N. C. (1969c). *Trans. Faraday Soc.* **65,** 2663–2672.

Blandamer, M. J., Hidden, N. J., and Symons, M. C. R. (1970). *Trans. Faraday Soc.* **66,** 316–320.

Bordi, S., and Vannel, F. (1958). *Ric. Sci.* **28,** 2039.

Bordi, S., and Vannel, F. (1962). *Gaz. Chim. Ital.* **92,** 82.

Brandts, J. F. (1967). *In* "Thermobiology" (A. H. Rose, ed.), pp. 25–72. Academic Press, New York.

Brandts, J. F. (1969). *In* "Structure and Stability of Biological Molecules" (S. N. Timasheff and G. D. Fasman, eds.), pp. 213–290. Marcel Dekker, New York.

Branton, D., and Park, R. B., eds. (1968). "Papers on Biological Membrane Structure." Little, Brown, Boston, Massachusetts.

Brock, T. D. (1967). *Science* **158,** 1012–1019.

Bruice, T. C., and Butler, A. R. (1965). *Fed. Proc., Fed. Amer. Soc. Exp. Biol.* **24,** S45–S49.

Buetow, D. E. (1962). *Exp. Cell Res.* **27,** 137.

Bullard, R. W., David, G., and Nichols, C. T. (1960). *In* "Mammalian Hibernation" (C. P. Lyman and A. R. Dawe, eds.), pp. 321–336.

Burk, D., and Woods, M. (1967). *Arch. Geschwulstforsch.* **28,** 305–319.

Calabrese, A. (1969). *Biol. Bull.* **137,** 417–428.

Catchpool, J. F. (1966). *Fed. Proc., Fed. Amer. Soc. Exp. Biol.* **25,** 979–989.

Cavaliere, R., Ciocatto, E. C., Giovanella, B. C., Heidelberger, C., Johnson, R. O., Margottini, M., Mondovi, B., Moricca, G., and Rossi-Fanelli, A. (1967). *Cancer* **20,** 1351–1381.

Cerbon, J. (1967). *Biochim. Biophys. Acta* **144,** 1–9.

Chapman, D. (1965). Unpublished observations.

Chapman, D. (1967). *Advan. Chem. Ser.* **63,** 157–166.

Chapman, D., ed. (1968). "Biological Membranes." Academic Press, New York.

Chapman, D., and McLauchlan, K. A. (1967). *Nature (London)* **215,** 391–392.

Chapman, D., and Salsbury, N. J. (1970). *Recent Progr. Surface Sci.* **3,** 121–168.

Chapman, D., and Wallach, D. F. H. (1968). *In* "Biological Membranes" (D. Chapman, ed.), pp. 125–202. Academic Press, New York.

Chapman, R. A. (1967). *Nature (London)* **213,** 1143–1144.

Chaudry, J. S., and Mishra, R. K. (1969). *Studia Biophys.,* Berlin, Band **17,** Heft 3, S. 99–108.

Cini, R., Loglio, G., and Ficalbi, A. (1969). *Nature (London)* **223,** 1148.

Clark, H. T., and Nachmansohn, D., eds. "Ion Transport Across Membranes." Academic Press, New York.

Clarke, K. U. (1967). *In* "Thermobiology" (A. H. Rose, ed.), pp. 293–352. Academic Press, New York.

Clifford, J., and Pethica, B. A. (1968). *In* "Hydrogen-Bonded Solvent Systems" (A. K. Covington and P. Jones, eds.), pp. 169–179. Taylor & Francis, London.

Clifford, J., Pethica, B. A., and Senior, W. A. (1965). *Ann. N. Y. Acad. Sci.* **125,** 458–470.

Coldman, M. F., and Good, W. (1967). *Comp. Biochem. Physiol.* **21,** 201–206.

Coldman, M. F., and Good, W. (1968a). *Biochim. Biophys. Acta* **150,** 194–205.

Coldman, M. F., and Good, W. (1968b). *Biochim. Biophys. Acta* **150**, 206–213.

Coldman, M. F., and Good, W. (1969). *Biochim. Biophys. Acta* **183**, 346–349.

Coldman, M. F., Gent, M., and Good, W. (1969a). *Comp. Biochem. Physiol.* **31**, 605–609.

Coldman, M. F., Good, W., and Swift, D. (1969b). *Biochim. Biophys. Acta* **173**, 62–70.

Cole, K. S. (1968). "Membranes, Ions and Impulses." Univ. of California Press, Berkeley, California.

Cope, F. W. (1967a) *Bull. Math. Biophys.* **29**, 583–596.

Cope, F. W. (1967b). *Bull. Math. Biophys.* **29**, 691–704.

Cope, F. W. (1967c). *J. Gen. Physiol.* **50**, 1353–1575.

Cope, F. W. (1969). *Biophys. J.* **9**, 303–319.

Cope, F. W. (1971). *Bull. Math. Biophys.* **33**, 39–47.

Coster, H. G. L., and Simons, R. (1970). *Biochim. Biophys. Acta* **203**, 17–27.

Covington, A. K., and Jones, P., eds. (1968). "Hydrogen-Bonded Solvent Systems." Taylor & Francis, London.

Cowey, C. B., Lush, I. E., and Knox, D. (1969). *Biochim. Biophys. Acta* **191**, 205–213.

Cyr, T. J. R., Janzen, W. R., and Dunnell, B. A. (1967). *Adv. Chem. Ser.* **63**, 13–25.

Dalton, T., and Snart, R. S. (1967). *Biochim. Biophys. Acta* **135**, 1059–1062.

Damadian, R. (1971). *Science* **171**, 1151–1153.

Damaschke, von K., and Becker, G. (1964). *Z. Naturforsch. B* **19**, 157–160.

Danielli, J. F., and Davson, H. (1935). *J. Cell. Comp. Physiol.* **5**, 495–508.

Danielli, J. F., Pankhurst, K. G. A., and Riddiford, A. C., eds. (1964). "Recent Progress in Surface Science," Vols. 1 and 2. Academic Press, New York.

Danielli, J. F., Riddiford, A. C., and Rosenberg, M. D., eds. (1970). "Recent Progress in Surface Science," Vol. 3. Academic Press, New York.

DasGupta, S., and Grewal, M. S. (1968). *Evolution* **22**, 87.

Davey, C. B., and Miller, R. J. (1964). *Soil Sci. Soc. Amer., Proc.* **28**, 1–289.

Davson, H., and Danielli, J. F. (1943). "Permeability of Natural Membranes." Cambridge Univ. Press, London and New York.

Dawson, R. M. C. (1968). *In* "Biological Membranes" (D. Chapman, ed.), pp. 203–232. Academic Press, New York.

De Bruijne, A. W., and Van Steveninck, J. (1970). *Biochim. Biophys. Acta* **196**, 45–52.

DeHaven, J. C., and Shapiro, N. Z. (1968). *Perspec. Biol. Med.* **12**, 31–59.

Del Bene, J., and Pople, J. A. (1969). *Chem. Phys. Lett.* **4**, 426–428.

Derjaguin, B. V. (1965). *Symp. Soc. Exp. Biol.* **19**, 55–60.

Desnoyers, J. E., and Jolicoeur, C. (1969). *In* "Modern Aspects of Electrochemistry" No. 5, pp. 1–89. Plenum Press, New York.

DeSylva, D. P. (1969). *In* "Biological Aspects of Thermal Pollution," pp. 229–293. Vanderbilt Univ. Press, Nashville, Tennessee.

Diamond, J. M., and Wright, E. M. (1969). *Proc. Roy. Soc., Ser. B* **172**, 273–316.

Dick, D. A. T. (1966). "Cell Water." Butterworth, London and Washington, D.C.

Dixon. M., and Webb, E. C. (1960). "Enzymes." Academic Press, New York.

Dodt, E., and Zotterman, Y. (1952a). *Acta Physiol. Scand.* **26**, 345–357.

Dodt, E., and Zotterman, Y. (1952b). *Acta Physiol. Scand.* **26**, 358–365.

Dreyer, G., Kahrig, E., Kirstein, D., Erpenbeck, J., and Lange, F. (1969). *Naturwissenschaften* **56**, 558–559.

Drost-Hansen, W. (1956). *Naturwissenschaften* **43**, 512.

Drost-Hansen, W. (1965a). *Ann. N. Y. Acad. Sci.* **125**, 471–501.

Drost-Hansen, W. (1965b). *Ind. Eng. Chem.* **57**, 18–37.

VI. STRUCTURE AND PROPERTIES OF WATER AT BIOLOGICAL INTERFACES 175

Drost-Hansen, W. (1966). Abstrt., *Proc. 2nd Int. Cong., Biophys., 1966,* Vienna, Austria.
Drost-Hansen, W. (1967a). *Advan. Chem. Ser.* **67,** 70–120.
Drost-Hansen, W. (1967b). *J. Colloid Interface Sci.* **25,** 131–160.
Drost-Hansen, W. (1967c). *Proc. Int. Symp. Water Desalination, 1st, 1965* Vol. 1, pp. 382–406.
Drost-Hansen, (1969a). *Chem. Phys. Lett.* **2,** 647–652.
Drost-Hansen, W. (1969b). *Ind. Eng. Chem.* **61,** 10–47.
Drost-Hansen, W. (1969c). *Chesapeake Sci.* **10,** 281–288.
Drost-Hansen, W. (1971). To be published.
Drost-Hansen, W., and Thorhaug, A. (1967). *Nature (London)* **215,** 506–508.
Ehrenberg, A., Malmstrom, B. G., and Vanngard, T. (1967). "Magnetic Resonance in Biological Systems." Pergamon Press, Oxford.
Eisenberg, D., and Kauzmann, W. (1969). "The Structure and Properties of Water." Oxford Univ. Press, London and New York.
Eisentraut, M. (1960). *In* "Mammalian Hibernation" (C. P. Lyman and A. R. Dawe, eds.), pp. 31–44.
Falk, M., Hartman, K., and Lord, R. C. (1962). *J. Am. Chem. Soc.* **84,** 3843–3847.
Falk, M., Hartman, K., and Lord, R. C. (1963). *J. Am. Chem. Soc.* **85,** 387–394.
Falk, M., Poole, A. G., and Goymour, C. G. (1970). *Can. J. Chem.* **48,** 1536–1542.
Faraday Society. (1956). *Discuss. Faraday Soc.* **21,** 1–288.
Farrell, J., and Rose, A. (1967). *Annu. Rev. Microbiol.* **21,** 101–120.
Flautt, T. J., and Lawson, K. D. (1967). *Advan. Chem. Ser.* **63,** 26–50.
Fogg, G. E. (1969). *Symp. Soc. Exp. Biol.* **23,** 123–142.
Forslind, E. (1966). *Sv. Naturvetenskap* **2,** 9–74.
Forslind, E. (1968). "The Mechanics of Liquids Containing Bubbles." Grenoble, France.
Frank, H. S. (1958). *Proc. Roy. Soc., Ser. A* **247,** 481.
Frank, H. S. (1963). *Nat. Acad. Sci.—Nat. Res. Counc., Publ.* **942,** 141.
Frank, H. S. (1965a). *Z. Phys. Chem. (Leipzig)* **228,** 364.
Frank, H. S. (1965b). *Fed. Proc., Fed. Amer. Soc. Exp. Biol.* **24,** Suppl. 15, S-1.
Frank, H. S. (1966). *In* "Chemical Physics of Ionic Solutions" (B. E. Conway and R. G. Barradas, eds.). Wiley, New York.
Frank, H. S. (1967). *Proc. Int. Symp. Water Desalination, 1st, 1965* Vol. 1.
Frank, H. S. (1970). *Science* **169,** 635–641.
Frank, H. S., and Evans, M. W. (1945). *J. Chem. Phys.* **13,** 507–532.
Frank, H. S., and Franks, F. (1968). *J. Chem. Phys.* **48,** 4746–4757.
Frank, H. S., and Quist, A. S. (1961). *J. Chem. Phys.* **34,** 604.
Frank, H. S., and Thompson, P. T. (1959a). *J. Chem. Phys.* **31,** 1086.
Frank, H. S., and Thompson, P. T. (1959b). *In* "The Structure of Electrolytic Solutions" (W. J. Hamer, ed.), Wiley, New York.
Frank, H. S., and Wen, W. Y. (1957). *Discuss. Faraday Soc.* **24,** 133.
Franks, F. (1966). *Nature (London)* **210,** 87–88.
Franks, F., ed. (1967). "Physico-Chemical Processes in Mixed Aqueous Solvents." Elsevier, Amsterdam.
Franks, F. (1968). *In* "Hydrogen-Bonded Solvent Systems" (A. K. Covington and P. Jones, eds.), pp. 31–47. Taylor & Francis, London.
Franks, F., and Ives, D. J. G. (1960). *J. Chem. Soc., London* p. 741.
Franks, F., and Ives, D. J. G. (1966). *Quart. Rev., Chem. Soc.* **20,** 1.
Franks, F., Ravenhill, J., Egelstaff, P. A., and Page, D. I. (1971). *Proc. Roy. Soc., Ser. A* (in press).
Franzen, J. S., Kuo, I., and Bobik, C. M. (1970). *Biochim. Biophys. Acta* **200,** 566–569.

VI. STRUCTURE AND PROPERTIES OF WATER AT BIOLOGICAL INTERFACES 175

Drost-Hansen, W. (1966). Abstrt., *Proc. 2nd Int. Cong., Biophys., 1966,* Vienna, Austria.

176   W. DROST-HANSEN

Frenkel, J. (1955). "Kinetic Theory of Liquids." Dover, New York.
Fritz, O. G., Jr., and Swift, T. J. (1967). *Biophys. J.* **7**, 675–687.
Fry, F. E. J. (1967). *In* "Thermobiology" (A. H. Rose, ed.), p. 775. Academic Press, New York.
Garvin, J. E. (1968). *In* "Conferences on Cellular Dynamics" (L. D. Peachey, ed.), pp. 278–316. N. Y. Acad. Sci., *Interdisciplinary Commun. Program*, New York.
Gary-Bobo, C. M., and Solomon, A. K. (1971). *J. Gen. Physiol.* **57**, 610–622.
Gary-Bobo, C. M., Lange, Y., and Rigaud, J. L. (1971). *Biochim. Biophys. Acta* **233**, 243–246.
Gawalek, G. (1969). "Einschlussverbindungen, Additionsverbindungen, Clathrate." Deut. Verlag Wiss., Berlin.
Gittens, G. J. (1969). *J. Colloid Interface Sci.* **30**, 406.
Glasel, J. A. (1968). *Nature (London)* **218**, 953–955.
Glasel, J. A. (1970a). *J. Amer. Chem. Soc.* **92**, 372–375.
Glasel, J. A. (1970b). *J. Amer. Chem. Soc.* **92**, 375–381.
Glew, D. N. (1962a). *J. Phys. Chem.* **66**, 605–609.
Glew, D. N. (1962b). *Nature (London)* **195**, 698.
Glew, D. N., Mak, H. D., and Rath, N. S. (1968). *In* "Hydrogen-Bonded Solvent Systems" (A. K. Covington and P. Jones, eds.), pp. 195–210. Taylor & Francis, London.
Goldup, A., Ohki, S., and Danielli, J. F. (1970). *Recent Progr. Surface Sci.* **3**, 193–261.
Good, W. (1960). *Biochim. Biophys. Acta* **44**, 130–143.
Good, W. (1961a). *Biochim. Biophys. Acta* **48**, 229–241.
Good, W. (1961b). *Biochim. Biophys. Acta* **49**, 397–399.
Good, W. (1961c). *Biochim. Biophys. Acta* **50**, 485–493.
Good, W. (1961d). *Biochim. Biophys. Acta* **52**, 545–551.
Good, W. (1967). *Nature (London)* **214**, 1250–1252.
Good, W., and Rose, S. M. (1968). *Biochim. Biophys. Acta* **163**, 483–493.
Gorter, E., and Grendal, F. (1925). *J. Exp. Med.* **41**, 439–443.
Graves, D. J., Sealock, R. W., and Wang, J. H. (1965). *Biochemistry* **4**, 290–296.
Green, K., and Otori, T. (1970). *J. Physiol. (London)* **207**, 93–102.
Gustafson, D. R. (1970). *Biophys. J.* **10**, 316–322.
Gutknecht, J. (1968). *Biochim. Biophys. Acta* **163**, 20–29.
Haase, R., and de Greiff, H. J. (1965). *Z. Phys. Chem. (Frankfurt am Main)* **44**, 301.
Haase, R., and Steiner, C. (1959). *Z. Phys. Chem. (Frankfurt am Main)* **21**, 270.
Hankins, D., Moskowitz, J. W., and Stillinger, F. H. (1970). *Chem. Phys. Lett.* **4**, 527–530.
Haskins, R. H. (1965). *Science* **150**, 1615–1616.
Hazlewood, C. F. (1971). *In* "Reversibility of Cellular Injury due to Inadequate Perfusion" (T. I. Malinin, ed.), C. C Thomas, Springfield, Illinois.
Hazlewood, C. F., and Nichols, B. L. (1968). *Johns Hopkins Med. J.* **123**, 198–203.
Hazlewood, C. F., and Nichols, B. L. (1969). *Physiologist* **12**, 251.
Hazlewood, C. F., Nichols, B. L., and Chamberlain, N. F. (1969). *Nature (London)* **222**, 747–750.
Hazlewood, C. F., Chang, D. C., Nichols, B. L., and Rorschach, H. E. (1971a). *Johns Hopkins Med. J.* **128**, 117–131.
Hazlewood, C. F. *et al.* (1971b). *J. Mol. Cell. Cardiol.* **2**, 51–53.
Hechter, O. (1965). *Ann. N. Y. Acad. Sci.* **125**, 625–646.
Henn, S. W., and Ackers, G. K. (1969). *Biochemistry* **8**, 3829–3838.
Henniker, J. C. (1949). *Rev. Mod. Phys.* **21**, 322–341.

Hensel, H. (1963). *In* "Temperature" (C. M. Herzfeld, ed.), Vol. 3, Part 3. Reinhold, New York.

Hensel, H., Iggo, A., and Witt, I. (1960). *J. Physiol.* (*London*) **153**, 113–126.

Hertz, H. G. (1970). *Angew. Chem., Int. Ed., Engl.* **9**, 124–138.

Hertz, H. G., Lindman, B., and Siepe, V. (1969). *Ber. Bunsenges. Phys. Chem.* **73**, 542.

Higasi, K.-i. (1955). "Studies on Bound Water," Monograph Ser. Res. Inst. Appl. Elec. No. 5, pp. 9–35.

Hildebrand, J. H. (1969). *Proc. Nat. Acad. Sci. U. S.* **64**, 1331–1334.

Hochachka, P. W., and Somero, G. N. (1968). *Comp. Biochem. Physiol.* **27**, 659–668.

Hori, T. (1960). "On the Super Cooling and Evaporation of Thin Water Films," U. S. Army Snow, Ice and Permafrost Res. Estab., Transl. No. 62.

Ilani, A., and Tzivoni, D. (1968). *Biochim. Biophys. Acta* **163**, 429–438.

Ingraham, J. L., and Maaløe, O. (1967). *In* "Molecular Mechanisms of Temperature Adaptation," Publ. No. 84, pp. 297–309. Am. Assoc. Advance Sci., Washington, D. C.

Jacobson, B. (1953). *Nature* (*London*) **172**, 666.

Jacobson, B. (1955). *Sv. Kem. Tidskr.* **67**, 1–7.

Järnefelt, J. (1968). "Regulatory Functions of Biological Membranes." Elsevier, Amsterdam.

Jeffrey, G. A. (1969). *Accounts Chem. Res.* **2**, 344–352.

Johnson, F. H., Eyring, H., and Polissar, M. J. (1954). "The Kinetic Basis of Molecular Biology." Wiley, New York.

Johnson, G. A., Lecchini, S. M. A., Smith, E. G., Clifford, J., and Pethica, B. A. (1966). *Discuss. Faraday Soc.* **42**, 120–142.

Johnson, S. M., and Bangham, A. D. (1969). *Biochim. Biophys. Acta* **193**, 92–104.

Kamb, B. (1968). *In* "Structural Chemistry and Molecular Biology" (A. Rich and N. Davidson, eds.), pp. 507–542. Freeman, San Francisco, California.

Karpovich, O. A. (1960). *In* "The Problem of Acute Hypothermia" (P. M. Starkov, ed.), pp. 32–43. Pergamon Press, Oxford.

Kavanau, J. L. (1964). "Water and Solute-Water Interactions." Holden-Day, San Francisco, California.

Kavanau, J. L. (1965). "Structure and Function in Biological Membranes," Vols. I and II. Holden-Day, San Francisco, California.

Kayushin, L. P., ed. (1969). "Water in Biological Systems." Consultants Bureau, New York.

Kemp, A., Groot, G. S. P., and Reitsma, H. F. (1969). *Biochim. Biophys. Acta* **180**, 28–34.

Kendrew, J., and Moelwyn-Hughes, E. A. (1940). *Proc. Roy. Soc., Ser. A* **176**, 352–367.

Kerr, J. E. (1970). Ph.D. Dissertation, University of Miami.

Kirsch, R., and Schmidt, D. (1966). *In* "Aktuelle Probleme aus dem Gebiet der Cancerologie" (W. Doerr, F. Linder, and G. Wagner, eds.). Springer-Verlag, Heidelberg.

Kistiakowsky, G. B., and Lumry, R. (1949). *J. Amer. Chem. Soc.* **71**, 2006.

Kleinzeller, A., and Kotyk, A., eds. (1961). "Membrane Transport and Metabolism." Academic Press, New York.

Klotz, I. M. (1958). *Science* **128**, 815–822.

Klotz, I. M. (1965). *Fed. Proc., Fed. Amer. Soc. Exp. Biol.* **24**, Part III, Suppl. 15, S24–S33.

Klykov, N. V. (1960). In "The Problem of Acute Hypothermia" (P. M. Starkov, ed.), pp. 82–92. Pergamon Press, Oxford.

Korson, L., Drost-Hansen, W., and Millero, F. J. (1969). J. Phys. Chem. 73, 34–39.

Krestov, G. A., Klopov, V. I., and Patsatsiya, K. M. (1969). J. Struct. Chem. 10, 343–347.

Kruyt, H. R., ed. (1949). "Colloid Science," Vol. II, Reversible Systems (see specifically pp. 684–685). Elsevier, Amsterdam.

Kruyt, H. R., ed. (1952). "Colloid Science," Vol. I. Irreversible Systems. Elsevier, Amsterdam.

Kuczenski, R. T., and Suelter, C. H. (1970). Biochemistry 9, 939–945.

Kushmerick, M. J., and Podolsky, R. J. (1969). Science 166, 1297.

Kuznetsova, Z. P. (1960). In "The Problem of Acute Hypothermia" (P. H. Starkov, ed.), pp. 93–106. Pergamon Press, Oxford.

Ladbrooke, B. D., and Chapman, D. (1969). Chem. Phys. Lipids 3, 304–367.

Laget, P., and Lundberg, A. (1949). Acta Physiol. Scand. 18, 121–138.

Lakshminarayanaiah, N. (1969). "Transport Phenomena in Membranes." Academic Press, New York.

Lang, J., and Zana, R. (1970). Trans. Faraday Soc. 66, 597–604.

Langridge, J., and McWilliam, J. R. (1967). In "Thermobiology" (A. H. Rose, ed.), Academic Press, New York.

Lehrer, G. M., and Barker, R. (1970). Biochemistry 9, 1533–1539.

Levinson, H. S., and Hyatt, M. T. (1970). J. Bacteriol. 101, 58–64.

Levitt, J. (1969). Symp. Soc. Exp. Biol. 23, 395–448.

Licht, P. (1967). In "Molecular Mechanisms of Temperature Adaptation," Publ. No. 84, pp. 131–145. Am. Assoc. Advance. Sci., Washington, D.C.

Ling, G. N. (1962). "A Physical Theory of the Living State: The Association-Induction Hypothesis." Ginn (Blaisdell), Boston, Massachusetts.

Ling, G. N. (1965). Ann. N. Y. Acad. Sci. 125, 401–417.

Lippold, O. C. J., Nicholls, J. G., and Redfearn, J. W. T. (1960). J. Physiol. (London) 153, 218–231.

Low, P. F. (1961). Advan. Agron. 13, 269–327.

Luck, W. (1964). Fortschr. Chem. Forsch. 4, 43–781.

Lumry, R., and Biltonen, R. (1969). In "Structure and Stability of Biological Macromolecules" (S. N. Timasheff and G. D. Fasman, eds.), pp. 65–212. Marcel Dekker, New York.

Lumry, R., and Rajender, S. (1971). "Enthalpy-Entropy Compensation Phenomena in Water Solutions of Proteins and Small Molecules." Wiley (Interscience), New York (in press).

Luzzati, V. (1968). In "Biological Membranes" (D. Chapman, ed.), pp. 71–123. Academic Press, New York.

Luzzati, V., Gulik-Krzywicki, T., Rivas, E., Reiss-Husson, F., and Rand, R. P. (1968). J. Gen. Physiol. 51, 37s–43s, Part 2.

Lyman, C. P., and Dawe, A. R. (1960). Bull Mus. Comp. Zool., Harvard Univ. 124, 1–549.

Lyman, C. P., and O'Brien, R. C. (1969). Bull. Mus. Comp. Zool., Harvard Univ. 124, 353–372.

Mak, T. C. W. (1965). J. Chem. Phys. 43, 2799–2805.

Mandell, L., Fontell, K., and Ekwall, P. (1967). Advan. Chem. Ser. 63, 89–124.

Markovitz, A., Klein, H. P., and Fischer, E. H. (1956). Biochim. Biophys. Acta 19, 267–273.

Martin-Löf, S., and Söremark, C. (1969a). *Sv. Träforskningsinsfitutet, Ser. B* No. 8, pp. 1–15.
Martin-Löf, S., and Söremark, C. (1969b). *Sv. Träforskiningsinstitutet, Ser. B* No. 1, 1–15.
Martin-Löf, S., and Söremark, C. (1969c). *Sv. Papperstidn.* **72**, 193–194.
Martin-Löf, S., and Söremark, C. (1970). Personal communication.
Massey, V., Curti, B., and Ganther, H. (1966). *J. Biol. Chem.* **241**, 2347–2357.
Mazur, P. (1966). *In* "Cryobiology" (H. T. Meryman, ed.), pp. 214–316. Academic Press, New York.
Mazur, P. (1970). *Science* **168**, 939–949.
Meryman, H. T., (1966a). *In* "Cryobiology" (H. T. Meryman, ed.), pp. 1–114. Academic Press, New York.
Meryman, H. T., ed. (1966b). "Cryobiology." Academic Press, New York.
Metsik, M. S., and Aidanova, O. S. (1966). *Res. Surface Forces, Proc. Conf., 2nd 1962* Vol. 2, pp. 169–175.
Meyer, H. H., Spahr, P. F., and Fischer, E. H. (1953). *Helv. Chim. Acta* **36**, 1924–1937.
Mikhailov, V. A. (1968). *J. Struct. Chem.* **9**, 332–339.
Miller, J. A., Jr. (1957). *In* "Influence of Temperature on Biological Systems" (F. H. Johnson, ed.), pp. 229–257. Am. Physiol. Soc., Washington, D. C.
Miller, R. J. (1968). Personal communication.
Miller, S. L. (1961). *Proc. Nat. Acad. Sci. U.S.*, **47**, 1515–1524.
Miller, S. L. (1968). *Fed. Proc., Fed. Amer. Soc. Exp. Biol.* **27**, 879–883.
Millero, F. J. (1970). *J. Phys. Chem.* **74**, 356–362.
Millero, F. J. (1971). *In* "Structure and Processes in Water and Aqueous Solutions" (R. A. Horne, ed.), Wiley, New York.
Mitchell, H. K., and Houlahan, M. B. (1946). *Amer. J. Bot.* **33**, 31.
Nelson and Blei (1966). "Model Membrane Studies Related to Ionic Transport in Biological Systems." U.S. Dept. Interior, OSW R & D Rpt. 221.
Nemethy, G. (1967). *Angew. Chem.*, **6**, 195–205.
Nemethy, G., and Scheraga, H. A. (1962a). *J. Chem. Phys.* **37**, 3382–3400.
Nemethy, G., and Scheraga, H. A. (1962b). *J. Chem. Phys.* **36**, 3401–3417.
Nemethy, G., and Scheraga, H. A. (1962c). *J. Phys. Chem.* **66**, 1773–1789.
Neuberger, A., and Tatum, E. L. (1967). eds. "Frontiers of Biology," Vol. 8. Wiley, New York.
New York Academy of Sciences. (1966). *Ann. N. Y. Acad. Sci.* **137**, 403–1048.
New York Heart Association. (1968). "Biological Interfaces: Flows and Exchanges." Little, Brown, Boston, Massachusetts.
Nishiyama, I. (1969). *Proc. Crop Sci. Soc. Jap.* **38**, 554–555.
Nishiyama, I. (1970). *Kagaku To Seibutsu* **8**, 14–20.
Odeblad, E. (1959). *Ann. New York Acad. Sci.* **83**, 189–207.
Ohki, S. (1970). *In* "Physical Principles of Biological Membranes" (F. Snell *et al.*, eds.), pp. 175–225. Gordon & Breach, New York.
Olmstead, E. G. (1966). "Mammalian Cell Water, Physiologic and Clinical Aspects." Lea & Febiger, Philadelphia, Pennsylvania.
Oppenheimer, C. H., and Drost-Hansen, W. (1960). *J. Bacteriol.* **80**, 21–24.
Pak, C. Y. C., and Gershfeld, N. L. (1967). *Nature (London)* **214**, 888–889.
Pauling, L. (1961). *Science* **134**, 15–21.
Peachey, L. D., ed. (1968). "Conferences on Cellular Dynamics." N. Y. Acad. Sci., Interdisciplinary Commun. Program, New York.

Peschel, G., and Adlfinger, K. H. (1967). *Naturwissenschaften* **54**, 614.
Peschel, G., and Adlfinger, K. H. (1969). *Naturwissenschaften* **56**, 558–559.
Peschel, G., and Adlfinger, K. H. (1970). Personal communication.
Pethica, B. A. (1961). *Exp. Cell Res. Suppl.* **8**, pp. 123–140.
Phillips, M. C., Ladbrooke, B. D., and Chapman, D. (1970). *Biochim. Biophys. Acta* **196**, 35–44.
Piccardi, G. (1962). "The Chemical Basis of Medical Climatology." C. Thomas, Springfield, Illinois.
Piguet, A., and Fischer, E. H. (1952). *Helv. Chim. Acta* **35**, 257–263.
Poland, D., and Scheraga, H. A. (1970). "Theory of Helix-Coil Transitions in Biopolymers." Academic Press, New York.
Porter, R. S., and Johnson, J. F. (1967). *Advan. Chem. Ser.* **63**, 1–332.
Powell, H. M. (1948). *J. Chem. Soc., London* pp. 61–73.
Prather, J. W., and Wright, E. M. (1970). *J. Membrane Biol.* **2**, 150–172.
Precht, H., Christophersen, J., and Hensel, H. (1955). "Temperatur und Leben." Springer, Berlin.
Privalov, P. L. (1958). *Biophysics (USSR)* **3**, 691–696.
Prokop'eva, E. M. (1960). *In* "The Problem of Acute Hypothermia" (P. M. Starkov, ed.), pp. 249–256. Pergamon Press, Oxford.
Prosser, C. L., ed. (1967). "Molecular Mechanisms of Temperature Adaptation," Publ. No. 84. Am. Assoc. Advance Sci., Washington, D. C.
Ramiah, M. V., and Goring, D. A. I. (1965). *J. Polym. Sci., Part C* **11**, 27–48.
Resing, H. A., and Neihof, R. A. (1970). *J. Colloid Interface Sci.* **34**, 480–487.
Richards, F. M. (1963). *Ann. Rev. Biochem.* **32**, 269–300.
Robertson, R. E., and Sugamori, S. E. (1969). *J. Amer. Chem. Soc.* **91**, 7254–7259.
Rosano, H. L., Duby, P., and Schulman, J. H. (1961). *J. Phys. Chem.* **65**, 1704.
Rose, A. H., ed. (1967). "Thermobiology." Academic Press, New York.
Rouser, G., Nelson, G. J., Fleischer, S., and Simon, G. (1968). *In* "Biological Membranes" (D. Chapman, ed.), pp. 5–70. Academic Press, New York.
Rushe, E. W., and Good, W. B. (1966). *J. Chem. Phys.* **45**, 4667.
S. E. B. (1965). *Symp. Soc. Exp. Biol.* **19**, 1–432.
Safford, G. J. (1966). *Cryobiology* **3**, 32–39.
Safford, G. J., and Leung, P. S. (1971). *In* "Techniques of Electrochemistry," Vol. II. Wiley, New York.
Samoilov, O. Ya. (1965). "Structure of Aqueous Electrolyte Solutions and the Hydration of Ions." Consultants Bureau, New York.
Schleich, T., and von Hippel, P. H. (1970). *Biochemistry* **9**, 1059–1066.
Schmidt, M. G., and Drost-Hansen, W. (1961). *Abstr., 140th Meet., Amer. Chem. Soc.*
Schoffeniels, E. (1967). "Cellular Aspects of Membrane Permeability." Pergamon Press, Oxford.
Schölgl, R. (1964). *Fortschr. Phys. Chem.* **9**, 1–123.
Schreiner, H. R. (1968). *Fed. Proc., Fed. Amer. Soc. Exp. Biol.* **27**, 872–883.
Schröder, W. (1968). *Naturwissenschaften* **55**, 542.
Schröder, W. (1969a). *Z. Naturforsch. B* **24**, 500–508.
Schröder, W. (1969b). Personal communication.
Schulman, J. H., and Teorell, T. (1938). *Trans. Faraday Soc.* **34**, 1337–1342.
Schultz, R. D., and Asunmaa, S. K. (1970). *Recent Progr. Surface Sci.* **3**, 291–332.
Schwan, H. P. (1965). *Ann. N. Y. Acad. Sci.* **125**, 344–354.
Seidell and Linke (1952). "Solubilities of Inorganic and Organic Compounds." Suppl. to 3rd Ed., Van Nostrand, New York.

Senghaphan, W., Zimmerman, G. O., and Chase, G. E. (1969). *J. Chem. Phys.* **51**, 2543–2545.

Shah, D. O. (1970). *Adv. Lipid Res.* **8**, 347–431.

Siegel, S. M. (1969). *Physiol. Plant.* **22**, 327–331.

Sitte, P. (1969). *Ber. Deutsch. Bot. Ges.* **82**, 329–383.

Skinner, F. A. (1968). *Proc. Roy. Soc., Ser. B* **171**, 77–89.

Small, D. M. (1970). *In* "Surface Chemistry of Biological Systems" (M. Blank, ed.), pp. 55–84. Plenum Press, New York.

Smith, E. B. (1969). *In* "The Physiology and Medicine of Diving and Compressed Air Work" (P. B. Bennett, ed.), pp. 183–192. Baillière, London.

Smith, M. W. (1967). *Biochem. J.* **105**, 65–71.

Snell, F., Wolken, J., Iverson, G., and Lam, J., eds. (1970). "Physical Principles of Biological Membranes." Gordon & Breach, New York.

Solomon, A. K. (1968). *J. Gen. Physiol.* **51**, 335.

Somero, G. N. (1969). *Biochem. J.* **114**, 237–241.

Spanner, D. C. (1954). *Sym. Soc. Exp. Biol.* **8**, 76–93.

Staal, G. E., and Veeger, C. (1969). *Biochim. Biophys. Acta* **185**, 49–62.

Stadelmann, E. (1970). *What's New in Plant Physiol.* **2**(5).

Starkov, P. M., ed. (1960). "The Problem of Acute Hypothermia." Pergamon Press, Oxford.

Stehli, F. G. (1957). *Amer. J. Sci.* **255**, 607–618.

Steim, J. M. (1968). *Advan. Chem. Ser.* **84**, 259–302.

Steim, J. M. (1969). *Biochem. Biophys. Res. Commun.* **34**, 434–440.

Steim, J. M., Edner, O. J., and Bargoot, F. G. (1968). *Science* **162**, 909–911.

Steim, J. M., Tourtellotte, M. E., Reinert, J. C., McElhaney, R. N., and Rader, R. L. (1969). *Proc. Nat. Acad. Sci. U.S.,* **63**, 104–109.

Stein, W. D. (1967). "The Movement of Molecules Across Cell Membranes." Academic Press, New York.

Stewart, G. W. (1931). *Phys. Rev.* **37**, 9–16.

Stillinger (1970). Personal communication.

Strumwasser, F. (1960). *In* "Mammalian Hibernation" (C. P. Lyman and A. R. Dawe, eds.), pp. 285–320.

Susi, H. (1969). *In* "Structure and Stability of Biological Macromolecules" (S. N. Timasheff and G. D. Fasman, eds.). Marcel Dekker, New York.

Symons, M. C. R., and Blandamer, M. J. (1968). *In* "Hydrogen-Bonded Solvent Systems" (A. K. Covington and P. Jones, eds.), pp. 211–220. Taylor & Francis, London.

Symposium on Cell Membrane Biophysics (1968). *J. Gen. Physiol.* **51**, Part 2.

Szent-Györgyi, A. (1957). "Bioenergetics." Academic Press, New York.

Szent-Györgyi, A. (1965). Personal communication.

Szent-Györgyi, A. (1971). *Perspec. Biol. & Med., Winter,* 239–250.

Tait, M. J., and Franks, F. (1971). *Nature* **230**, 91–94.

Takahashi, T., and Ohsaka, A. (1970). *Biochim. Biophys. Acta* **198**, 293–307.

Tappel, A. L. (1966). *In* "Cryobiology" (H. T. Meryman, ed.), pp. 163–177. Academic Press, New York.

Thompson, P. A. (1969). *Hort Res.* **9**, 130–138.

Thompson, P. A. (1970a). *Ann. Bot.* (*London*) [N. S.] **34**, 427–449.

Thompson, P. A. (1970b). *Nature* (*London*) **225**, 827–831.

Thompson, T. E. (1964). *In* "Cellular Membranes in Development" (M. Locke, ed.), pp. 83–96. Academic Press, New York.

Thorhaug, A. (1967). Unpublished results.

Ting, H. P., Bertrand, G. L., and Sears. D. F. (1966). *Biophys. J.* **6,** 813–823.

Ting, H. P., Huemoeller, W. A., Lalitha, S., Diana, A. L., and Tien, H. T. (1968). *Biochim. Biophys. Acta* **163,** 439–450.

Tracey, M. V. (1968). *Proc. Roy. Soc., Ser. B* **171,** 59–65.

Ushakov, B. P. (1968). *Mar. Biol.* **1,** 153–160.

Uzelac, B. M., and Cussler, E. L. (1970). *J. Colloid Interface Sci.* **32,** 487–491.

van der Waals, J. H., and Platteeuw, J. C. (1959). *Advan. Chem. Phys.* **2,** 1–57.

Vaslow, F. (1963). *J. Phys. Chem.* **67,** 2773.

Vieira, F. L., Schaafi, R. I., and Solomon, A. K. (1970). *J. Gen. Physiol.* **55,** 451–466.

von Ardenne, M. (1965). *Naturwissenschaften* **52,** 419.

von Ardenne, M., and Reitnauer, P. G. (1966b). *Z. Naturforsch. B* **21,** 841–848.

von Ardenne, M., Elsner, J., Kruger, W., Reitnauer, P. G., and Rieger, F. (1966a). *Klin. Wochenschr.* **44,** 503–511.

von Hippel, P. H., and Schleich, T. (1969). *In* "Structure and Stability of Biological Macromolecules" (S. N. Timasheff, and G. D. Fasman, eds.). pp. 417–574. Marcel Dekker, New York.

von Hippel, A., and Farrell, E. F. (1971). "I. A Molecular Interpretation of the Phase Diagram of Ice." Technical Report 9 MIT (New Series) (to Office of Naval Research, Washington, D.C.).

Walker, J. M. (1969). *Soil. Sci. Soc. Amer., Proc.* **33,** 729–736.

Warner, D. T. (1965). *Ann. N. Y. Acad. Sci.* **125,** 605–624.

Wang, J. H. (1965). *J. Phys. Chem.* **69,** 4412.

Wersuhn, G. (1967). *Naturwissenschaften* **54,** 27.

Wetlaufer, D. B., and Lovrien, R. (1964a). *J. Biol. Chem.* **239,** 596–608.

Wetlaufer, D. B., Malik, S. K., Stoller, L., and Coffin, R. L. (1964b). *J. Amer. Chem. Soc.* **86,** 508.

Wetzel, R., Zirwer, D., and Becker, M. (1969). *Biopolymers* **8,** 391–401.

Wishnia, A. (1962). *Proc. Nat. Acad. Sci. U. S.* **48,** 2200–2204.

Woessner, D. E. (1963). *J. Chem. Phys.* **39,** 2783–2787.

Woessner, D. E. (1966). *J. Phys. Chem.* **70,** 1217–1230.

Woessner, D. E. (1971). Personal communication.

Woessner, D. E., and Zimmerman, Y. (1963). *J. Phys. Chem.* **67,** 1590.

Woessner, D. E., and Snowden, B. S., Jr. (1968). *J. Colloid Interface Sci.* **26,** 297–305.

Woessner, D. E., and Snowden, B. S., Jr. (1969a). *J. Chem. Phys.* **50,** 1516–1523.

Woessner, D. E., and Snowden, B. S., Jr. (1969b). *J. Colloid Interface Sci.* **30,** 54–67.

Woessner, D. E., Snowden, B. S., Jr., and Chiu, Y.-C. (1970a). *J. Colloid Interface Sci.* **34,** 283–289.

Woessner, D. E., Snowden, B. S., Jr., and Meyer, G. H. (1970b). *J. Colloid Interface Sci.* **34,** 43–52.

Wright, E. M., and Diamond, J. M. (1969). *Proc. Roy. Soc., Ser. B* **172,** 227–271.

Wright, E. M., and Prather, J. W. (1970). *J. Membrane Biol.* **2,** 127–149.

Yamashita, S., and Sato, M. (1965). *J. Cell. Comp. Physiol.* **66,** 1–18.

Yastremskii, P. S. (1963). *J. Struct. Chem. (Russian)* **4,** 161–164.

Zhilenkov, A. P. (1963). Abstract of Doctor's Dissertation, Institute of Physical Chemistry of the Academy of Sciences of the U.S.S.R.

Zotterman, Y. (1959). *In* "Handbook of Physiology: Neurophysiology" (J. Field, ed.), p. 431. Amer. Physiol. Soc., Washington, D.C.

Zundel, G. (1969). "Hydration and Intermolecular Interaction: Infrared Investigations with Polyelectrolyte Membranes." Academic Press, New York.

SUPPLEMENTARY REFERENCES

Adolph, E. F. (1963). How Do Infant Mammals Tolerate Deep Hypothermia? *In* "Temperature: Its Measurement and Control in Science and Industry" (C. M. Herzfeld, ed.), Vol. 3. Reinhold, New York.

Berlin, E., Kliman, P. G., and Pallansch, M. J. (1970). Changes in State of Water in Proteinaceous Systems. *J. Colloid and Interface Sci.* **34**, 488–494.

Blanchard, K. C. (1940). Water, Free and Bound. *Cold Spring Harbor Symp. Quant. Biol.* **8**, 1–8.

Burt, D. H., and Green, J. W. (1971). The Sodium Permeability of Butanol-Treated Erythrocytes—the Role of Calcium. *Biochim. Biophys. Acta* **225**, 46–55.

Child, T. F., Pryce, N. G., Tait, M. J., and Ablett, S. (1970). Proton and Deuteron Magnetic Resonance Studies of Aqueous Polysaccharides. *Chem. Comm.,* pp. 1214–1215.

Cuthbert, A. W., and Dunant, Y. (1970). Diffusion of Drugs through Stationary Water Layers as the Rate Limiting Process in Their Action at Membrane Receptors. *Br. J. Pharmac.* **40**, 508–521.

Friedenberg, R. M. (1967). "The Electrostatics of Biological Cell Membranes: Frontiers of Biology," Vol. 8. North-Holland Publishing Co., Amsterdam, Holland and Wiley, New York.

Gilbert, J. C., Gray, P., and Heaton, G. M. (1971). Anticonvulsant Drugs and Brain Glucose. *Biochem. Pharmac.* **20**, 240–243.

Grigera, J. R., and Creijido, M. (1971). The State of Water in the Outer Barrier of the Isolated Frog Skin. *J. Memb. Biol.* **4**, 148–155.

Hadzi, D., ed. (1969). "Hydrogen Bonding." Pergamon Press, London.

Harrison, B. D. (1956). Studies on the Effect of Temperature on Virus Multiplication in Inoculated Leaves. *Ann. Appl. Biol.* **44**, 215–226.

Henrikson, K. (1970). Observations by Nuclear Magnetic Resonance of the Interactions of Water with Lecithin Micelles in Carbon Tetrachloride Solution. *Biochim. Biophys. Acta* **203**, 228–232.

Hogg, J., Williams, E. J., and Johnston, R. J. (1968). The Temperature Dependence of the Membrane Potential and Resistance in *Nitella translucens*. *Biochim. Biophys. Acta* **150**, 640–648.

Horne, R. A., Almeida, J. P., Day, A. F., and Yu, N-T. (1971). Macromolecule Hydration and the Effect of Solutes on the Cloud Point of Aqueous Solutions of Polyvinyl Methyl Ether: A Possible Model for Protein Denaturation and Temperature Control in Homeothermic Animals. *J. Colloid and Interface Sci.* **35**, 77–84.

Iampietro, P. F. (1971). Use of Skin Temperature to Predict Tolerance to Thermal Environments. *Aerospace Med., Apr.* 396–399.

Ladbrooke, B. D., Williams, R. M., and Chapman, D. (1968). Studies on Lecithin–Cholesterol–Water Interactions by Differential Scanning Calorimetry and X-Ray Diffraction. *Biochim. Biophys. Acta* **150**, 333–340.

Lecuyer, H., and Dervichian, D. G. (1969). *J. Mol. Biol.* **45**, 39.

Ling, G. N. (1970). Diphosphoglycerate and Inosine Hexaphosphate Control of Oxygen Binding by Hemoglobin: A Theoretical Interpretation of Experimental Data. *Proc. Natl. Acad. Sci.* **67**, 296–301.

Ling, G. N., and Negendank, W. (1970). The Physical State of Water in Frog Muscles. *Physiology, Chemistry and Physics* **2**, 15–33.

Miller, R. J., and Davey, C. B. (1967). The Apparent Effect of Water Structure on K Uptake by Plants. *Soil Sci. Soc. Am. Proc.* **31**, 286–287.

184   W. DROST-HANSEN

Monnier, A. M. (1968). Experimental and Theoretical Data on Excitable Artificial Lipidic Membranes. *J. Gen. Physiol.* **51**, 26s–36s, Part 2.

Morimoto, H., and Halvorson, H. O. (1971). Characterization of Mitochondrial Ribosomes from Yeast. *Proc. Natl. Acad. Sci.* **68**, 324–328.

Murthy, A. S. N., and Rao, C. N. R. (1970). Recent Theoretical Studies of the Hydrogen Bond. *J. Mol. Struct.* **6**, 253–282.

Peschel, G. and Adlfinger, K. H. (1971). Thermodynamic Investigations of Thin Layers Between Solid Surfaces. *Zeit. Naturforsch.* 26A, 705–715.

Pimentel, G. C., and McClellan, A. L. (1960). "The Hydrogen Bond." Freeman, San Francisco, California.

Rahman, A., and Stillinger, F. H. (1971). Molecular Dynamics Study of Liquid Water. (Submitted: *Journal Chem. Phys.*)

Rotunno, C. A., Kowalewski, V., and Cereijido, M. (1970). Nuclear Spin Resonance Evidence for Complexing of Sodium in Frog Skin. *Biochim. Biophys. Acta* **135**, 170.

Spier, H. L., and van Senden, K. G. (1965). *Steroids* **6**, 871.

Sugimoto, S., and Nosoh, Y. (1971). Thermal Properties of Fructose-1,6-diphosphate Aldolase from Thermophilic Bacteria. *Biochim. Biophys. Acta* **235**, 210–221.

Weiss, L. (1967). "The Cell Periphery, Metastasis and Other Contact Phenomena" In Frantiers of Biology, Vol. 7. North-Holland Publishing Co. Amsterdam, Holland and Wiley, New York.

Whittam, R. (1964). Transport and Diffusion in Red Blood Cells. *In* "Monographs of the Physiological Society." Williams & Wilkins, Baltimore.

CHAPTER VII

Matrix-Supported Enzymes

HARRY DARROW BROWN AND FRANK X. HASSELBERGER

I. Prologue. The Nature of the Biological Model and Its Application to the Study of Membrane-Borne Enzymes

The relationship between the biological membrane and the enzyme which, *in vivo*, is bound to it—or part of it in fact—and the effect of this

185

relationship upon the catalytic activity is of pervading theoretical importance. It seems very likely indeed that a large percentage of the physiologically active catalysts of the living cell are not free in solution and that even those of the cytosol are in environments which are far different from the high-dilution conditions conventionally used for their study in the laboratory. Because of the complexity of enzyme–membrane binding even in isolated systems, attempts have been made to produce synthetic enzyme–membrane systems for which the binding situations are defined. The adsorption of enzymes at lipid interfaces and the physical changes in the adsorbed proteins were studied by Langmuir and his associates in the early 1930's. This extended the literature which was begun as early as 1840 by Ascherson who observed that a delineated surface formed spontaneously around oil droplets which were suspended in egg albumin solutions. He compared characteristics of this model system to characteristics of the living cell. Devaux (1903) and Ramsden (1903) early studied proteins at air–water interfaces and a number of systems (egg albumin, oxyhemoglobin, and methemoglobin) were described in 1927 by Wu and Ling. Throughout this literature the intent to design biologically germane models has been implicit. Tien and James (Chapter IV, Part A, this work), have reviewed the literature of enzymes on lipid surfaces, and many matters germane to this subject have been considered in the review articles of Bull (1947), Rothen (1947), Cheesman and Davies (1954), Kauzmann (1954, 1956), Fraser (1957), and James and Augenstein (1966).

It seems a reasonable *prima facie* assumption that when a particle-bound enzyme is solubilized for study in the conventional biochemical system, two major categories of change are brought about. The first is conformational since many studies have illustrated that the three-dimensional form of these proteins is at least in part dependent upon its interaction with other membrane components. Second, the solubilization of a membrane enzyme must bring about changes in the local environment of the moieties composing the protein. Figure 1 is a schematic representation of the possible change in conformation which may accompany protein solubilization while changes in the local environment are also stressed. This thesis is illustrated in its simplest form by Fig. 2.

The concepts of membrane–enzyme interaction can be extrapolated to structured systems such as those of the endoplasmic reticulum or the mitochondria. The frequent suggestion that, for example, the components of endoplasmic reticulum mixed-function oxidation are rigidly organized as a single structural entity would certainly bring into question the use of solution systems for studying the chemical reactions involved. In such a system local environment must refer not only to a charge interaction, the first factor to receive attention, but also to spatial interrelationships which, in fact, preclude the random character of the reactions in ideal solutions.

Efforts in many enzymology investigations have been expended to

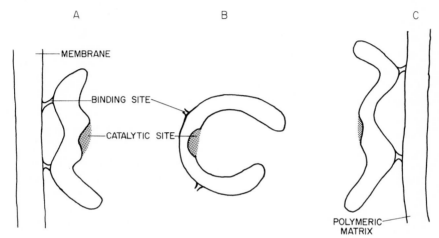

Fig. 1. Schematic representation of the possible change in conformation which may accompany protein solubilization.

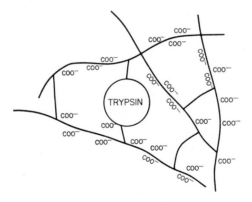

Fig. 2. Schematic presentation of an insoluble maleic acid–ethylene trypsin particle in suspension. [Redrawn from Goldstein *et al.* (1964), by permission.]

evaluate the contribution of these several factors to the behavior of enzymes. The matrix-supported enzyme model has provided an approach to the question of the role of conformation and local environment in *in vitro* systems which, in theory we are free to suggest, may relate to the native situation. From the standpoint of a biological model which would serve as the *in vitro* equivalent of the catalyst *in situ*, the lipid support because it appears most like the membrane itself had offered the first promise. However, difficulties have beset scholars using this tool. Generally adsorption of enzymes to lipid surfaces has resulted in denaturation. Reports of irreversible protein denaturation by Langmuir and Schaefer (1938) were corroborated and made the basis of theoretical studies by Barnett and Bull (1959a,b). Too, the lipid–protein interface models have proven to be very difficult to duplicate with precision from experiment to experiment, and physical manipulation alone has often been a tedious and frustrating exercise. Because of these factors others have chosen adsorption, entrapment, or covalent bonding to matrices that were not lipids as alternative approaches but based upon the same rationale.

Complete success in the construction of a biologically meaningful model would imply that the matrix-supported enzyme, *in vitro*, should have a conformation and local environment (and perhaps other characteristics of importance, too) identical with that of the same enzyme in its native state within the living cell. To be sure, this goal has not been realized; it has not at this point been approached. For the present we must employ essentially "shotgun" attempts to produce models which resemble the native membrane-bound enzyme. The criteria are the conventional parameters of the enzymologist; that is, when a matrix-supported enzyme system is studied, it can be compared to the isolated membrane system and, in very limited ways, to the same enzyme in the cell. Ideally, we should have a much more insightful understanding of the factors that affect enzyme function in the native but active state and compare these with the *in vitro* matrix-supported model system. Certainly not the ultimate in critical comparison, but still eluding us is the actual conformation of the model as compared with the isolated, fragmented membrane–enzyme systems. Even this appears difficult beyond present abilities.

Since adsorption and lattice entrapment offer less control over conformation of the entrapped or adsorbed protein, these methods do not seem the most likely to provide ultimately the greatest hope of yielding a useful biological model. Covalent bonding does allow us a choice of the reactive binding points along our protein. Up to now the number of chemical methods has been limited, primarily by the necessity for reactions through functional groups of the protein which will react under very mild conditions while at the same time avoiding hindrance of the catalytic sites themselves. As the studies of enzymes have progressed over the years, the essentiality for what we generally describe as mild conditions has become less apparent. Indeed, it may be possible to find chemical conditions which, although not considered mild in a normal sense, are nevertheless appropriate for very useful derivitization of the proteins and which will not irreversibly denature the molecule. This is clearly a difficult direction in which to proceed and more of the methodology—perhaps with significant and unique modification—of the synthetic chemist must be brought to bear on the problem if we are to overcome what at present seems to be significant limitations. The elucidation of the models and of the progress which has been made for the ultimate biological goal will be further illustrated in discussion of specific enzymes.

II. Introduction

The present chapter constitutes a review of the literature of matrix-supported enzymes. The emphasis is that of the authors and reflects their specialized interests.* The literature was searched using *Current Contents* through March 1970. A review was published by Silman and Katchalski (1966). Manecke (1962a) and Katchalski (1962a,b, 1968) have reviewed much of their own work. Guilbault (1966, 1968) has included sections on matrix-supported enzymes in his reviews of "Use of Enzymes in Analytical Chemistry." Foreign language reviews have been written by Manecke (1964) [German] and by Chibata and Tosa (1966) [Japanese]. Microencapsulation, antigen–antibody precipitations, and the activity of enzymes at aqueous–organic interfaces are areas not covered in this chapter. Many appropriate techniques have seen parallel development in studies of affinity chromatography (Lerman, 1953; Arsenis and McCormick, 1964, 1966; McCormick, 1965; Cuatrecases *et al.*, 1968). Throughout the chapter, under each subheading, the discussion has been organized to accord with the numerical classification of Enzyme Commission nomenclature (Table I).

The application of matrix-supported enzymes to the development of biological models has been discussed above. Much of the literature, however, has had an orientation which has been based either on a substrate conversion goal or one related to the purification of a complex system. The matrix-supported enzymes have been prepared as soluble materials, suspensions, in column distributions, in sheets, and as membranes. The advantage of long-term stability which often—though not inevitably— follows these derivatizations has made it possible to develop substrate conversion systems in which the catalysts are essentially used for long periods of time. Purification procedures particularly have been applied to

---

* Help of workers in the field was directly solicited, and we thank those who have provided us with sets of reprints.

TABLE I

Chemical Methods Used for Matrix Support of Enzyme

E.C. number[a]	Enzyme	Enzyme formal name	Ref.
		Adsorption or Ion Exchange	
1.11.1.6	Catalase	Hydrogen peroxide: hydrogen peroxide oxidoreductase	Mitz (1956); Mitz and Yanari (1956)
2.7.7.-	Ribonuclease	Ribonucleate nucleotido-2'-transferase (cyclizing)	Barnett and Bull (1959a,b); Mkrtumova and Deborin (1962)
3.1.1.8	Cholinesterase	Acylcholine acyl hydrolase	Bauman et al. (1965); Guilbault and Kramer (1965)
3.2.1.1, 3.2.1.2	Diastase (mixture of α and β amylases)	α-I,4-Glucan 4-glucanohydrolase and α-I,4-glucan maltohydrolase	Usami and Taketomi (1965)
3.2.1.17	Lysozyme	Mucopeptide N-acetyl muramylhydrolase	McLaren (1954a,b); L. G. Augenstine et al. (1958); Estermann et al. (1959); James and Hilborn (1968)
3.2.1.21	β-Glucosidase	β-D-Glucoside glucohydrolase	Reese and Mandels (1958)
3.2.1.26	Invertase	β-D-Fructofuranoside fructohydrolase	Reese and Mandels (1958); W. N. Arnold (1966); Suzuki et al. (1966)
3.4.4.1	Pepsin	—	McLaren (1954a); L. G. Augenstine et al. (1958); Mitz and Schlueter (1959)
3.4.4.4	Trypsin	—	Hartman et al. (1953); McLaren and Estermann (1956); L. G. Augenstine et al. (1958); Mitz and Schlueter (1959); E. S. Augenstine et al. (1960); Kobamoto et al. (1966); Lofroth and Augenstein (1967); Haynes and Walsh (1969)
3.4.4.5	α-Chymotrypsin	—	McLaren (1954a,b, 1957); McLaren and Estermann (1956, 1957); Erlanger (1958); L. G. Augenstine et al. (1958); Mitz and Schlueter (1959); Ghosh and Bull (1962); Haynes and Walsh (1969)
3.4.4.9	Cathepsin C	—	Mitz and Yanari (1956)
3.4.4.10	Papain	—	Maneeke and Gunzel (1967); Messing (1969)
3.5.1.1	Asparaginase	L-Asparagine amidohydrolase	Nikolaev and Mardashev (1961); Nikolaev (1962)
3.5.1.5	Urease	Urea amidohydrolase	Sundaram and Crook (1967); Guilbault and Montalvo (1969)
3.5.1.14	Aminoacylase	N-Acylamino acid amidohydrolase	Tosa et al. (1966a,b, 1967a,b, 1969a,b)
3.5.4.-	Adenosine triphosphate deaminase		Chung et al. (1968)
4.1.2.10	Aldehyde lyase (D-oxynitrilase)	Mandelonitrile benzaldehyde lyase	Becker et al. (1965); Becker and Pfeil (1966)

190

Entrapment/Polyacrylamide Gel/Silastic

1.1.1.27	Lactate dehydrogenase	L-Lactate: nicotinamide adenine dinucleotide oxidoreductase	Hicks and Updike (1966)
1.1.3.4	Glucose oxidase	β-D-Glucose: oxygen oxidoreductase	Hicks and Updike (1966); Updike and Hicks (1967); Pennington et al. (1968b,c)
1.11.1.7	Peroxidase	Donor: hydrogen peroxide oxidoreductase	Van der Ploeg and Van Duijn (1964); Pennington et al. (1968b,c)
2.7.1.1	Hexokinase	Adenosine triphosphate: D-hexose 6-phosphotransferase (glycolytic enzymes)	Brown et al. (1968e)
2.7.5.3	Phosphoglycerate mutase		Bernfeld et al. (1969)
2.7.7.–	Ribonuclease (RNase)	Ribonucleate nucleotido-2'-transferase (cyclizing)	Bernfeld and Wan (1963)
3.1.1.7	Acetylcholinesterase	Acetylcholine hydrolase	Pennington et al. (1968b)
3.1.1.8	Cholinesterase	Acylcholine acyl hydrolase	Pennington et al. (1968a)
3.1.3.1	Alkaline phosphatase	Orthophosphoric monoester phosphohydrolase	Van Duijn et al. (1967); Van der Ploeg and Van Duijn (1968)
3.1.3.2	Acid phosphatase	Orthophosphoric monoester phosphohydrolase	Brederoo et al. (1968 a,b)
3.2.1.1	α-Amylase	α-1,4-Glucan 4-glucanohydrolase	Bernfeld and Wan (1963)
3.2.1.2	β-Amylase	α-1,4-Glucan maltohydrolase	Bernfeld and Wan (1963)
3.4.4.4	Trypsin	—	Bernfeld and Wan (1963); Mosbach and Mosbach (1966); Brown et al. (1968a)
3.4.4.5	α-Chymotrypsin	—	Bernfeld and Wan (1963); Brown et al. (1968a)
3.4.4.10	Papain		Bernfeld and Wan (1963)
3.5.1.5	Urease	Urea amidohydrolase	Guilbault and Montalvo (1969)
3.6.1.5	Apyrase	Adenosine triphosphate diphosphohydrolase	Brown et al. (1968a)
4.1.1.–	Orsellinic acid decarboxylase	Orsellinic acid decarboxylase	Mosbach and Mosbach (1966)
4.1.2.13	Aldolase	Fructose 1,6-diphosphate D-glyceraldehyde-3-phosphate lyase	Bernfeld and Wan (1963); Bernfeld et al. (1968); Brown et al. (1968e)
4.2.1.11	Enolase	2-Phospho-D-glycerate hydrolyase	Bernfeld and Bieber (1969)
5.3.1.9	Phosphoglucoisomerase (glucose-phosphate isomerase)	D-Glucose-6-phosphate ketol isomerase (glycolytic enzymes)	Brown et al. (1968e)

Cross-linking Enzyme

1.1.1.27	Lactate dehydrogenase	L-Lactate: nicotinamide adenine dinucleotide oxidoreductase	Casu and Avrameas (1969)
1.1.3.4	Glucose oxidase	β-D-Glucose: oxygen oxidoreductase	Selegny et al. (1968); Avrameas et al. (1969); Broun et al. (1969)
1.2.1.9	Glyceraldehyde 3-phosphate dehydrogenase	D-Glyceraldehyde 3-phosphate: nicotinamide adenine dinucleotide phosphate oxidoreductase	Lynn and Falb (1969)

TABLE I (Continued)

Cross-linking Enzyme (Continued)

E.C. number	Enzyme	Enzyme formal name	Ref.
1.11.1.7	Peroxidase	Donor: hydrogen peroxide oxidoreductase	Nakane and Pierce (1967); Avrameas and Ternynck (1967, 1969); Avrameas et al. (1969)
2.4.1.1	Glycogen phosphorylase B	Glucan phosphorylase; glycogen phosphorylase	Wang and Tu (1969)
2.6.1.1	Glutamic aspartic transaminase	L-Aspartate: 2-oxoglutarate aminotransferase	Patramani et al. (1969)
2.7.7.–	Ribonuclease (RNase)	Ribonucleate nucleotido-2'-transferase (cyclizing)	Mowbray and Scholand (1966); Avrameas and Ternynck (1969); Davis et al. (1969)
2.7.7.16	Ribonuclease (RNase)	Ribonucleate pyrimidine nucleotide-2'-transferase (cyclizing)	Marfey et al. (1965)
3.1.1.7	Acetylcholinesterase	Acetylcholine hydrolase	A. B. Patel et al. (1969b)
3.1.3.11	Fructose-1,6-diphosphatase	D-Fructose 1,6-diphosphate	Lynn and Falb (1969)
3.1.4.5	Deoxyribonuclease	Deoxyribonucleate oligonucleotidylhydrolase	A. B. Patel et al. (1969b)
3.2.1.17	Lysozyme	Mucopeptide N-acetyl muramylhydrolase	Herzig et al. (1962, 1964); Hiremath and Day (1964); Avrameas and Ternynck (1967); Stevens and Long (1969)
3.4.2.1	Carboxypeptidase A	Peptidyl L-amino acid hydrolase	Quiocho and Richards (1964, 1966; Quiocho et al., 1967; Bishop et al. (1966); Doyle et al. (1968)
3.4.4.3	Rennin	—	Green and Crutchfield (1969)
3.4.4.4	Trypsin	—	Goldstein et al. (1964); Levin et al. (1964); Ong et al. (1966); Rimon et al. (1966a); Habeeb (1967); Katchalski (1968); Lowey et al. (1967); Avrameas and Ternynck (1969); Haynes and Walsh (1969)
3.4.4.5	α-Chymotrypsin	—	Katchalski (1968); Haynes and Walsh (1969); Jansen and Olson (1969)
3.4.4.10	Papain	—	Goldman et al. (1965, 1968); Silman et al. (1966); Katchalski (1968); Jansen and Olson (1969)
3.4.4.16	Subtilisin (subtilopeptidase A)		Ogata et al. (1968)
3.6.1.3	Adenosinetriphosphatase	Adenosine triphosphate phosphohydrolase	Brown et al. (1968c)
3.6.1.5	Apyrase	Adenosine triphosphate diphosphohydrolase	Brown et al. (1968c); A. B. Patel et al. (1969b)
4.1.2.13	Aldolase	Fructose 1,6-diphosphate D-glyceraldehyde-3-phosphate lyase	Lynn and Falb (1969)

Chemical (Covalent) Binding

Azide

EC	Enzyme	Systematic name	References
2.7.3.2	Adenosine triphosphate phospho-transferase (creatire kinase)	Adenosine triphosphate: creatine phosphotransferase	Hornby et al. (1968)
2.7.7.-	Ribonuclease (RNase)	Ribonucleate nucleotido-2'-transferase (cyclizing)	Epstein and Anfinsen (1962); Lilly et al. (1965)
3.1.1.8	Cholinesterase	Acylcholine acyl hydrolase	A. B. Patel et al. (1969b)
3.1.4.5	Deoxyribonuclease	Deoxyribonucleate oligonucleotidylhydrolase	A. B. Patel et al. (1969b)
3.2.1.1, 3.2.1.2	Diastase; mixture of α and β amylase	α-I,4-Glucan 4-glucanohydrolase and α-I,4-glucan maltohydrolase	Manecke and Foerster (1966)
3.2.1.26	Invertase (β-D-fructosidase)	β-D-Fructofuranoside fructohydrolase	Manecke and Singer (1960b)
3.4.4.-	Pronase	—	A. B. Patel et al. (1971)
3.4.4.4	Trypsin	—	Mitz and Summaria (1961); Epstein and Anfinsen (1962); Takami (1968)
3.4.4.5	α-Chymotrypsin	—	Micheel and Ewers (1949); Mitz and Summaria (1961); Lilly et al. (1965); Lilly and Sharp (1968); Takami (1968)
3.4.4.10	Papain	—	Manecke and Foerster (1966)
3.4.4.12	Ficin	—	Lilly et al. (1965, 1966); Hornby and Lilly (1966)
3.4.4.24	Bromelain	—	Wharton et al. (1968a,b)
3.5.1.1	Asparaginase	L-Asparagine amidohydrolase	Hasselberger et al. (1970a)
3.6.1.3	Adenosinetriphosphatase	Adenosine triphosphate phosphohydrolase	Brown et al. (1966a,b, 1967, 1968c); Wheeler et al. (1969)
3.6.1.5	Apyrase	Adenosine triphosphate diphosphohydrolase	Brown et al. (1966c, 1968c); Whittam et al. (1968); A. B. Patel et al. (1969b); Wheeler et al. (1969)
3.6.1.8	Adenosinetriphosphatase	Adenosine triphosphate pyrophosphohydrolase	Brown et al. (1964a,b, 1967, 1968c)

Carbodiimide

EC	Enzyme	Systematic name	References
1.1.3.4	Glucose oxidase	β-D-Glucose: oxygen oxidoreductase	Weliky and Weetall (1965)
1.11.1.7	Peroxidase	Donor: hydrogen peroxide oxidoreductase	Weliky and Weetall (1965); Weetall and Weliky (1966); Nakane and Pierce (1967); Weliky et al. (1969)
2.7.7.-	Ribonuclease	Ribonucleate nucleotido-2'-transferase (cyclizing)	Weliky and Weetall (1965); Riehm and Scheraga (1966)
3.1.3.1	Alkaline phosphatase	Orthophosphoric monoester phosphohydrolase	Weliky and Weetall (1965)
3.1.3.2	Acid phosphatase	Orthophosphoric monoester phosphohydrolase	Weliky and Weetall (1965)
3.1.4.5	Deoxyribonuclease	Deoxyribonucleate oligonucleotidylhydrolase	Weliky and Weetall (1965)
3.2.1.17	Lysozyme	Mucopeptide N-acetyl muramylhydrolase	Hasselberger et al. (1970b)
3.2.1.31	β-Glucuronidase	β-D-Glucuronide glucuronohydrolase	Weliky and Weetall (1965)
3.4.4.-	Panprotease	—	Weliky and Weetall (1965)
3.4.4.4	Trypsin	—	Weliky and Weetall (1965)
3.4.4.5	α-Chymotrypsin	—	Weliky and Weetall (1965); Banks et al. (1969)
3.4.4.16	Subtilisin (subtilopeptidase A)	—	Weliky and Weetall (1965)
3.5.1.1	Asparaginase	L-Asparagine amidohydrolase	Hasselberger et al. (1970a)

TABLE I (Continued)

E.C. number	Enzyme	Enzyme formal name	Ref.
		Chemical (Covalent) Binding (Continued)	
Cyanogen bromide			
3.1.1.7	Acetylcholinesterase	Acetylcholine hydrolase	Axen et al. (1969)
3.1.1.8	Cholinesterase	Acylcholine acyl hydrolase	Axen et al. (1969)
3.2.1.17	Lysozyme	Mucopeptide N-acetyl muramylhydrolase	Hasselberger et al. (1970b)
3.4.4.–	Pronase	—	A. B. Patel et al. (1971)
3.4.4.3	Rennin	—	Green and Crutchfield (1969)
3.4.4.5	α-Chymotrypsin	—	Axen et al. (1967); Porath et al. (1967); Green and Crutchfield (1969)
3.5.1.1	Asparaginase	L-Asparagine amidohydrolase	Hasselberger et al. (1970a)
Cyanuric chloride/Procian Brilliant Orange			
1.1.1.27	Lactate dehydrogenase	L-Lactate: nicotinamide adenine dinucleotide oxidoreductase	Kay et al. (1968); Wilson et al. (1968a,b)
1.1.3.4	Glucose oxidase	β-D-Glucose: oxygen oxidoreductase	Avrameas et al. (1969)
1.11.1.6	Catalase	Hydrogen peroxide: hydrogen peroxide oxidoreductase	Surinov and Manoilov (1966)
1.11.1.7	Peroxidase	Donor: hydrogen peroxide oxidoreductase	Avrameas et al. (1969)
2.7.1.40	Pyruvate kinase	Adenosine triphosphate: pyruvate phosphotransferase	Kay et al. (1968); Wilson et al. (1968a,b)
2.7.3.2	Creatine kinase	Adenosine triphosphate: creatine phosphotransferase	Kay et al. (1968)
2.7.7.–	Ribonuclease (RNase)	Ribonucleate nucleotido-2'-transferase (cyclizing)	Surinov and Manoilov (1966)
3.1.1.7	Acetylcholinesterase	Acetylcholine hydrolase	Stasiw et al. (1970)
3.1.1.8	Cholinesterase	Acylcholine acyl hydrolase	A. B. Patel et al. (1969a)
3.2.1.21	β-Glucosidase	β-D-Glucoside glucohydrolase	Sharp et al. (1969)
3.2.1.23	β-Galactosidase	β-D-Galactoside galactohydrolase	Kay et al. (1968)
3.4.4.4	Trypsin	—	Surinov and Manoilov (1966)
3.4.4.5	α-Chymotrypsin	—	Surinov and Manoilov (1966); Kay and Crook (1967); Kay et al. (1968); Kay and Lilly (1970)
3.6.1.5	Apyrase	Adenosine triphosphate diphosphohydrolase	Wheeler et al. (1969)
Diazonium coupling			
1.1.3.4	Glucose oxidase	β-D-Glucose: oxygen oxidoreductase	Avrameas et al. (1969); Selegny et al. (1968)
1.11.1.6	Catalase	Hydrogen peroxide: hydrogen peroxide oxidoreductase	Surinov and Manoilov (1966)
1.11.1.7	Peroxidase	Donor: hydrogen peroxide oxidoreductase	Avrameas et al. (1969)
2.7.7.–	Ribonuclease	Ribonucleate nucleotido-2'-transferase (cyclizing)	Grubhofer and Schleith (1953); Mitz and Summaria (1961); Silman et al. (1963); Lilly et al. (1965); Surinov and Manoilov (1966); Davis et al. (1969)

Diazonium coupling (*Continued*)

EC no.	Enzyme	Systematic name	References
3.1.3.1	Alkaline phosphatase	Orthophosphoric monoester phosphohydrolase	Weetall (1969a)
3.2.1.1, 3.2.1.2	Diastase (mixture of α and β amylases)	α-1,4-Glucan 4-glucanohydrolase and α-1,4-glucan malto-hydrolase	Grubhofer and Schleith (1953, 1954)
3.2.1.17	Lysozyme	Mucopeptide N-acetyl muramylhydrolase	Hasselberger et al. (1970b)
3.4.2.1	Carboxypeptidase A	Peptidyl-L-amino acid hydrolase	Grubhofer and Schleith (1953)
3.4.4.-	Pronase	—	Cresswell and Sanderson (1969)
3.4.4.4	Trypsin	—	Bar-Eli and Katchalski (1960, 1963); Glazer et al. (1962); Alexander et al. (1965); Engel and Alexander (1965); Rimon et al. (1966a); Surinov and Manoilov (1966); Katchalski (1968); Weetall (1969b)
3.4.4.5	α-Chymotrypsin	—	Mitz and Summaria (1961); Lilly et al. (1965); Surinov and Manoilov (1966); Katchalski (1968)
3.4.4.10	Papain	—	Grubhofer and Schleith (1953); Cebra et al. (1961, 1962); Cebra (1964); Goldman et al. (1965, 1968); Jaquet and Cebra (1965); Silman et al. (1966); Katchalski (1968); Weetall (1969b)
3.4.4.12	Ficin	—	Lilly et al. (1965); Hornby et al. (1968)
3.4.4.14	Streptokinase (fibrinolysin); plasminokinase	—	Rimon et al. (1963, 1966b); Gutman and Rimon (1964)
3.5.1.5	Urease	Urea amidohydrolase	Riesel and Katchalski (1964)
3.6.1.5	Apyrase	Adenosine triphosphate diphosphohydrolase	Brown et al. (1968b,d); Weetall and Hersh (1969)

Isothiocyanate method

EC no.	Enzyme	Systematic name	References
3.2.1.2	β-Amylase	α-1,4-Glucan maltohydrolase	Axen and Porath (1966)
3.4.4.4	Trypsin		Axen and Porath (1966)
3.4.4.5	α-Chymotrypsin	—	Axen and Porath (1966)

Woodward reagent K

EC no.	Enzyme	Systematic name	References
3.1.1.7	Acetylcholinesterase	Acetylcholine hydrolase	A. B. Patel et al. (1969b)
3.1.1.8	Cholinesterase	Acylcholine acyl hydrolase	A. B. Patel et al. (1969b)
3.1.4.5	Deoxyribonuclease	Deoxyribonucleate oligonucleotidylhydrolase	A. B. Patel et al. (1969b)
3.4.4.4	Trypsin		Wagner et al. (1968)
3.4.4.5	α-Chymotrypsin	—	Patel and Price (1967); R. P. Patel et al. (1967); Wagner et al. (1968); A. B. Patel et al. (1969b)
3.5.1.1	Asparaginase	L-Asparagine amidohydrolase	Hasselberger et al. (1970a)
3.6.1.5	Apyrase	Adenosine triphosphate diphosphohydrolase	A. B. Patel et al. (1969b)

TABLE I (*Continued*)

E.C. number	Enzyme	Enzyme formal name	Ref.
		Chemical (Covalent) Binding (*Continued*)	
Ethylene–maleic anhydride copolymer method			
3.1.1.7	Acetylcholinesterase	Acetylcholine hydrolase	A. B. Patel *et al.* (1969b)
3.1.4.5	Deoxyribonuclease	Deoxyribonucleate oligonucleotidylhydrolase	A. B. Patel *et al.* (1969b)
3.2.1.17	Lysozyme	Mucopeptide *N*-acetyl muramylhydrolase	Hasselberger *et al.* (1970b)
3.4.4.–	Pronase	—	A. B. Patel *et al.* (1971)
3.4.4.4	Trypsin	—	Goldstein *et al.* (1964); Alexander *et al.* (1965); Engel and Alexander (1965); Ong *et al.* (1966); Katchalski (1968); Westman (1969)
3.4.4.5	α-Chymotrypsin	—	Katchalski (1968); Goldstein *et al.* (1967a)
3.6.1.5	Apyrase	Adenosine triphosphate diphosphohydrolase	A. B. Patel *et al.* (1969b)
Fluorine replacement (nitrated copolymer of methacrylic acid and methacrylic acid *m*-fluoroanilide; polystyrene nitrofluorobenzene sulfonyl copolymer)			
3.2.1.1, 3.2.1.2	Diastase (mixture of α and β amylases)	α-1,4-Glucan 4-glucanohydrolase and α-1,4-glucan maltohydrolase	Manecke and Gunzel (1962); Manecke and Foerster (1966)
3.2.1.26	Invertase	β-D-Fructofuranoside fructohydrolase	Manecke and Singer (1960b); Manecke (1962a, b)
3.4.4.1	Pepsin	—	Manecke (1962a, b); Manecke and Gunzel (1962)
3.4.4.10	Papain	—	Manecke and Foerster (1966)

Bridging: Enzymes bound through "bridging compounds" have been considered under appropriate headings above

[a] E.C. number = Enzyme Commission number.

196

antigens and antibodies; though this literature is very interesting and promising in providing a tool for enzyme purification, the application has not extensively been made. Erlanger (1958) used chymotrypsin, chymotrypsinogen, trypsin, trypsinogen, pepsin, and ribonuclease (RNase) in the development of a method for purification using the specificity of enzyme substrate interaction. More of the specific applications will be presented within the text below.

The methods which have been used most frequently for the solid support of biocatalysts include diazonium coupling, the acylazide reaction, and the reaction of the ethylene–maleic anhydride copolymer. The cyanogen bromide and carbodiimide reactions are being used with increasing frequency, and cyanuric chloride-related methods are being developed in England (Kay and Crook, 1967; Kay et al., 1968; Kay and Lilly, 1970). Activation of acid-containing polymers by Woodward's reagent K and bonding to isothiocyanates are less frequently used methods. Manecke and Gunzel (1962) have incorporated 2,4-dinitrofluorobenzene into polymers and used these as reactive matrices.

All of these methods involve nucleophilic reactions. Most are nucleophilic substitutions, but the isothiocyanate method uses a nucleophilic addition (Drobnica and Augustin, 1965a,b,c). (The mechanism of the cyanogen bromide reaction is not known, but the coupling is almost certainly a nucleophilic reaction.)

The principal nucleophilic moieties present in proteins are the amino, hydroxyl and sulfhydryl functions. The usual order of reactivity for these groups is $-NH_2 > -OH > -SH$. When working with large molecules such as proteins, steric factors may block reaction of more nucleophilic groups making less active nucleophiles the reaction site. The pH may influence the relative reactivities; thus, Drobnica and Augustin (1965c) found that sulfhydryl groups added to isothiocyanates much more rapidly than amino groups when these reactions were carried out at pH 9.8. This can be explained by the observation that the sulfhydryl groups ionize before reacting. Thus, in this instance, the reactive species is the mercaptide anion rather than the sulfhydryl group. If pH is maintained close to neutrality, dissociation of the sulfhydryl group is not extensive and reaction at more active sites is favored. Nevertheless, protection of the sulfhydryl groups during the coupling reaction is still advisable if these groups are involved in the active site of the enzyme.

The carbodiimide and Woodward's reagent K methods have been developed for use in organic synthesis, for the formation of amide bonds. For matrix support, these methods are sometimes used in a two-step procedure: the activating reagent is removed before the protein coupling reaction is carried forward. Both reagents can be used to polymerize proteins

(Sheehan and Hlavka, 1957; R. P. Patel and Price, 1967); this will occur as a side reaction (presumably unwanted) if the reagents are not removed before coupling the proteins to the activated matrices. Even if this precaution is taken, nucleophilic moieties on the protein, other than that chosen as the reactive site, can react with the activated matrix. Banks and coworkers (1969) have illustrated this potentially significant problem in their description of the inactivation of α-chymotrypsin by a water-soluble carbodiimide. The enzyme was inactivated because a serine moiety at the catalytic site added to carbodiimide-activated carboxymethyl (CM) cellulose matrix.

The azide and ethylene–maleic anhydride methods lead to matrices which are negatively charged in alkaline solution. This is often an undesirable condition since substrates may also be negatively charged under these conditions. This unfavorable electrostatic situation may be reflected in a high apparent K_m value (Hornby et al., 1968).

A method developed by Kay and co-workers (Kay et al., 1968; Sharp et al., 1969) leads to formation of a positively charged matrix regardless of the assay conditions. This method which uses cellulose, cyanuric chloride, and N-(3-aminopropyl) diethanolamine should result in a favorable electrostatic situation when substrates are negatively charged.

The diazonium coupling reaction is usually described as an electrophilic reaction since the diazo group carries a positive ionic charge. It should be recognized, however, that this electrophilic group reacts with electron-rich (nucleophilic) groups of the protein. Thus, the phenolic, imidazolyl, amino, sulfhydryl, and indolyl groups of proteins have all been found to react with diazonium compounds (Howard and Wild, 1957).

III. Adsorbed and Ionically Bonded Enzymes

Adsorption has been used to insolubilize a number of enzymes. The method has been used to model a great diversity of phenomena ranging from intracellular reactions to complex systems involving enzymes at soil–water interfaces. Much of the methodology per se has been described originally in the immunology literature (e.g., Campbell et al., 1951; Isliker, 1953; Nezlin, 1961).

Study of the kinetics of adsorbed enzyme-catalyzed reactions is complicated by the fact that a fairly rapid desorption of the enzyme occurs under assay conditions. Despite difficulty in evaluation of adsorbed systems, they may nonetheless very strongly reflect the nature of the biological materials. If this is true, then a considerable departure from conventional enzyme kinetic theory of the steady state must be introduced; diffusion of the substrate or some motile force of the enzyme must be considered. Conformational or even more extreme (chemical) changes may occur as result of the reorganization of the interface at the time the adsorption occurs. To this point, it is known that the reactivity of enzymes under such circumstances (oil–water interface) is strongly intensified (Fraser *et al.*, 1955). Too, since the model systems almost always imply single rather than multiple interfacial layers, the simplicity of even the best developed of these models may defeat interpretation in terms of the biological system. Kinetics of enzymes at interfaces is the subject of Chapter V, Part A (this work), by Laidler and Sundaram.

A. Oxidoreductases

An insoluble catalase (1.11.1.6) was prepared by Mitz (1956) by adsorbing an enzyme purified from liver to a cellulose ion exchanger. Unlike many adsorbed enzymes the catalase preparation was quite active as a catalyst, retaining 70% of the activity level of the soluble form. It was not removed from the adsorbent by washing with water, but, predictably, its solubility was affected by a shift in pH or increase in ionic strength of the surrounding medium. Other proteins displaced the adsorbed enzyme but the substrate, hydrogen peroxide, did not free the catalase from the cellulose. In Mitz's experiments, peroxide solutions (3–6%) were passed through this enzyme column for several days during which time the enzyme remained unchanged. At the end of the period, the enzyme was rapidly desorbed by the addition of a salt solution.

Interestingly, Mitz speculates that some insoluble enzymes, in the cell, function in a manner analogous to his catalase column. He suggests that enzymes such as catalase though electrostatically bound in the cell are active only when in true solution. He envisages that structural proteins, mucopolysaccharides or phospholipids, may serve as ionic exchange agents which with change in pH or ionic strength of the medium or in the presence of charged large molecules release the enzymes and so, in fact, transform them from inactive into physiologically active catalysts (Fig. 3).

B. Transferases

Ribonuclease (2.7.7.–) was adsorbed by Barnett and Bull (1955a,b) to Dowex 50 cation- and Dowex 2 anion-exchange resins. Both were active after washing but only the Dowex-50 preparation remained active after incubation with substrate. Drying the protein on the adsorbent did not alter its properties and the free solution pH optimum (pH 8.0) of the ribonuclease remained unchanged when adsorbed. Studies using the Tiselius moving boundary technique, however, showed there were significant differences in electrophoretic mobility of adsorbed and free solution preparations. Adsorbed protein, the workers postulated, was distorted at the interface. This rationalization may be carried further to suppose that significant electrostatic interaction between the protein and the adsorbent may result in a decrease in the isoelectric point pH of the

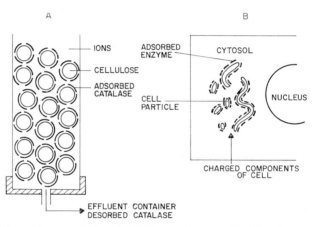

Fig. 3. Schematic presentation of Mitz's model, comparing his column experiments using catalase (A) to a situation in the biological cell (B) where charged molecules of the sol are equivalent to the displacing ions, and the cell particulates are analogous to the medium of the column.

adsorbed protein on a positively charged particle and an increase on a negatively charged particle. Data for ribonuclease adsorbed on Pyrex glass, on Dowex 50 resin, on Nujol, and on paraffin in tris buffer are consistent with the idea of electrostatic interaction between the adsorbed protein and the adsorbing surface. The results obtained when RNase was adsorbed to Dowex 2 resin and to paraffin in Michaelis buffers, however, are contrary to the notion of electrostatic interaction. Hence alternative explanations are pertinent: close packing of the protein molecules may result in a significant shift in pK of the ionizing groups; the protein undergoes extreme structural changes, i.e., a denatured form is produced which has a vastly different array of surface charges and hence an altered isoelectric point. Barnett and Bull assume that some combination of these rationalizations must be employed.

Mkrtumova and Deborin (1962) adsorbed ribonuclease on "sulfo resin SBS-4" to compare its enzymic activity in the dissolved and adsorbed states. More than half of the activity was lost by the adsorbed enzyme. When the enzyme was desorbed, 99% of the original enzymic activity was regained. It was concluded that on adsorption, the active centers of the enzyme are affected, but its structure is not disturbed.

C. HYDROLASES

McLaren (1963; McLaren and Babcock, 1959), although not immediately concerned with enzymes bound to subcellular structure, has employed techniques and made kinetic analyses which are germane to subcellular events. Estermann et al. (1959) studied monolayers of lysozyme (3.2.1.17) adsorbed on kaolinite, bentonite, and lignin. Chymotryptic digests were obtained after formation of reactive enzyme–substrate–adsorbent preparations. Substantial differences in the rates of digestion were found when lignin rather than clay was the adsorbent. Slower digestion of lysozyme adsorbed to lignin may result from physical adsorption of the protein within the gel structure of the lignin.

James and Hilborn (1968) studied the adsorption of hen's egg white lysozyme at the air–water interface. Their results were consistent with the view that lysozyme is a compact protein which does not readily denature at surfaces though activity was altered by moderate reversible unfolding.

Using adsorbed β-glucuronidase (3.2.1.21) and invertase (β-fructofuranosidase; 3.2.1.26), Reese and Mandel (1958) described a continuous two-phase column for partition chromatography. New substrate entered constantly from the solvent phase, and products were removed so that the enzyme action was continuous. The advantages of this approach are common to many substrate conversion systems using insolubilized enzymes: there is neither inhibition by reaction products nor reduction in rate due to depletion of substrate; the column is useful in obtaining short-lived intermediates. In these experiments, the substrates were passed through the columns in alcoholic solution and the enzymic reaction resulted in transfers of sugar moieties from the substrate to the alcohol molecule, yielding a glycoside.

W. N. Arnold (1966) used an insoluble β-fructofuranosidase which was bonded to sedimented cellular debris as a natural occurrence of isolation and purification. This material was used to fill void spaces in a glass bead column. The author suggested that the method was appropriate for characterization of such coarse cellular fractions.

Water-insoluble yeast invertase was prepared by binding to diethyl-aminoethyl (DEAE) cellulose (Suzuki *et al.*, 1966). The bound invertase was about half as active as the free form at pH 3.4. The apparent pH optimum of the bound enzyme was shifted toward the more acid by about 2 pH units. Contrary to expectation, thermal stability of the bound invertase was slightly less than the solution enzyme (at pH 5.2). In their batch reaction system, the invertase reactions could be repeated about 10 times before the product yield dropped to about half that of the initial level.

The proteolytic enzymes, pepsin (3.4.4.1) and trypsin (3.4.4.4) represent, respectively, the complementary properties of acidic and basic proteins. Mitz and Schlueter (1959) related this to data obtained when the enzymes were bound to cellulose. DEAE cellulose was effective for acidic and neutral enzymes and CMC phosphate and cellulose citrate for basic proteins. Use of these cellulose derivatives to isolate proteolytic enzymes from solution varied in efficiency from 1 to 50%, but the procedure did offer considerable storage stability. The desorbed enzyme was favorable in activity when compared with the same enzyme in the original solution. In the same investigation acylase, trypsin, chymotrypsin, and the proteases from kidney and spleen were studied. Pepsin, chymotrypsin (3.4.4.5) (McLaren, 1954a,b), and trypsin (McLaren and Estermann, 1956) were inactive when adsorbed to kaolinite. But when the enzymes were added to the bulk medium in which the same clay was the adsorbent for lysozyme, it was found that the proteolysis of the adsorbed substrate followed the preliminary formation of a reactive kaolinite–enzyme–substrate complex. With chymotrypsin and trypsin, the clay–enzyme substrate complexes were stable at low pH (about pH 4.5) and became reactive at pH 5 and above. If chymotrypsin was adsorbed on kaolinite before its substrate, then the rate of reaction was much slower than if the enzyme was added to substrate adsorbed on kaolinite. McLaren and Estermann interpreted this in terms of the effect of surface on reaction rate.

Kobamoto *et al.* (1966) studied trypsin adsorbed to cellulose, glass, and quartz. The rate and amount of adsorption were a function of protein concentration, concentration of solution ions, pH, and temperature. The pH of maximum adsorption differed for trypsin and RNase. This observation suggested that the extent of the adsorption is determined by pK of interacting groups of the protein and the adsorbent. Active and inactive forms of trypsin adsorbed and desorbed at different rates indicating that conformational changes are involved.

Compression characteristics and their effect upon recoverable tryptic activity were studied by L. G. Augenstine *et al.* (1958). Trypsin films were formed on the surface of ammonium sulfate, and reactions followed in the pH range of 2.4 to 9. Data indicate that this model offers promise for understanding the relationship of activity to conformation. Effects upon films of the addition of folded and unfolded populations were considered. Adsorption of trypsin onto glass and quartz was studied by Lofroth and Augenstein (1967) as a function of radiation. They found that radiation sensitivity of trypsin at these air–solid interfaces is the same as or less than that of trypsin in solution. The 228 nm absorbance of trypsin is 1.7 times that of trypsin in solution. This they interpreted to indicate either a preferential orientation of dipoles relative to the interface or conformational changes. The extent of change in inactivation yield depended upon the polarization of the light relative to the interface.

Haynes and Walsh (1969) found that the adsorption of trypsin as a monolayer onto colloidal silica particles was independent of pH values between 6.8 and 8.7. Their further study of trypsin monolayers on colloidal silica employed glutaraldehyde cross-linking (discussed in Section V, this chapter). In a study of chymotrypsin, McLaren (1957) interpreted his data with reference to subcellular phenomena. The pH activity curves of some enzyme reactions in intact cells and mitochondria are similar to those in solution. It has been suggested that these enzymes must be peripherally located. This conclusion requires the assumption that the internal pH of the cell is independent of the external pH of the ambient buffer and that the permeability of the membrane is independent of pH. However, if the surface carrying an enzyme is charged, one cannot expect that the pH optimum of the enzyme will be the same as the enzyme in solution. The charged surface, of course, will either attract or repel hydrogen ions. Hence, an enzyme acting at a surface will be exposed to and in equilibrium with hydrogen ion activity which differs from that of the ambient buffer. The point was illustrated by comparing the action of chymotrypsin on a substrate protein in solution with substrate on the surface of 1-μ kaolinite particles. Optimum pH for the enzyme adsorbed to clay was higher (Fig. 4), and McLaren calculated that the hydrogen ion activity at the surface is 100 times greater than in solution. This concept of local environment effect upon enzyme activity has been developed by other authors as well, particularly as illustrated by covalently bound enzymes, and will be discussed by reference to appropriate examples.

Chymotrypsin as well as chymotrypsinogen, trypsin, trypsinogen, pepsin, and RNase were used by Erlanger (1958) in the development of a method for purification that depended on the specificity of the enzyme–substrate interaction. Similar work has been carried forth by others in the purification of antigen and antibodies (cited in the review of Silman and Katchalski, 1966). The approach is based upon the thesis that the enzyme will be selectively adsorbed by substrates or by competitive inhibitors.

Ghosh and Bull (1962) adsorbed chymotrypsin on particles of N-octadecane in phosphate buffer. The adsorbed enzyme was inactivated, and, after desorption, its specific activity was decreased presumably in consequence of autolysis. The adsorbed enzyme had an isoelectric point of pH 6.15 as compared with pH 7.3 for the same enzyme dissolved in 0.01 M phosphate buffer. The mobility of the octadecane particles cov-

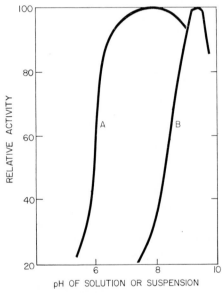

Fig. 4. Effect of pH on chymotrypsin activity in solution (A) or adsorbed on kaolinate (B). [Redrawn from McLaren, *Science* **125**, (1957), by permission.]

ered with chymotrypsin was identical whether or not the inhibitor N-acetyl tryptophan was present. Thus it was apparent that the inhibitor did not desorb the enzyme. Interestingly, this observation appeared to indicate that the negative charge of chymotrypsin on octadecane particles is due to orientation of the enzyme molecules on the surface.

Ghosh and Bull believed that the adsorption of the chymotrypsin involved two distinct processes: diffusion which brings the protein molecules to the water–hydrocarbon interface and then spreads the protein into a film on the particle surface. As the protein concentration in solution is increased, the rate of arrival of the protein at the interface exceeds the rate at which the molecules can spread completely. Hence at very high concentrations, the protein molecules at the interface are practically undistorted and, perhaps, in a native configuration. If the rate of surface denaturation is sufficiently rapid, then a denatured monolayer will form whatever the magnitude of the protein concentration may be. This may result in an underlying denatured film with an adsorbed second layer of more or less native molecules.

The preparation and stability of papain (3.4.4.10) included in porous glass was described by Messing (1969). Papain adsorbed to glass retained activity over extended periods of time.

Nikolaev and Mardashev (1961) used CM cellulose to support asparaginase (3.5.1.1). Their asparaginase was obtained as an acetone powder from *Bacillus cadaveris*. The procedure was technically useful in that approximately 40% of the asparaginase, but only 1% of the total protein of the acetone powder, was adsorbed onto the CM cellulose. The enzyme column had remarkable long-term stability. D-Asparagine was not hydrolyzed by this preparation and had no effect upon the rate of hydrolysis of the L-asparagine.

Nikolaev (1962) also studied asparaginase on DEAE cellulose and reported this preparation superior to that using CM cellulose. The DEAE cellulose was preferable for asparaginase because the optimum pH range of the enzyme could be maintained and, hence, it produced a more active complex. The rate of hydrolysis relative to substrate concentration followed an anomalous, nonlinear course which made it impossible to determine a Michaelis constant. However, the experimental points fit a straight line if the reciprocal of the reaction rates is plotted against the square of the reciprocal of the substrate concentrations. The reaction constant determined from this plot is $1.8 \times 10^{-6} M^2$ for both the soluble enzyme and for the DEAE cellulose preparation. The rate of heating did not affect the thermal stability of the DEAE cellulose enzyme. Loss in activity which resulted from heating within the range 20–60°C was recovered after cooling. Temperatures higher than 60°C resulted in a loss of activity presumably as a result of denaturization. Energies of activation plotted for the native and adsorbed preparations were 6.1 and 6.0 kcal/mole, respectively.

Tosa and associates (1966a,b, 1967a,b, 1969a,b) have studied ionically bound and physically adsorbed preparations of aminoacylase (3.5.1.14). Derivatives of dextran (Sephadex) and cellulose were used as the supports. Although the development of continuous substrate conversion systems appropriate to industrial applications motivated the study, observations of these authors are of substance to biochemical model interpretation for aminoacylase and other enzymes. They derived a rate equation applicable to the aminoacylase column (Tosa *et al.*, 1969a):

$$\frac{(S_0 - (S_h))}{(S_0)} = \frac{k(E_0)}{(S_0)V_s}$$

where the k value in the equation is specific activity of the enzyme column; S_0 is the initial concentration of substrate; S_h is the concentration
of substrate in effluent; E_0 is the enzyme concentration; and V_s is the
space velocity. Reaction rates are independent of the column height.

Adenosine triphosphate (ATP) deaminase (3.5.4.–) was studied on
DEAE cellulose by Chung et al. (1968). Optimum pH was shifted toward
the acid side by about 2 pH units compared with the free solution enzyme.
Activity of bound deaminase to ATP corresponded to about one-fourth
of free deaminase at pH 3.0.

D. LYASES

The flavoprotein enzyme, D-oxynitrilase (aldehyde lyase; D-hydroxy-
nitrile lyase, 4.1.2.10) was bound to the ion exchanger ECTEOLA
cellulose by Becker and Pfeil (1965, 1966) to form a stable catalyst for
continuous column synthesis of D-α-hydroxynitriles from aldehydes and
hydrocyanic acid. The method is useful for the synthesis of D-(+)-man-
delonitrile and can be used for conversion of many aliphatic, aromatic,
and heteroaromatic aldehydes into D-α-hydroxynitriles which can be
readily transformed chemically into optically active D-α-hydroxycar-
boxylic acid, substituted ethanolamines, or acyloins. Reaction rate and
solubility varied with respect to the aldehyde used.

IV. Lattice-Entrapped Enzymes

Enzymes have been entrapped within starch gel (Bauman et al., 1965;
Guilbault and Kramer, 1965), within polyacrylamide gels (numerous
references, Table I), and within Silastic (Brown et al., 1968a). In general,
most of the studies employing entrapment of enzymes have been motivated toward the use of the material either as an analytical tool or as a
useful substrate conversion device. Hence, two goals have been para-
mount: that the enzyme activity be stable and that the reagent form be
reproducible. There have, however, been exceptions provided by studies
based upon other goals, e.g., Van der Ploeg and Van Duijn (1964) used
peroxidase (1.11.1.7) entrapped within a polyacrylamide film as a
model system to test a hypothesis dealing with the mechanism of the
3,4-dihydroxyphenylalanine (dopa) reaction. They had entertained the

hypothesis that peroxidase in the presence of hydrogen peroxide accelerated the conversion of 5,6-dihydroxyindole into indole-5,6-quinone and, hence, that 5,6-dihydroxyindole was the actual substrate in the dopa reaction, indole-5,6-quinone being transformed into insoluble dopa melanin at the site of the enzyme molecules in the cells. To test this thesis by obtaining the end products of this reaction, a model was used in which peroxidases were incorporated into polyacrylamide films.

In their system, peroxidase caused rapid melanin formation in the films during incubation in solutions containing 5,6-dihydroxyindole and hydrogen peroxide. The authors compared these findings with what they considered to be the analogous situation in leukocytes and erythrocytes. They concluded that membranes in which enzymes are incorporated can be used as models to study histochemical reactions quantitatively. The use of the entrapped enzyme as a model system for the histochemist and electron microscope histochemist is in itself a major unexplored area. From the standpoint of those of us interested in the properties of restricted enzymes, it is to be hoped that the electron microscope fraternity will find interest here.

The same group of workers (Van Duijn et al., 1967; Van der Ploeg and Van Duijn, 1968; Brederoo et al., 1968a,b) used alkaline phosphatase (3.1.3.1) and acid phosphatase (3.1.3.2) in a series of studies technically similar to the cytochemical model of entrapped enzymes but with substantially different intent. The enzymes were restricted to the gel interstices, and substrate was allowed to diffuse into the gel. In this way kinetics of the reaction under conditions believed by the authors to be analogous to those used in cytochemical procedures could be studied. Findings with respect to the polyacrylamide–phosphatase model system were compared to the conditions employed in microchemical preparations using tissue sections. Under the conditions conventional to the cytochemist the enzyme is constrained within a cell structure into which the substrate must diffuse and the product must be trapped or converted by an additional agent into an insoluble final product (the stained object). That there are several distinct phenomena complicates rate determinations of reactions since diffusion of the substrate toward the site of the enzyme molecule and the rate of formation of the final product may influence the amounts of the final product formed. These authors have used the polyacrylamide enzyme to model these conditions and have evaluated the model mathematically.

Lattice-entrapped enzymes were used in still another type of model by Brown et al. (1968e) which bears upon substrate conversion technology but also has reference to a subcellular biological system. They isolated four enzymes of the glycolytic sequence in a polyacrylamide column which served to allow isolation of the end product of the sequence of the reactions as a column eluate. Hexokinase (2.7.1.7), phosphoglucoisomerase (5.3.1.9), phosphofructokinase (2.7.1.11), and aldolase (4.1.2.7) were entrapped separately and packed into a column in the

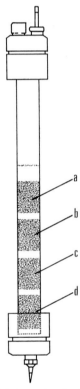

Fɪɢ. 5. Arrangement of the multiphase glycolytic enzyme column. The column was packed with layers of polyacrylamide gel, each containing an entrapped enzyme. At the bottom of the column and between the layers, 2-cm spacers of the same gel granules were used. A 4-cm band of granules was placed on top of the column. (a) Hexokinase, (b) phosphoglucoisomerase, (c) phosphofructokinase, (d) aldolase. [From Brown et al., J. Chromatog. **35**, 103–105 (1968e), by permission.]

same order as that in which the reactions occur in nature. This arrangement is illustrated in Fig. 5. Substrate solution placed on the column contained glucose, ATP, and $MgCl_2$ in a pH 9 buffer. The column was eluted with the same buffer. One hundred and sixteen micromoles of glucose were consumed; 12 μmoles of glyceraldehyde 3-phosphate were obtained in the product. In view of the adversity of conditions under which these four enzymes functioned in isolated systems, the degree of efficiency represented was probably as great as could be expected. It is quite apparent that the ability of the cell's architecture to maintain chemical events separate and ideally poised has not yet been achieved in this substrate conversion system, but the column does represent a "visualization" in which enzymes are isolated and the reactions are restricted to a chosen sequence analogous to a known sequence of intermediary metabolism. There appeared to be no interaction between components of the system, and thus the authors suggested that it might be possible to study effects of hormones, drugs, etc., in this manner without enzyme interactions.

A. Oxidoreductases

Lactate dehydrogenase (1.1.1.27) and glucose oxidase (1.1.3.4) were entrapped in cross-linked polyacrylamide by Hicks and Updike (1966). These workers undertook to determine the most appropriate gel composition for enzymes stabilized by entrapment. In addition to the primary models, catalase, amino oxidase, and glutamic dehydrogenase were entrapped. In general, with the exception of thermal stability, the properties of the entrapped enzymes were not greatly different from those of free solution preparations. Figure 6 indicates the stability of lactic dehydrogenase (LDH) in gel as compared with the enzyme in solution. These authors favor the thesis that a stable fraction of the LDH population is selected by entrapment rather than the conventional assumption that change in thermal stability results as a function of the entrapment itself.

Glucose oxidase was entrapped in cross-linked polyacrylamide and used by Updike and Hicks (1967) in an "enzyme electrode." In this promising and novel device the immobilized glucose oxidase was supported on a polarographic oxygen electrode in such a way that glucose in the surrounding medium served as a substrate to produce a proportionate yield of oxygen which, in turn, was then measured by the conventional polarographic method (Fig. 7).

Coupled glucose oxidase–peroxidase (1.11.1.7) were entrapped in a Silastic matrix by Pennington et al. (1968b,c). Although this menstruum served as a convenient carrier for these enzymes, properties of the system were, with the exception of a higher apparent K_m, much like those of the same materials in solution.

B. Transferases

Bernfeld and Wan (1963) in a description of methodology for the entrapment of proteins in polyacrylamide gels studied RNase (2.7.7.–), trypsin, α-chymotrypsin, papain, α-amylase, β-amylase, and aldolase. Enzymes retained about 1% of their original activity. There was no indication that the properties of the insoluble preparations were in any way different from the corresponding soluble ones. In particular, the pH optima of the various enzymes were unchanged. Contact with the substrate did not release the entrapped enzyme. Bernfeld and Wan believed that 2 to 6% of the original enzyme may have been entrapped but did not exhibit biological activity because of steric hindrances.

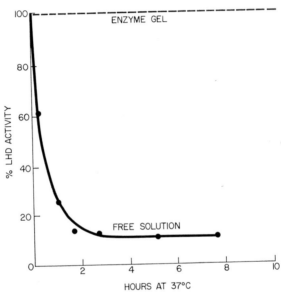

FIG. 6. Thermal stability of lactate dehydrogenase (LDH) in gel and in solution. [Redrawn from Hicks and Updike, *Anal. Chem.* **38,** 726–730 (1966), by permission.]

Brown *et al.* (1968e) entrapped the transferases hexokinase (2.7.1.7) and phosphofructokinase (2.7.1.11) together with other enzymes of a glycolytic sequence in polyacrylamide gels as a multiphase reaction system.

Insoluble phosphoglycerate mutase (2.7.5.3) from rabbit muscle was immobilized in cross-linked polyacrylamide by Bernfeld *et al.* (1969). Kinetic behavior of the soluble and insoluble forms of this enzyme was identical in many respects (enzyme concentration, substrate concentration, and temperature). However, pH optimum of the insoluble form of mutase was one unit lower than that of the water-soluble form, in spite of the fact that the carrier of the insoluble enzyme was electrostatically neutral. Insoluble mutase was more sensitive to activation by 2,3-diphos-phoglycerate than the soluble form.

Comparison between the insoluble forms of phosphoglycerate mutase and enolase, prepared by the same procedure, indicated that differences between soluble and insoluble forms of these enzymes depend more on the nature of the enzyme than on that of the entrapment medium.

C. HYDROLASES

Bauman *et al.* (1965) and Guilbault and Kramer (1965) used cholines-terase (3.1.1.8) in the development of an organic phosphorus-detecting

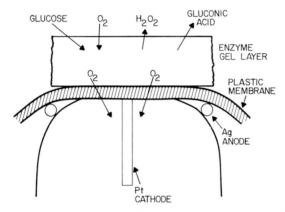

FIG. 7. Principle of enzyme electrode. [Redrawn from Updike and Hicks, *Nature* **214**, 986–988 (1967), by permission.]

device. In their studies the enzyme was within a starch gel which was then formed in a loosely woven polyurethan pad. The impregnated pad was used for the electrochemical detection of anticholinergic agents. The reaction was measured electrochemically. In this technique butyrylthio-choline iodide was hydrolyzed to thiocholine iodide with a resulting electrochemical potential. In a continuously monitored system, the anti-cholinergic agent is detected because its presence, as an enzyme inhibi-tor, results in the absence of an easily oxidized substance. The potential across the electrochemical cell will rise markedly in contrast to that which is characteristic of the iodide–iodine complex. In the study of Guilbault and Kramer (1965), the immobilized cholinesterase was used to hydrolyze napthol esters to form free napthol, a fluorescent molecule. Cholinesterase inhibitors could then be detected fluorometrically.

Acetylcholinesterase was entrapped in a Silastic matrix by Pennington *et al.* (1968a). The enzyme was dramatically stabilized, and acetylcholinesterase in Silastic had a lower K_m value than it did in solution. Silastic as a menstruum for the entrapment of the enzymes offers interesting possibilities in practical application. The suggestion has been made (Brown *et al.*, 1968a; Pennington *et al.*, 1968b) that biomedical materials having enzymic activity may be prepared in this way. The thermal stability properties achieved are, in some systems, useful, and it may be that the material will have utility as an analytic tool. It is important to draw attention to the small, but significant, solubility of Silastic in water which may present anomalies when the material is to be maintained over a very long term in aqueous environment (for example, a cardiovascular implant).

Alkaline and acid phosphatases (3.1.3.1, 3.1.3.2) have been studied in entrapped systems (Van Duijn *et al.*, 1967; Van der Ploeg and Van Duijn, 1968; Brederoo *et al.*, 1968a,b). Trypsin (3.4.4.4), orsellinic acid decarboxylase, and entire microorganisms were entrapped within crosslinked polyacrylamide gels by Mosbach and Mosbach (1966). The gels were granulated and then assayed in a continuously eluted column. About 2% of the activity was retained in the granules (compared with the free solution enzyme) and leakage of the enzyme from the matrix was negligible during a 2-month test period.

More novel than the purified enzyme columns were lichen columns which were prepared in the same way and then used as a source of esterase and decarboxylase activity. This complex material remained active after 3 months of storage at 20°C. Mosbach and Mosbach suggest that entrapment of an entire organism should prove of practical advantage, assuming that substrate penetration of the cell wall does not affect the results, by overcoming the necessity for laborious and difficult steps involved in enzyme preparation for such column conversion systems.

In a study (Brown *et al.*, 1968a) to characterize enzymes under the restricted conditions of lattice entrapment, trypsin and chymotrypsin were insolubilized in polyacrylamide as well as in hydrophobic Silastic matrices. Esterase activity of trypsin and chymotrypsin was measured using three substrates. The entrapped trypsin had 62% of the activity of the free enzyme when the substrate was *p*-tosyl-L-arginine methyl ester hydrochloride and 40% of the activity of free enzyme when α-*N*-benzoyl-L-arginine ethyl ester hydrochloride was the substrate. Neither the free enzyme nor the entrapped enzyme was catalytically active when *N*-acetyl-L-tyrosine ethyl ester was the substrate. Entrapped chymotrypsin had 48% of the activity of the native enzyme toward *p*-tosyl-L-arginine methyl ester hydrochloride and about 50% of the activity of the native enzymes toward N-acetyl-L-tyrosine ethyl ester.

Urease (3.5.1.5) was employed by Guilbault and Montalvo (1969) in a polyacrylamide gel in a device not unlike the enzyme electrode of Hicks and Updike. In their "urea transducer" membrane, immobilized urease was used as surface for an ammonium electrode. Ammonia produced by hydrolysis of the substrate was measured potentiometrically.

Characteristics of entrapped apyrase (3.6.1.5) were compared with the free solution preparation (Brown et al., 1968a). Calcium activation optima of apyrase and acrylamide-entrapped apyrase were identical though the extent of stimulation was different. Too, high concentrations of calcium inhibited the entrapped enzyme. The activity of the polyacrylamide-entrapped enzyme differs as a function of substrate, presumably because of the gel structure which limited approachability of the ester to the protein surface. Thermal stability of polyacrylamide-entrapped apyrase was greater than that of the native enzyme. The lattice preparation not only survived 60°C for 60 minutes but actually showed enhanced apyrase activity. The authors suggested, however, that this may be due to the effect of heat upon the gel structure which allowed greater access of substrate to the enzyme.

D. Lyases

Aldolase (4.1.2.7) has been included in a number of lattice-entrapped enzyme studies (Bernfeld and Wan, 1963; Bernfeld et al., 1968; Brown et al., 1968e). Radioactive aldolase was used by Bernfeld et al. (1968) to study distribution of enzyme protein vs. activity in insoluble carrier and aqueous phase after polyacrylamide polymerization. All of the radioactivity was accounted for, and 44.5% of the total enzyme activity was recovered. The ratio of enzyme activity to radioactivity in the aqueous phase remaining after the completion of the polymerization was about 25% lower than the original soluble aldolase. This was interpreted to indicate that the polymerization catalysts caused no more than a 25% enzyme inactivation. Ratio of enzyme activity to protein in the gel indicated that four times as much protein was present as was indicated by the catalytic activity. Bernfeld and his associates rationalized this by suggesting that only the aldolase at or near the surface of the carrier particle is able to react. That the enzyme had not been rendered insoluble by the entrapment was demonstrated by ultrasonically disrupting the carrier particles to release some of the entrapped enzyme. When solubilized in this way the aldolase was restored to activity.

Inclusion of aldolase in a column containing enzyme representative of a portion of glycolytic sequence has been discussed above. Bernfeld and Bieber (1969) entrapped enolase (4.2.1.11) in cross-linked polyacrylamide, and the kinetics of the reaction catalyzed by the enzyme were compared with the kinetics of the reaction catalyzed by enolase in solution. There were no differences between the two forms with regard to the influences of pH, enzyme, or substrate concentration. Michaelis constants and maximum velocity as well as the turnover number were identical. Both

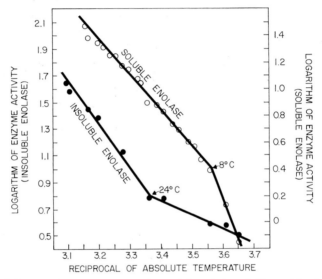

FIG. 8. Influence of temperature on the activity of the soluble and insoluble forms of rabbit muscle enolase. Plots of the logarithm of activity versus the reciprocal of absolute temperature × 10^3. [Redrawn from Bernfeld and Bieber, *Arch. Biochem. Biophys.* **131**, 587–595 (1969), by permission.]

forms required magnesium for maximum activity, but the insoluble form was not inhibited by excess magnesium as was soluble enolase. Interesting differences were noted between the two forms in their reaction temperature (Fig. 8). Between 1° and 24°C the activity of the insoluble form increased much more slowly than that of the soluble form. Optimum activity was reached at 44°C for the soluble and at 50°C for the insoluble forms. The soluble enzyme was inactive at 50°C (the optimum temperature for the insoluble form), but insoluble enolase retained 23% of its activity at 56°C. Activation energies were similar for both forms in the temperature range above 24°C (11,800 and 15,300 cal/mole for the soluble- and insoluble forms, respectively). The extremely low value of 4600 cal/mole for insoluble enolase below 24°C indicates that this form of enzyme is a more efficient catalyst at low temperatures than is the soluble enzyme. This observation [similar to those of Hicks and Updike (1966) for several enzymes] is certainly worthy of further thought both from theoretical and practical (low-temperature analysis) points of view.

V. Cross-Linked Enzymes

Figure 9 illustrates chemical methods of cross-linking which have been used for water-insoluble enzyme preparations. A schematic conceptualization of cross-linking, as used here, is shown in Fig. 10.

Although the number of bifunctional reagents used in this way has been relatively small, many others presumably would be appropriate. We shall consider cross-linking to be essentially a complex of the enzyme molecules themselves or the formation by the protein of a tight-fitting "skin" around an inert matrix. The covalent binding of an enzyme to a matrix through a bifunctional reagent is considered in Section VII, and pertinent concepts are discussed by Habeeb in Chapter VIII, this volume.

Although the development of methodology has scarcely begun, there does exist a base for technological development and for the construction of appropriate biological models. Cross-linking can be used in many of those applications in which it is impractical or undesirable to use a reactive matrix. Although published works have dealt with enzymes superficial to the matrix, there is no apparent reason that fine pores could not be lined with protein. The cross-linking techniques lend themselves to virtually any configurational requirements.

Reactions of bifunctional cross-linking agents with proteins were described in a series of studies by Zahn and associates (1956; Zahn and Waschka, 1955; Zahn and Meienhofer, 1958) using the agents bisdiazohexane, difluorodinitrobenzene, and p,p'-difluoro-m,m'-dinitrophenyl sulfone with keratin, fibroin, and insulin. These reagents reacted with α- and ϵ-amino, hydroxyl, and phenolic groups as well as others. Wold (1961) used difluorodinitrophenyl sulfone to cross-link the serines of albumin and described the intramolecular cross-linking which occurred. The same reagent was used in experiments with RNase, but the authors reported that under the conditions of their experiments no cross-linking occurred (Broomfield and Scheraga, 1961).

Herzig et al. (1962, 1964) used disulfonyl chlorides in experiments with lysozyme. They considered the reagents phenol-2,4-disulfonyl chloride and α-napthol-2,4-disulfonyl chloride to be suited for the determination

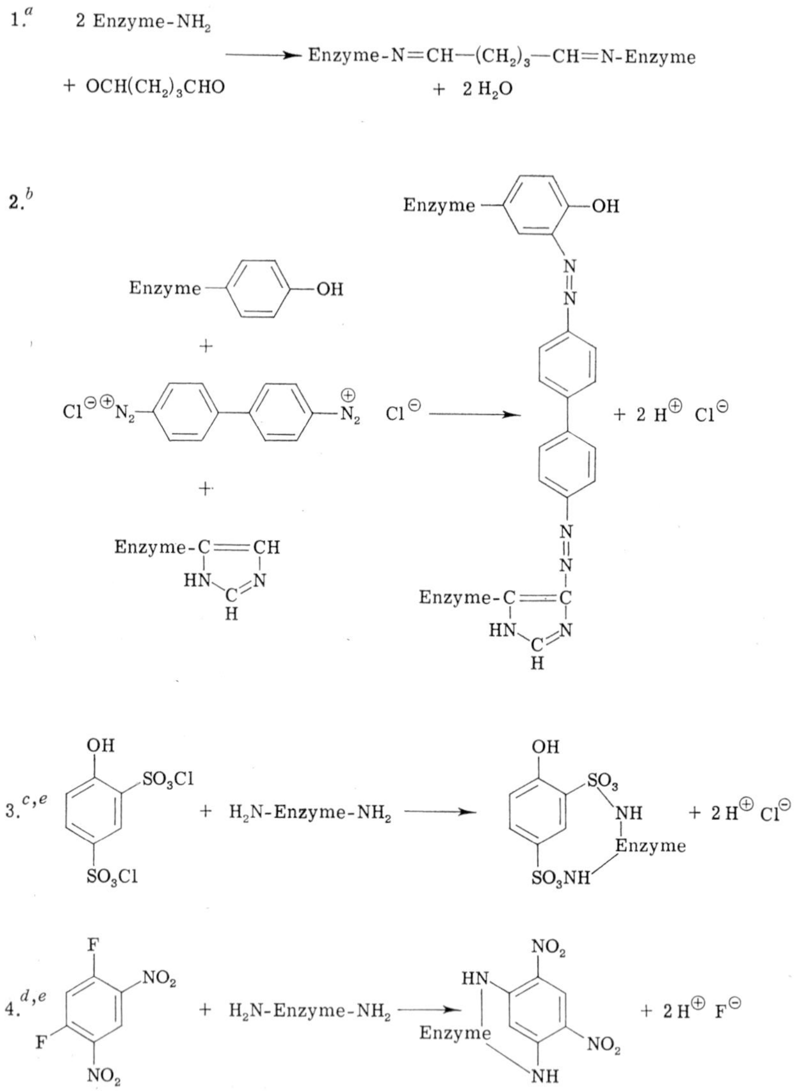

1.[a]   2 Enzyme-NH$_2$

$\longrightarrow$ Enzyme-N=CH—(CH$_2$)$_3$—CH=N-Enzyme

+ OCH(CH$_2$)$_3$CHO   + 2 H$_2$O

2.[b]

Enzyme—⟨benzene⟩—OH

+

Cl$^{\ominus}$ $^{\oplus}$N$_2$—⟨biphenyl⟩—N$_2$ $^{\oplus}$ Cl$^{\ominus}$

+

Enzyme-C=CH
     | |
    HN   N
       ⟍C⟋
        H

$\longrightarrow$

Enzyme—⟨benzene⟩—OH
                      N
                      ‖
                      N
                  ⟨biphenyl⟩
                      N
                      ‖
                      N
Enzyme-C=C
     | |
    HN   N
       ⟍C⟋
        H

+ 2 H$^{\oplus}$ Cl$^{\ominus}$

3.[c,e]   ⟨benzene with OH, SO$_3$Cl, SO$_3$Cl⟩ + H$_2$N-Enzyme-NH$_2$ $\longrightarrow$ ⟨benzene with OH, SO$_3$-NH-Enzyme, SO$_3$NH⟩ + 2 H$^{\oplus}$ Cl$^{\ominus}$

4.[d,e]   ⟨benzene with F, NO$_2$, F, NO$_2$⟩ + H$_2$N-Enzyme-NH$_2$ $\longrightarrow$ ⟨benzene with NO$_2$, HN-Enzyme, NO$_2$, NH⟩ + 2 H$^{\oplus}$ F$^{\ominus}$

FIG. 9. Representative cross-linking reactions. (a) Quiocho and Richards (1964). Quiocho and Richards (1966) suggested that the reaction goes beyond the Schiff's base formation which is illustrated. (b) Mowbray and Scholand (1966). Coupling through tyrosyl and histidyl residues is illustrated, but other moieties can react. See discussion of diazonium coupling. (c) Herzig et al. (1964). (d) Marfey et al. (1965). (e) Intramolecular link.

of protein conformation because at proper pH they react preferentially with the amino group of lysine forming stable sulfonamide bonds; aromatic sulfonyl chlorides react with primary and secondary groups under basic conditions; p-iodobenzenesulfonyl chloride and p-toluenesulfonyl chloride react with α- and ϵ-amino groups in proteins; and earlier investigators (Geschwind and Li, 1957) had shown that none of the lysines in lysozyme were at the active site.

By partial analysis of the peptides, Herzig *et al.* showed that cross-linking between ϵ-amino groups of lysine residues had occurred. Too, definite deletions and additions to the chains were shown on peptide maps after tryptic digestion. This observation is cogent to consideration of the

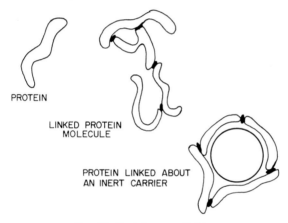

FIG. 10. Protein cross-linking.

entire literature since we, in most applications, assume that such changes are not inherent in cross-linking procedures. Lysozyme was stabilized by cross-linking.

Other theoretical implications and studies of cross-linking procedures have been reviewed by Habeeb in Chapter VIII, this volume.

A. OXIDOREDUCTASES

Broun *et al.* (1969) prepared cellophane glucose oxidase (1.1.3.4) membranes using glutaraldehyde as the cross-linking reagent. Cellophane-supported preparations retained 12% of the specific activity of the enzyme. There was no change in the pH optimum. Thermal dependence of membrane-borne catalyst was greater at its optimum than was thermal dependence of glucose oxidase in solution, though the insoluble preparation

218 HARRY D. BROWN AND FRANK X. HASSELBERGER

was thermally more stable than its solution counterpart. Both forms had the same Michaelis constant, and the substrate specificities were identical. These authors report application of the same techniques to trypsin, chymotrypsin, urease, carbonic anhydrase, and urate oxidase.

B. TRANSFERASES

Wang and Tu (1969) modified glycogen phosphorylase *b* (2.4.1.1) with glutaraldehyde. Sixty percent inactivation was seen after incubation of the enzyme with 0.05% reagent for 10 minutes. Inactivated enzyme showed multiple protein bands on polyacrylamide gel after electrophoresis. The various protein species were differentially susceptible to thermal denaturation. A modified and heat-treated enzyme was considerably more resistant than native phosphorylase *b* toward heat, cold, or urea denaturation. The glutaraldehyde-treated contained 7 to 11 fewer free amino groups than did native phosphorylase *b*. Modified enzyme contained 10% fewer lysyl residues than native phosphorylase *b;* specific activity of the modified and heat-treated phosphorylase *b* was 65% that of the free solution enzyme. Although affinities of this enzyme species toward glucose 1-phosphate and glycogen were unchanged, the homotropic interaction of adenosine monophosphate (AMP) of "soluble" enzyme could not be demonstrated with the cross-linked phosphorylase *b*.

Marfey *et al.* (1965) used the bifunctional reagent, 1,5-difluoro-2,4-dinitrobenzene to cross-link pancreatic RNase A (2.7.7.16). They demonstrated that the cross-linked preparations were monomeric (molecular weight 14,000) and that the cross-links were between the side chains of lysine residues. Three different RNase A derivatives were separated by ionic exchange chromatography, and their spectra were interpreted to indicate that they contained a single cross-link per mole of RNase A. With the substrate cytidine 2′,3′-cyclic phosphate, enzymic activity was 15 to 49% of the native purified bovine pancreatic RNase A.

C. HYDROLASES

Carboxypeptidase A (3.4.2.1), a relatively insoluble crystal, has been compared to a larger particle prepared by cross-linking (Quiocho and Richards, 1964, 1966; Bishop *et al.*, 1966; Quiocho *et al.*, 1967). Carboxypeptidase crystals are somewhat soluble at high ionic strengths. The solubility of the native crystal is very low at low ionic strength, and enzymic activity measured under these conditions is attributable to the crystal itself rather than to a small solubilized population.

Crystals were treated with 1% aqueous glutaraldehyde to form a cross-linked particle which was totally insoluble in 1 M NaCl. Glutaraldehyde cross-linked carboxypeptidase crystals were remarkably stable, functioning under column conditions at room temperature for several months.

Mechanical properties of the crystal were changed by cross-linking. Generally, large single protein crystals are fragile and easily cracked or chipped during manipulation. After glutaraldehyde treatment, it was not possible to fragment the crystals even by vigorous agitation. Despite the mechanical changes, however, the diffraction patterns of the treated and untreated crystals were similar, though not identical. There was a small decrease in average intensity in the higher orders indicating the absence of marked disordering effects of the glutaraldehyde reaction; hence, apparently there was little change in the molecular structure or orientation. The authors suggested that the decreased enzymic activity—the cross-linked crystals retained slightly better than 30% of the activity of the native enzymes—could be due either to the alteration of functional groups directly affecting the catalytic activity or to an increase in the restriction of diffusion caused by a decrease in the effective pore size of the channels. Treatment of highly cross-linked carboxypeptidase with 8 M urea caused complete loss of crystalline reflections though without evidence of solubilization of the protein.

Bishop and associates (1966) used cross-linked carboxypeptidase to study the relationship of enzyme activity to metals. One gram atom of zinc per mole of enzyme was measured and the metal was not removed by treatment with aqueous phenanthroline, a chelating agent. Phenanthroline in nitrate solution, however, did remove zinc. This was interpreted as an indication that the chelating agent is capable of entering the crystal lattice. Cobalt solution caused a marked activation, an effect reversed by washing with water. Addition of zinc had no further effect.

If zinc content of a column of cross-linked crystals was lowered to a barely detectable level and the enzyme was equilibrated with cobalt until the preparation had about 1 atom per protein molecule and a zinc cobalt solution was passed over the crystals, the zinc content increased to 2 gram atoms per mole of protein while at least 1 gram atom of cobalt remained. This observation was interpreted by these authors to indicate that there were a number of metal-binding sites in addition to the catalytic site itself.

With nitrate as the principal anion, substantial amounts of mercury could be bound to the carboxypeptidase crystals without displacement of zinc. When the mercury concentration was increased to 10^{-2} M (and the pH lowered to 5.2), however, the zinc was displaced. The activity of this column was about 50% of the level of the original material. These crystals contained barely detectable amounts of zinc and 4 atoms of mercury per molecule of enzyme. By exposure to a zinc solution, an atom of zinc was added and an atom of mercury removed. After this, the assay indicated 140% activity. When the zinc content was lowered to 1 mole and mercury to about 1.5 moles, the activity was 100% that of the original preparation. Whether this phenomenon of "extra metal sites" (more than 1 mole of a metal bound per mole of protein) was an intrinsic property of the protein or a peculiarity of the crystal lattice was not determined. Interestingly, the enhanced or decreased activities observed to be related to these extra sites indicate that they do affect catalytic behavior of the protein.

Properties of the cross-linked carboxypeptidase have been considered by Quiocho and Richards in an attempt to determine whether the cross-linking reaction involved the entire crystal lattice or merely its surface. Aliquots of the preparation were removed during the reaction, crushed, and extracted. When the supernatant was assayed it was inactive; the pelleting crystals were enzymically active. They concluded, therefore, that the entire crystal is affected by the reaction to form a true three-dimensional net. Amino acid analyses indicated that only lysine content was altered in the cross-linked material.

In a study of the bond formation it was shown that Schiff bases involving analine and arylaldehyde were rapidly attacked by semicarbazide to yield a semicarbazone and the free amine. Attempts to reverse the reaction, however, were unsuccessful, and it was concluded that the reaction of glutaraldehyde and the ϵ-amino group of lysine proceeded beyond the stage of simple Schiff base formation. Other bifunctional reagents, 1,3-difluoro-4,6-dinitrobenzene and biphenyl-4,4'-bisdiazonium dichloride, were, like glutaraldehyde, found to be effective cross-linking agents for carboxypeptidase.

The effects of inhibitors on glutaraldehyde cross-linked carboxypeptidase were studied. The block copolymer of glutamic acid and tyrosine (ratio 95:5) is a potent inhibitor of carboxypeptidase in low ionic strength solution. Cross-linked crystals and native carboxypeptidase were almost completely inhibited, but no inhibition was observed when the enzyme was in the form of an amorphous cross-linked mass. Phenylpropionic acid, a substrate analog, effectively inhibited carboxypeptidase in solution and in a cross-linked amorphous preparation. In the crystalline state, the inhibition constant was increased by an order of magnitude. The basis for the differences in the effects of these inhibitors is not known, although further studies, particularly those using the inhibitor β-phenylpropionate, were interpreted in terms of the presence of modifier sites. Inhibition of peptidase activity by chelating agents, such as phenanthroline and hydroxyquinoline sulfonic acid, appears to result from competition with apoenzyme for its atom of zinc. Quiocho and Richards' study indicated that the zinc atom may be removed and reinserted in the amorphous enzyme and in soluble preparations with corresponding activity changes. Resistance of cross-linked crystals to inhibition by the chelating agent was interpreted as an indication that the zinc atoms were not removed from enzyme molecules within the crystal lattice. Some inhibition of the enzyme by the chelating agent without removal of the metal ion was observed, and it was proposed that a ternary complex among the apoenzyme, the metal ion, and the chelating agent existed.

Activity of amorphous cross-linked enzyme was less subject to ionic strength than were the cross-linked crystals, although unit cell dimensions of crystals did not vary significantly with ionic strength. Activation energy of the crystal was calculated to be 10.7 kcal; the same enzyme in solution has activation energy of 9.6 kcal. Quiocho and Richards (1964, 1966) hypothesized that the difference in specific activity between soluble and crystalline forms can be attributed almost entirely to entropy of activation.

Habeeb (1967) described cross-linked trypsin (3.4.4.4) prepared by a treatment of the enzyme with glutaraldehyde. The insoluble material was highly reactive retaining 66 to 88% of its solution level. Activity was retained after repeated digestions of casein. Dioxane at 50% concentration destroyed the activity of the glutaraldehyde-insoluble cross-linked trypsin.

Haynes and Walsh (1969) formed glutaraldehyde cross-linked trypsin monolayers on colloidal silica and applied the method to other enzymes including α-chymotrypsin (3.4.4.5).

Papain (3.4.4.10) was used in a study by Silman et al. (1966). The enzyme was cross-linked with bisdiazobenzidine. This preparation retained much of the activity of the native enzyme on low and high molecular weight substrates. A graded series of cross-linked papains, ranging from partially insoluble to completely insoluble, could be obtained with this agent by using varying amounts of the cross-linking reagent. Jansen and Olson (1969) were able to retain a cross-linked papain with glutaraldehyde in aqueous solution.

An enzymically active membrane was prepared by Goldman et al. (1965, 1968) by impregnating a collodion membrane with crystalline papain using the cross-linking agent diazobenzidine 3,3'-sulfonic acid. The papain membrane retained much of its enzymic activity when activated with sulfhydryl compounds. There was no evidence for desorption from the membrane either in the presence of substrate or after 4 months of storage under water at 4°C. Enzyme membranes with two- or three-layer structures were prepared. The three-layer system had alternate layers of enzyme sandwiching an inner layer of membrane. The papain membrane was as permeable to sucrose and to benzoyl-L-arginine ethyl ester (BAEE) as was the unmodified collodion membrane from which it was derived. At pH 6 the papain membrane had about 5% of the activity of an equivalent amount of the crystalline enzyme on BAEE and 40% of the activity on benzoyl-L-arginine amide.

Behavior of the membrane enzyme to BAEE is markedly different from the behavior of the native enzyme, and activity appears to increase with increasing pH (to pH 9.6). Decrease in activity at low pH is attributed by Goldman to a lowering of the internal pH of the membrane during reaction. Additional evidence is advanced to support the belief that there is a change in internal pH, and calculations are presented which indicate internal pH values differing from those of the external pH to the extent of several pH units.

The data demonstrate that an enzyme imbedded in a membrane can by its own action change its environment markedly and thereby alter its own activity. Such effects may play a part in feedback and control systems at the cellular and intracellular level. It is conceivable that a wave reaction may be propagated in the plane of a membrane as the result of acid released by one enzyme molecule serving to trigger the action of an adjacent molecule which might even in the presence of substrate be initially active. It is scarcely necessary to comment upon the interest of Goldman's membrane enzyme model as it may relate to the present renaissance of interest in Peter Mitchell's (1966a,b) hypothesis in which pH differences at membranes are related to ATP synthesis and to the other phenomena of oxidative phosphorylation. The membrane enzyme model may also find use in the development of illustrative systems relating to ion transport itself in cells. Certainly Goldman's membrane is a very direct demonstration, however limited, that the papain membrane itself may be a useful enzyme-bearing membrane for the construction of pertinent physiologically related models.

Water-insoluble apyrase (3.6.1.3) was prepared by Brown *et al.* (1968c) by treating the enzyme with glutaraldehyde or with glutaric acid dihydrazide. In the same studies, these authors employed a metal-sensitive ATPase derived from heart muscle membranes. Glutaraldehyde cross-linked apyrase was less active than the native enzyme although apyrase reacted with glutaric acid dihydrazide had enhanced activity. This was tentatively attributed to altered configurations of the complexes which made catalytic sites either more or less accessible to the substrate. The data were discussed by the authors in terms of changed enzyme conformations and microenvironment.

Lysozyme (3.2.1.17) was studied in cross-linked preparations by Herzig *et al.* (1964).

VI. Covalently Bound Enzymes

A. Introduction

The covalent coupling of physiologically active proteins to a matrix must be accomplished by reactions with groups which are sufficiently distal to the catalytic site to avoid functional inactivation. Although it has been conventional to state that the reaction may be carried forth only under very mild conditions, it is probably better to take the more pragmatic view that the reaction conditions must not remove irreversibly catalytic activity of the protein. A number of authors have discussed the conditions appropriate to protein derivatization (Habeeb, Chapter VIII of this volume; Fraenkel-Conrat, 1959; Sri Ram et al., 1962). Manecke (1962a), citing Herriot (1947), has outlined the protein groups that are available for reaction with matrix polymers: free carboxyl groups (acid amino acids or terminal carboxy groups of amino acids); free amino groups (basic amino acids or terminal amino groups of amino acids); the phenol group of tyrosine, which couples with diazonium salts; the imidazole group of histidine; the imino group of tryptophan, free thiol groups (cystine); the disulfide link in cystine; the guanidino structure in arginine; aliphatic hydroxyl groups (serine, threonine, oxyproline, oxyglycine, oxyglutamic acid), the methylmercapto group of methionine; the phenol group in phenylalanine; amide groups (asparagine, glutamine); and in some cases peptide links.

B. Methods Involving Amide Linkages

Use of N-ethyl-5-phenylisoxazolium-3′-sulfonate (Woodward's reagent K) to convert a carboxylate group into a reactive ester group by a very rapid and smooth reaction, under exceptionally mild conditions, was reported by Woodward and Olofson (1961). In an accompanying article, Woodward et al. (1961) reported the application of this reaction as the carboxyl-activating step to the synthesis of peptides.

R. P. Patel and Price (1967) prepared enzymically active polymers of α-chymotrypsin and also matrix derivatives of this enzyme using polyacrylic acid, CM cellulose, and poly-L-glutamic acid as the matrices and Woodward's reagent K as the condensing reagent. The chemistry of these reactions has been outlined schematically by Patel and co-workers (Fig. 11).

Sheehan and Hess (1955) reported the preparation of peptide (or amide) bonds by using 1,3-dicyclohexylcarbodiimide. Water-soluble carbodiimides were later prepared by Sheehan and co-workers (Sheehan and Hlavka, 1956; Sheehan *et al.*, 1961) and shown to be useful in forming

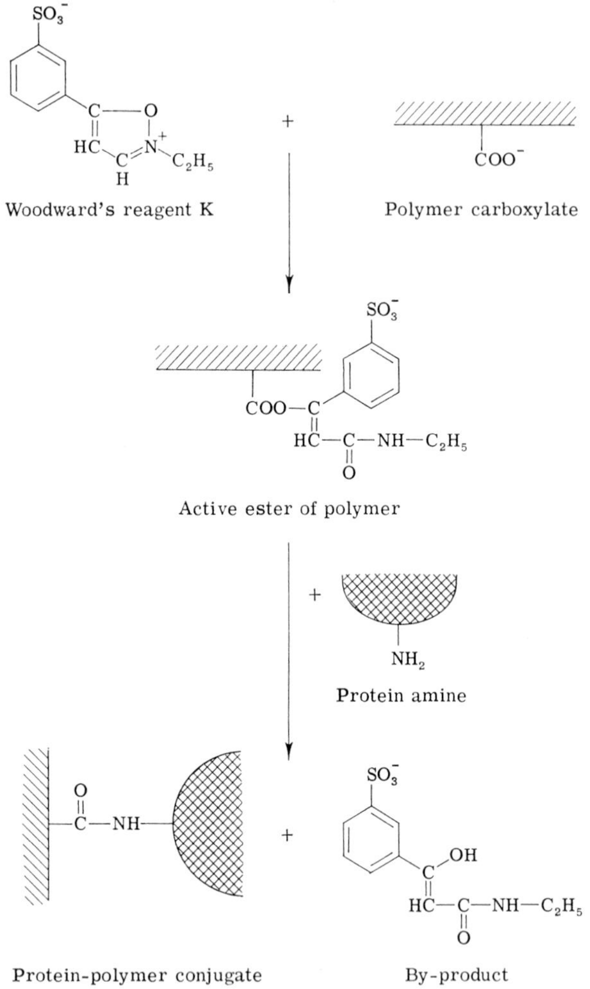

Woodward's reagent K   Polymer carboxylate

Active ester of polymer

Protein amine

Protein-polymer conjugate   By-product

Fig. 11. Schematic representation of conjugation of protein with carboxyl-containing polymer by using *N*-ethyl-5-phenylisoxazolium-3'-sulfonate. [Redrawn from R. P. Patel *et al.*, *Biopolymers* **5**, 577–582 (1967), by permission.]

amide bonds. This latter development was significant to protein chemistry since it permitted the enzyme-supporting reaction to be carried out in aqueous solution without requirement of an organic cosolvent.

Experience in our laboratory has indicated that using a water-soluble carbodiimide in acid solution as an activating agent for the matrix and washing the matrix before coupling the enzyme yields a more active preparation than does the mixing of matrix, enzyme, and carbodiimide in a single step. This may result from the fact that in the one-step procedure the enzyme can cross-link with itself (Sheehan and Hlavka, 1957) as well as undergo the desired reaction.

It is important to recognize that the carbodiimide reaction may inactivate an enzyme if reaction occurs at the catalytic site. Thus, Banks and co-workers (1969) have shown that α-chymotrypsin is inactivated by a water-soluble carbodiimide because a seryl residue at the active site reacts with the matrix–carbodiimide adduct to form an inactive isoureylene derivative of the enzyme. These results should be contrasted with those obtained by R. P. Patel and Price (1967) and R. P. Patel et al. (1967) with N-ethyl-5-phenylisoxazolium-3′-sulfonate. They prepared enzymically active polymerized α-chymotrypsin and CM cellulose supported α-chymotrypsin using this reagent.

In solid-support studies, N,N'-disubstituted carbodiimides have usually been used with CM cellulose (Weliky and Weetall, 1965; Weetall and Weliky, 1966; Hasselberger et al., 1970b). It should be recognized that, although the carbodiimide reagent shows a strong preference for carboxyl groups, other groups are also reactive toward this reagent (see, for instance, Hoare and Koshland, 1967; Carraway and Koshland, 1968), and they too can be used as the support matrices reactive site. Thus, we have found it possible to activate unsubstituted cellulose with a water-soluble carbodiimide and then to form enzyme derivatives of the activated cellulose.

Nevertheless, the CM cellulose reaction is historically the important one. The mechanism for this reaction is thought to be

$$R-\underset{\underset{O}{\|}}{C}-OH \;+\; H^{\oplus} \;+\; \underset{\underset{\underset{R''}{|}}{\overset{\|}{N}}}{\overset{R'}{\underset{\|}{N}}}\overset{\displaystyle R'}{\underset{\overset{\|}{C}}{|}} \;\longrightarrow\; R-\underset{\underset{O}{\|}}{C}-O-\underset{\underset{R}{|}}{\overset{NH}{\overset{|}{\underset{NH^{\oplus}}{\|}}}}\overset{R'}{\underset{}{}} \tag{1}$$

$$\text{Protein-NH}_2 \;+\; R-\underset{\underset{O}{\|}}{C}-O-\underset{\underset{R''}{|}}{\overset{\overset{R'}{|}}{\overset{NH}{\overset{|}{\underset{NH^{\oplus}}{\|}}}}} \;\longrightarrow\; \text{Protein-NH}-\underset{\underset{O}{\|}}{C}-R \;+\; O=\underset{\underset{R''}{\underset{+\,H^{\oplus}}{|}}}{\overset{\overset{R'}{|}}{\overset{N-H}{\overset{|}{\underset{NH}{}}}}} \tag{2}$$

In the first step (1), the carboxyl group undergoes an acid-catalyzed addition to the cumulated double-bond system of the carbodiimide. Insoluble CM cellulose–carbodiimide adducts may be isolated and excess carbodiimide washed out prior to enzyme coupling. This latter step (2) involves a nucleophilic attack by a free amino group of the enzyme on the carbonyl group of the adduct. The proton catalyst, regenerated in the protein coupling step, would doubtless be bonded to one of the many nucleophilic sites available, just as it is in the intermediate.

The mechanisms of carbodiimide reactions have been studied by DeTar and co-workers (DeTar and Silverstein, 1966; DeTar et al., 1966). Reviews of carbodiimide chemistry have been presented by Khorana (1953) and more recently by Kurzer and Douraghi-Zadeh (1967). Several hundred references dealing with carbodiimide chemistry can be found in these articles.

Woodward's reagent K method and the carbodiimide method complement each other. The latter depends upon an acid-catalyzed activation step, whereas the former is carried out in basic solution. This should be considered in planning the preparation of matrix-supported enzymes. (See discussion above of carbodiimide for reference to inactivation of α-chymotrypsin by a water-soluble carbodiimide.)

The ethylene–maleic anhydride (EMA) copolymer is useful in solid-support chemistry because of the reactive nature of the carboxylic acid anhydride function. It is possible to draw "resonance structures" having charge separation in the carbonyl groups. Thus, the extreme forms would be

Such moieties are susceptible to attack by nucleophilic groups on the protein molecule. That is, groups which have an atom having an unshared pair of electrons will attack the "partially positive" carbon of the carbonyl. Such attack is illustrated for the amino functions:

Free hydroxyl and presumably free sulfhydryl groups can give analogous reactions. Thus, the protein could be bonded to the matrix via amide, ester, or thioester bonds.

This method has been used for the preparation of both soluble and insoluble enzyme derivatives. Thus, Levin and co-workers (1964) have prepared insoluble trypsin derivatives by coupling the enzyme to an EMA sample having an average molecular weight of 40,000. Hexamethylene diamine was used as a cross-linking reagent in preparing this insoluble derivative. Westman (1969) intentionally omitted cross-linking reagents and isolated a soluble EMA derivative of the same enzyme on a matrix having an average molecular weight of 30,000.

As indicated in the above illustration, when the nucleophile attacks the anhydride function, the ring is opened and a negative charge is developed on the part of the anhydride not involved in binding the protein. In addition, groups not reacting with protein will be hydrolyzed by the weakly basic medium usually used (or even by water). This results in the formation of two negative charges for each anhydride group. [This is illustrated schematically in the paper by Levin et al. (1964).] The protein is, therefore, held in an environment of highly negative charge. This should favor reaction with positively charged substrates and should hinder reaction with negatively charged substrates, which would be repulsed by the negative charge on the matrix.

Weetall and Weliky (1966) developed peroxidase (1.11.1.7) paper for use in a semiquantitative "spot test." The material was capable of detecting peroxide in small volumes at concentrations as low as 10^{-6} M. The enzyme paper retained activity after refrigerated storage for 2 months. It was prepared by bonding horseradish peroxidase to CM cellulose paper strips in the presence of N,N'-dicyclohexylcarbodiimide. The procedure makes use of the color change resulting from the oxidation of benzidene. Chymotrypsin, trypsin, panprotease, subtilisin, nagarase, alkaline phosphatase, acid phosphatase, β-glucuronidase, deoxyribonuclease (DNase), RNase, and glucose oxidase were also linked to matrices by the carbodiimide method (Weliky and Weetall, 1965).

Weliky et al. (1969) used dicyclohexylcarbodiimide to bind peroxidase to CM cellulose. The pH optimum of the supported enzyme had a much narrower range than that of solution enzyme, and its behavior was changed after freeze-drying and rehydrating compared with the soluble material handled in the same way. Oddly enough, stability of the solid-supported peroxidase was less than that of the solution material though the matrix-supported enzyme was resistant to azide inhibition. These authors also described differences in the shape of the curve of guaiacol oxidation for the two preparations—an observation which they rationalized in terms of the time required to obtain a steady-state concentration of free radicals.

They calculated the activation energies of the peroxidase to be 6.5 kcal/ mole for the free enzyme and 7.2 kcal/mole for the bound enzyme.

Wagner *et al.* (1968) coupled trypsin (3.4.4.4) and chymotrypsin (3.4.4.5) to a random copolymer of L-alanine and L-glutamic acid. The copolymer was preswollen in acetonitrile and then reacted with Wood-

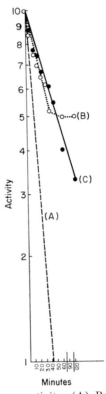

Fig. 12. Plot of log of enzyme activity. (A) Pronase in solution; (B) cyanogen bromide-activated preparation; (C) ethylene maleic anhydride preparation.

ward's reagent K (*N*-ethyl-5-phenylisoxazolium-3'-sulfonate). After this, the activated copolymer was resuspended in the buffer which contained the enzyme to be coupled. In each of the reactions, over 70% of the enzyme was bonded. Trypsin retained 43% and chymotrypsin 38% of their free solution activities.

A. B. Patel *et al.* (1971) have contrasted the activity of pronase bound to an EMA copolymer with the same enzyme bound to CM cellulose by the azide method and to cyanogen bromide-activated microcrystalline cellulose matrices. Figure 12 is a plot of log activity versus time. The insolubilized enzymes showed first-order rate constants, whereas, as expected, pronase in solution deviated from this. The authors suggested that the autolytic changes of the solution pronase were not present in the matrix-supported forms but that the changes which did occur might rather represent thermal denaturation. These data indicated that thermal stability properties of water-insolubilized pronase are a function of the specific character of the complex itself.

Both the carbodiimide method and the method employing Woodward's reagent K were used by Hasselberger et al. (1970a) to bind an asparaginase (3.5.1.1) from Escherichia coli to CM cellulose. This material was used in a miniature column as an asparagine filter designed to lower blood level of asparagine in the therapy of neoplastic disease.

Westman (1969) prepared a water-soluble trypsin by reacting the enzyme with EMA copolymer of 30,000 average molecular weight. It was purified by repetitive chromatography using a dextran molecular sieve. At high ionic strength, unbonded protein was removed from the soluble EMA–trypsin adduct. Substantial activity was present in the adduct solution: about 65% of the initial activity remained after 17 days in solution (compared to the original trypsin preparation which lost most of its activity within 2 to 3 days). The pH profile of soluble EMA–trypsin mimicked that of insoluble EMA–trypsin (which employed as matrix a higher molecular weight polymer) with pH optima shifted toward a higher pH and a tailing of activity into higher pH ranges.

Westman found by end-group analysis and digestion of oxidized RNase that the soluble EMA–trypsin was capable of hydrolyzing fewer bonds than was the free solution trypsin preparation. If tryptic catalysis represents the activity of a single molecular site, then the change in specificity of the site which results from binding of the enzyme to a short synthetic polymer invites a reexamination of some of the enzymologists' "truths" about the nature of this catalytic mechanism.

Goldstein and associates (1967a) described the preparations of poly-L-ornithyl chymotrypsin, poly-L-glutamyl chymotrypsin, and EMA–chymotrypsin. Activity profiles as a function of pH at low ionic strengths of the EMA–chymotrypsin and poly-L-glutamyl chymotrypsin were displaced about 1 pH unit toward the more alkaline as compared to the crystalline chymotrypsin in solution. The turnover number at pH 9.5 was higher by about 30% than that of the native enzyme at its pH optimum (pH 8.5). Interestingly, the activity profile of polyornithyl chymotrypsin was displaced by 1 pH unit toward the more acidic, and the turnover value was lower by about 30% at pH 8.0 than that of the native enzyme. These effects were abolished at high ionic strength. The authors stated that the Donnan-type electrostatic model could not explain these experimental findings, and they developed, therefore, a new theoretical model. They assumed that the hydrolysis of esters by serine-containing proteases was preceded by a general base catalysis mechanism possibly involving the histidine residue. A linear free energy relationship of the Bronsted type would correspond to the dependence on the rate constant, the turnover number and the apparent basic dissociation constant of the group participating in the general base catalysis.

C. Azide Method

The azide method (Mitz and Summaria, 1961) is useful in that the matrix can easily be prepared in large quantities and can be stored (desiccated) for relatively long periods with little loss in protein bonding power. The starting material is usually commercially available CM cellulose. This is esterified and the ester is subjected to hydrazinolysis. The resulting hydrazide reacts with nitrous acid to give the azide (Fig. 13).

Micheel and Ewers (1949) reported that the azide reacts with free amino, hydroxyl, and sufhydryl groups of the protein forming amide, ester, and thioester bonds. Since sufhydryl groups are frequently a part of the active sites of enzymes, protective groups are sometimes added before subjecting the enzyme to the azide-containing matrix (Brown et al., 1968c). The protective groups are removed from the matrix-supported derivative to restore enzymic activity.

These reactions presumably involve nucleophilic attack by the nitrogen, oxygen, and sulfur fractions on the carbonyl carbon, resulting in replacement of the azide. This is the same type of mechanism as that for EMA.

Both soluble and insoluble CM cellulose enzyme derivatives have been prepared by the azide method (Mitz and Summaria, 1961). The enzymic activities of soluble derivatives were greater than the activities of the free enzymes.

Adenosine triphosphate–creatine phosphotransferase (2.7.3.2) was covalently bonded to CM cellulose by the azide method and to p-aminobenzyl cellulose by diazo coupling (Hornby et al., 1968). Hornby and coworkers considered apparent K_m when an ionic substrate is attacked by an enzyme bound to either a polyanionic support (CM cellulose) or an uncharged support (p-aminobenzyl cellulose). The design of the experiment was conceived in terms of the thesis (discussed later in this section) that attachment of an enzyme to a charged support matrix provided a microenvironment that is partially determined by the nature of the charged groups of the support. Hence, there is an unequal distribution of ions, as compared with that which would exist if the enzyme were free in solution. This inequality, which is assignable to electrostatic interaction between the charged field of the support and the ionic molecules in the system, may result in an alteration of the pH activity profile and the apparent K_m of the bound enzyme.

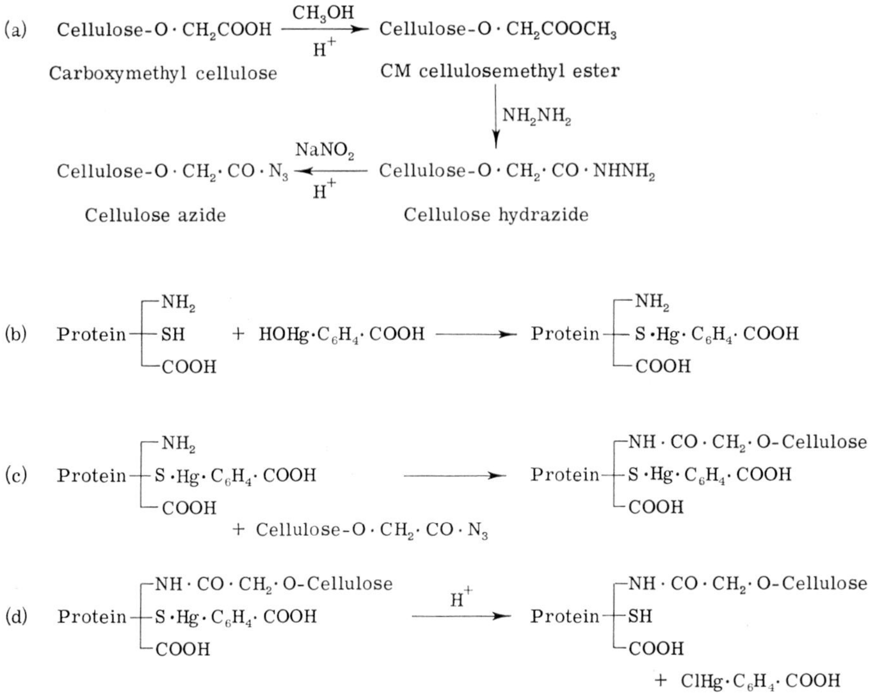

(a) Cellulose-O·CH$_2$COOH $\xrightarrow[\text{H}^+]{\text{CH}_3\text{OH}}$ Cellulose-O·CH$_2$COOCH$_3$

    Carboxymethyl cellulose   CM cellulosemethyl ester

$\downarrow$ NH$_2$NH$_2$

Cellulose-O·CH$_2$·CO·N$_3$ $\xleftarrow[\text{H}^+]{\text{NaNO}_2}$ Cellulose-O·CH$_2$·CO·NHNH$_2$

   Cellulose azide   Cellulose hydrazide

(b) Protein $\left[\begin{array}{l}\text{—NH}_2 \\ \text{—SH} \\ \text{—COOH}\end{array}\right.$ + HOHg·C$_6$H$_4$·COOH $\longrightarrow$ Protein $\left[\begin{array}{l}\text{—NH}_2 \\ \text{—S·Hg·C}_6\text{H}_4\text{·COOH} \\ \text{—COOH}\end{array}\right.$

(c) Protein $\left[\begin{array}{l}\text{—NH}_2 \\ \text{—S·Hg·C}_6\text{H}_4\text{·COOH} \\ \text{—COOH}\end{array}\right.$ $\longrightarrow$ Protein $\left[\begin{array}{l}\text{—NH·CO·CH}_2\text{·O-Cellulose} \\ \text{—S·Hg·C}_6\text{H}_4\text{·COOH} \\ \text{—COOH}\end{array}\right.$

            + Cellulose-O·CH$_2$·CO·N$_3$

(d) Protein $\left[\begin{array}{l}\text{—NH·CO·CH}_2\text{·O-Cellulose} \\ \text{—S·Hg·C}_6\text{H}_4\text{·COOH} \\ \text{—COOH}\end{array}\right.$ $\xrightarrow{\text{H}^+}$ Protein $\left[\begin{array}{l}\text{—NH·CO·CH}_2\text{·O-Cellulose} \\ \text{—SH} \\ \text{—COOH}\end{array}\right.$

                               + ClHg·C$_6$H$_4$·COOH

FIG. 13. (a) Cellulose azide, to serve as the reactive matrix, is prepared as shown. (b) In some instances sulfhydryl groups of the enzyme may be protected by reaction with hydroxymercuribenzoate. This presumably serves to prevent binding to residual carboxymethyl groups of the matrix. (c) The enzyme mercuribenzoate complex is reacted with the cellulose azide. (d) The mercuribenzoate is hydrolyzed in the presence of HCl. The sulfhydryl protection steps are not required for many enzymes. [From Brown et al., Biochem. Biophys. Res. Commun. **25**, 304–308 (1966b), by permission.]

The K_m of ATP creatine phosphotransferase increased tenfold with respect to ATP and to Mg^{2+} when attached to CM cellulose but only 23% when attached to p-aminobenzyl cellulose. For enzyme derivatives of CM cellulose, the extent of the reaction with the thiol-binding agent, 5,5′-dithiobis-2-nitrobenzoic (DTNB) acid, was dependent upon ionic strength. With similar derivatives of p-aminobenzyl cellulose, the extent of this reaction was independent of ionic strength. Data for this enzyme and for ficin as well support the electrostatic interaction (microenvironment) thesis (Table II). These generalized concepts of electrostatic interaction in such systems provide a basis for prediction and, hence, are worthy of consideration of other matrix-supported enzymes.

The thesis predicts that unlike charges on the substrate and support tend to enhance the substrate concentration in the microenvironment of the enzyme relative to the bulk of the suspending medium. Conversely, like charges on both substrate and support result in a decrease in substrate concentration in the enzyme's environment relative to that of the bulk

TABLE II

ENZYME SUBSTRATE AFFINITY AS IT RELATES TO CHARGE-CHARGE INTERACTION[a,b]

E. C. number	Enzyme	Support	Charge	Substrate	Charge	K_m	Ref.
2.7.3.2	ATP creatine phosphotransferase	None		ATP	−	6.5×10^{-4}	
		p-Aminobenzyl cellulose	0	ATP	−	8.0×10^{-4}	Hornby et al. (1968)
		CM cellulose-90	−	ATP	−	7.0×10^{-3}	
3.4.4.4	Trypsin	None		BAA	+	6.8×10^{-3}	
		Maleic acid–ethylene copolymer	−	BAA	+	2.0×10^{-4}	Goldstein et al. (1964)
3.4.4.5	Chymotrypsin	None		ATEE	0	2.7×10^{-4}	C. Money and E.
		CM cellulose-70	−	ATEE	0	5.6×10^{-4}	M. Crook (in Hornby et al., 1968)
3.4.4.10	Papain	None		BAEE	+	1.9×10^{-2}	Silman et al.
		p-Aminophenylalanine-L-leucine copolymer	0	BAEE	+	'No change from free enzyme'	(1966)
3.4.4.12	Ficin	None		BAEE	+	2.0×10^{-2}	
		CM cellulose-70	−	BAEE	+	2.0×10^{-3}	Hornby et al. (1966)

[a] Data derived from Hornby et al. (1968).

[b] Abbreviations: ATP—adenosine triphosphate; CM—carboxymethyl; BAA—benzoyl-L-arginine amide; E.C. number—Enzyme Commission number; BAEE—benzoyl-L-arginine ethyl ester; ATEE—acetyl-L-ethyl ester.

solution. If the substrate or support or both were uncharged, then there would be no difference between the substrate concentration in the two environments. One would expect to find an apparent decrease in K_m when unlike charges were on the substrate and support, an apparent increase in K_m when substrate and support carried like charges, and little or no effect on the apparent K_m when one or both of the components were uncharged.

Results of thiol titration experiments using DTNB can also be interpreted upon the basis of electrostatic interaction among enzyme, support, and substrate. At pH 8, DTNB, thiol groups, and the CM cellulose support are all negatively charged. As expected, increase in ionic strength promoted interaction, presumably by screening interacting similar charges. The DTNB titrations indicate that the effect of microenvironment extends to molecules other than substrate and, hence, may very well be pertinent to questions of enzyme activity control in the biological cell.

Examples are presented in the data of Hornby *et al.* (1968) in which charge-charge interactions between substrate and support are absent but which, nevertheless, show increases in the apparent K_m. The data are considered significant by the authors. They state that this could result from conformation changes of the protein, although the alternative explanation of a diffusion-limiting layer was put forward.

In terms, then, of a modified environment hypothesis involving both enzyme–matrix–substrate, charge-charge interactions and the existence of a diffusion-limiting layer, these authors have derived an equation* to describe the equilibrium distribution of charges between substrate and enzyme. Reasoning employed to develop the mathematical statement was based upon electrical and diffusion effects, from the Nernst-Planck equation, in a manner similar to that used to describe movement of ions in an ion-exchange system (Helfferich, 1962). The equation is given in a Michaelis–Menten form with K_m replaced by K'_m which is a function of both charge-charge interaction and diffusion. The form of the function is such that like charges on substrate and support increase K'_m and unlike charges decrease K'_m compared with K_m. The effect of their diffusion term is always to increase K'_m. The magnitude of the increase is dependent upon the reaction velocity of the system, the thickness of the diffusion layer, and the magnitude of the diffusion coefficient of the substrate.

In contrast to the thesis that activity of bound enzyme is almost completely determined by charge-charge interactions modified by a diffusion-limiting layer, Epstein and Anfinsen (1962) considered conformational effects upon catalytic activity. In their experiments, RNase (2.7.7.–) and trypsin (3.4.4.4) were subjected to 8 M urea. Columns of CM cellulose-supported enzymes were reactivated after urea denaturation by treatment with 0.2% β-mercaptoethanol. Since the extent of reactivation was much greater than would be expected, on theoretical grounds, for a random re-

---

* Hornby *et al.* (1968) gives the following equilibrium distribution:

$$K'_m = \left(K_m + \frac{x \cdot V}{D}\right)\left(\frac{R \cdot T}{R \cdot T - z \cdot x \cdot F \cdot \text{grad } \psi}\right)$$

x = thickness of diffusion layer; grad ψ = gradient of electrical potential; D = diffusion constant of the substrate; K_m = Michaelis constant; K'_m = apparent Michaelis constant; R = gas constant; T = absolute temperature; V = maximum reaction velocity of the enzyme; F = Faraday constant; Z = electrochemical valence of the substrate.

oxidation process, they concluded that the three-dimensional configuration of at least the catalytically active portion of the protein is determined
by the amino acid sequence alone. (See Fig. 14.)

Ribonuclease as well as chymotrypsin and ficin were supported on cellulose matrices by the azide method in a study by Lilly *et al.* (1965). They
interpreted their results to indicate that the bonding was through the enzyme's free amino group. Chymotrypsin and RNase were also bonded to

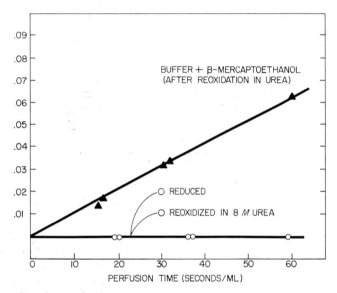

PERFUSION TIME (SECONDS/ML)

Fig. 14. Complete reduction of carboxymethyl cellulose–trypsin columns with 8
M urea and β-mercaptoethanol, followed by reoxidation at pH 8.3, resulted in a recovery of 4% of the original activity. Reoxidation in 8 *M* urea produced no return
in activity, but the enzyme could then be activated by treatment with 0.2% β-
mercaptoethanol. [From Epstein and Anfinsen, *J. Biol. Chem.* **237**, 2175–2179 (1962),
by permission.]

the benzoyl derivatives of cellulose by diazo coupling which involves the
tyrosyl and histidyl residues of the enzyme. In their comparison, the azo
coupling method gave less active preparations than did the azide method.
These authors pointed to the importance of matrix swelling as a condition
to be considered in experimentation involving supported enzymes. Chymotrypsin supported by a matrix which had been preswollen was five
times as active as a control preparation when small molecule substrates
were used.

Acetylcholinesterase (3.1.1.8), DNase (3.1.4.5), and other enzymes
were studied by A. B. Patel *et al.* (1969b). A number of matrices, includ-

ing those employing the azide method, and various physical parameters were compared by these authors.

Manecke and Foerster (1966) studied diastase (3.2.1.1 and 3.2.1.2) bound to a styrene–divinylbenzene copolymer, 4- and 3-fluorostyrenes, a 4-fluorostyrene–methacrylic acid–divinylbenzene copolymer, and a 3-fluorostyrene–methacrylic acid–divinylbenzene copolymer. In each instance the enzyme–resin complexes were formed, and the authors concluded that these matrices were suitable as reactive hydrophobic polymers for biologically active substances.

Pronase (3.4.4.–) (*Streptomyces griseus*-derived proteolytic enzyme complex) was bound to CM cellulose azide and to other polymers (A. B. Patel *et al.*, 1971). The activity of the enzyme was studied as a function of the support matrix; thermal stability was enhanced.

Manecke and Singer (1960b) used a copolymerized methacrylic acid–*m*-fluoroanilide copolymer for the support of invertase (3.2.1.26) to produce an enzymically reactive, supported preparation which had altered stability properties as compared with free solution enzyme. They considered the effects of cross-linking within the copolymer, time dependence of the invertase binding, and other characteristics of the matrices as they related to the provision of hydrophilic properties in solution of varying pH.

Trypsin and chymotrypsin were coupled to CM cellulose by the azide method in the now classic work of Mitz and Summaria (1961). *p*-Diazobenzyl cellulose matrix was used also for chymotrypsin and RNase. Mitz and Summaria found that all couplings were unaffected by water, salt solution, and acid buffer, and they were dried by lyophilization without loss in activity. Derivatives of the *p*-diazobenzyl cellulose were insoluble; the CM cellulose derivatives varied from insoluble to water soluble, depending upon the chain length of the matrix. Soluble derivatives had activity comparable to that of the free enzyme; the insoluble preparations were less active. Activity of the derivatives was related to the degree of hydration of the product rather than to the amount of enzyme coupled. Both insoluble and soluble CM cellulose derivatives of chymotrypsin were more heat-stable than the free solution enzyme—the insoluble derivatives were the most stable. There was no shift in the pH optimum of the bound chymotrypsin as compared with the solution enzyme. The insoluble derivatives of trypsin and chymotrypsin were more active on small substrates than on larger ones, a fact since observed for a number of proteolytic enzyme preparations and apparently to be interpreted in terms of the spatial limitations imposed.

The study of Hornby *et al.* (1966) included ficin (3.4.4.12). The enzyme was reduced with cysteine, separated from the cysteine by gel filtration on a G-25 dextran Sephadex column, and then blocked with *p*-chloromercuribenzoate. No residual activity toward BAEE was detected, although after treatment with cysteine the original activity was restored. This material was used in a study of the effect of reaction conditions upon the CM cellulose–ficin derivatives. Protein content and enzymic activity were found to be dependent upon the degree of CM substitution in the original cellulose. The pH optimum of CM cellulose–ficin was identical with that of the native enzyme. The enzyme–matrix suspension was stored in water, pH 5 at 2°C, and remained stable for periods up to 4 months with only a minor (15%) decrease in activity. The insoluble enzyme was found to be less dependent upon cysteine for activation than was the free enzyme.

It is worth special note that the activity levels were different for the esterase as compared with the proteolytic activity per se of ficin. Supported enzyme retained 9 to 12% of its original esterase activity but only 4 to 5% of its proteolytic activity. Results using the mercury-inhibitor, p-chloromercuribenzoate, indicated that loss in activity was not due to the formation of a thiol ester between the reactive thiol group of ficin and the azide group of the cellulose. Steric hindrance of the reactive center by the cellulose matrix was indicated by greater activity of the derivatives toward smaller molecules rather than toward larger substrates. Restriction in conformation must also be considered a contributing factor.

K_m values of derivatives were 10 to 12% of that of the free enzyme, presumably indicating increased substrate concentration around the polymer because of electrostatic interaction between positively charged substrate and carboxyl groups of the polymer.

Ficin derivatives were examined with respect to reaction kinetics and flow characteristics in a packed column (Lilly et al., 1966). The apparent Michaelis constant (K'_m) of these preparations was dependent upon the flow rate at low rates of perfusion.

Wharton et al. (1968a,b) attached purified bromelain (3.4.4.24) to a CM cellulose matrix using the azide method. Wharton has considered bound protein to be an imprecise parameter for activity reference and has used DTNB to titrate reactive thiol groups. When the concentration of thiol groups was used as criterion of the concentration of active sites, CM cellulose–bromelain was 77% as active as bromelain in free solution. Wharton took this to mean either that some CM cellulose–bromelain molecules which have a reactive thiol group are completely inactive and the rates characteristic for catalysis by the supported and by the free enzyme solutions are the same, or that, for at least some of the CM cellulose–bromelain, the magnitude of the rate constant, which determines the value of K_{cat}, changes when bromelain is attached to the matrix. The esterase (BAEE) K_m values of bromelain, as a function of ionic strength of the medium, were evaluated. The equation was modeled for a system in which diffusion was neglected. The study supported the concept that perturbation of the K_m of the free-solution enzyme catalysis by attachment of the enzyme to a CM cellulose matrix is explicable largely in terms of charge-charge interactions.

Hasselberger et al. (1970a) have used four methods, including the azide technique, to prepare a matrix-supported asparaginase. The intent of the study was to produce a derivatized form of the enzyme useful as an antineoplastic agent. CM cellulose was the matrices. Each of the methods produced preparations with considerably enhanced thermal stability, and the authors suggested using the insolubilized enzyme in an extracorporeal filtering material.

Brown et al. (1967) contrasted properties of ATPase derived from barley root as membrane fragments, detergent-solubilized, and matrix-supported enzyme. Apparent rarity in higher plants of the Na^+ vs. K^+ active transport common to animals has not encouraged a widespread search for a transport ATPase in plants. Motivation of this study was to consider the individual properties associated with the so-called transport enzyme. The relationship of membrane enzyme, a detergent-solubilized preparation, and matrix-supported solubilized material to the cardiac glycoside ouabain (a specific inhibitor of the transport enzyme) and to Na^+, K^+, and Mg^{2+} has been considered. It was observed that the greatest sensitivity to ouabain was exhibited by membrane fragments and the least sensitivity by the matrix-supported enzyme. Potassium stimulated the membrane preparation only slightly; the effect upon the matrix-supported enzyme was considerable. Although no single preparation from barley endoplasmic reticulum met precisely the criteria of the transport enzyme, most of the individual characteristics (Skou, 1957) identified with this activity could be demonstrated in one or more of the several preparations. The authors have discussed the implications of these data to the transport ATPase literature.

Carboxymethyl cellulose–azide was used as a matrix for apyrase (commercial preparation from potato; Brown et al., 1966c) and rabbit heart sarcoplasmic reticulum ATPase (3.6.1.3) (Brown et al., 1966a,b, 1968c). The authors observed that heart sarcoplasmic reticulum ATPase linked to CM cellulose was inactive unless protected during the reaction sequence by binding p-hydroxymercuribenzoate to the sulfhydryl group. Apyrase exhibited a bimodal pH optimum. Binding to CM cellulose–azide resulted in a shift to the more alkaline. Whittam et al. (1968) described a single pH optimum peak using an apyrase obtained from another commercial source. Brown and associates described changes in the activating metal requirements of the bound apyrases compared with the free solution material. Interpretation of these data was based upon conformational change and electrostatic interaction.

D. Diazonium Coupling

Historically, diazonium coupling has been the method most frequently used for derivatizing enzymes with application to wide variety of matrices. Among the matrices were synthetic polypeptides (Cebra et al., 1961), p-aminostyrene (Grubhofer and Schleith, 1953), p-aminobenzyl cellulose (Mitz and Summaria, 1961), collodion (Goldman et al., 1965), cellophane (Selegny et al., 1968), and glass (Weetall, 1969a,b). Diazonium coupling has been used to bind enzymes to other proteins and to cross-link molecules of the same species (Mowbray and Scholand, 1966).

Pauly (1915) reported that both tryosine and histidine can couple with diazonium salts in the ratio 1:2. Boyd and co-workers (Boyd and Hooker, 1934; Boyd and Mover, 1935) indicated that protein moieties, other than tryosine and histidine, can react with diazonium salts. Gelewitz et al. (1954) studied the reaction of diazonium salts with serum albumin. They found that far fewer groups were introduced than had been anticipated and that only about a third of these were attached to the protein by an —N=N— bridge (the mode of bonding of the diazonium coupling reaction). None of the histidine or tyrosine residues contained two attached azobenzene groups. Higgins and Harrington (1959) reacted simple aromatic diazonium compounds with various proteins and found that it was possible to form bistyrosyl and bishistidyl derivatives. They confirmed that a spectral adsorption, previously reported by Gelewitz et al. (1954), was due to the reaction of the diazonium salt with the ϵ-amino groups of lysine on the proteins. It was stressed that reaction conditions are important and that these must be chosen on the basis of "trial and experience." Figure 15 gives reactions representative of the diazonium coupling.

Howard and Wild (1957) studied the reaction of diazonium compounds with amino acids and compounds containing groupings present in proteins. To summarize their findings, the phenol group of tyrosine, the imidazole group of histidine, and the ϵ-amino group of lysine react with 2 moles of diazonium compound; the sulfhydryl group of cysteine and the indolyl group of tryptophan react with 1 mole; the guanidino group of arginine reacts with at least 1, and probably 3 moles; the amino group of glycine reacts with 2 moles of diazonium salt when the glycine is a terminal residue; amino groups of other amino acids present are deaminated.

Tabachnick and Sobotka (1960) studied the coupling of diazotized arsanilic acid with various proteins. With bovine serum albumin it was found that azotyrosine and azohistidine account for about 50% of the protein-bound arsenic. When the ε-amino groups of the albumin were blocked by acetylation, practically all of the bound arsenic was found as

1.[a,c,d]

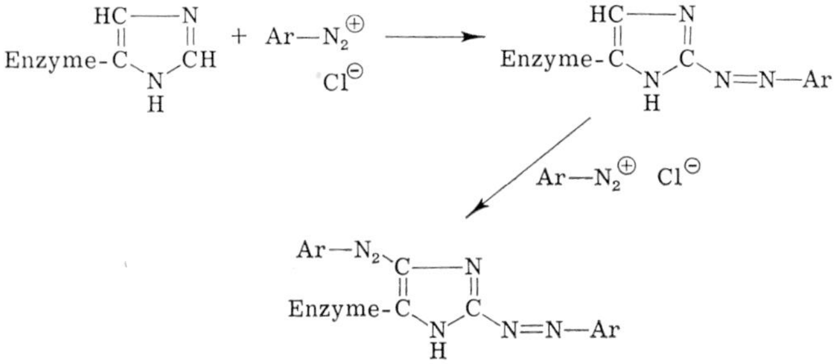

2.[a,c,d]

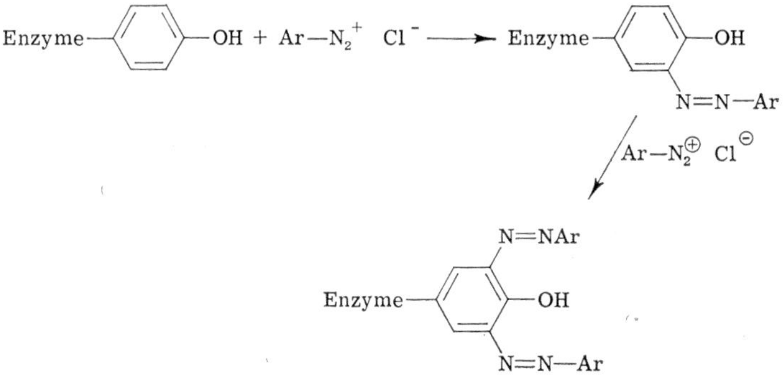

3.[b,c]

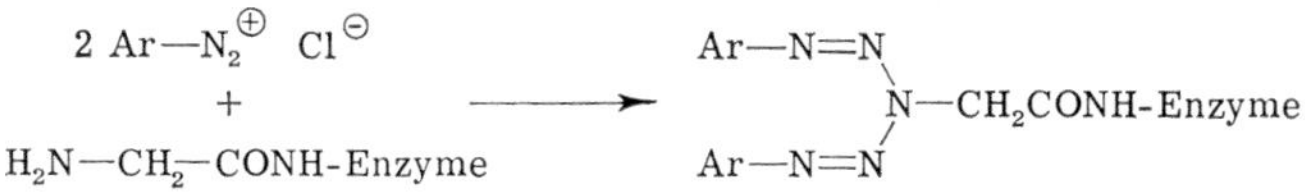

Fig. 15. Reactions representative of the diazonium coupling method. (a) Higgins and Fraser (1952). (b) Busch et al. (1934). Howard and Wild (1957) suggested that the ε-amino group of lysine reacts similarly. (c) The Ar indicates an aryl group, which may be free or linked to a polymeric matrix. (d) Formation of mono- and bishistidyl and mono- and bistyrosyl derivatives is illustrated.

azotyrosine and azohistidine. This demonstrated that appreciable quantities of diazonium compounds can be bound by free ε-amino groups of the lysine moieties of proteins.

The relatively promiscuous nature of diazonium salts should be considered when they are used to bond biologically active proteins. It may be necessary to protect essential groups during the coupling reaction. The coupling of proteins by diazonium coupling is generally favored electro-statically. The diazonium salt has a full ionic positive charge, whereas most proteins are expected to be negatively charged under the alkaline conditions most frequently used for diazonium coupling.

Selegny *et al.* (1968) prepared glucose oxidase (1.1.3.4)-surfaced mem-branes on paper or cellulose by using the condensing agent bisdiazotized *o*-dianisidine. This technique results in cross-linking of free amino func-tions of the polypeptide chains of the enzyme. These membranes, after equilibration in buffer, were employed in a diffusion experiment. Flow of substrate through the membrane was measured and proved to be in accord with theoretical prediction for such diffusion phenomena.

Hornby *et al.* (1968) used diazo coupling to bind ATP–creatine–phos-photransferase (2.7.3.2) to *p*-aminobenzyl cellulose. Ribonuclease (2.7.7.–) was bound to diazotized polyaminostyrene by Grubhofer and Schleith (1953) and to *p*-diazobenzyl cellulose by Mitz and Summaria (1961). Silman *et al.* (1963) used a diazotized polymer of *p*-aminophenyl-alanine–leucine as a matrix for RNase derived from bovine pancreas. In order to obtain preparations that were more active, the enzyme was coupled to the copolymer in the presence of cytidilic acid, an inhibitor and substrate analog of RNase. The preparation obtained in this way possessed 8% of the activity of the free solution enzyme toward high molecular weight RNA and 35% of the original activity toward cytidine 2',3'-cyclic phos-phate. Lilly *et al.* (1965) used diazo coupling to bind RNase and chymo-trypsin to benzoyl amino derivatives of cellulose.

Weetall (1969a) used silanized porous glass to couple alkaline phospha-tase (3.1.3.1) using diazotization. Diastase (3.2.1.–) was bound by Grubhofer and Schleith (1953) to a diazotized polyaminostyrene. They ex-tended the methodology also to carboxypeptidase (3.4.2.1). Hasselberger *et al.* (1970b) included diazotization as a method in a study of bound lysozyme (3.2.2.17).

Cresswell and Sanderson (1969) used a diazotized copolymer of *p*-ami-nophenylalanine–leucine copolymer to bind pronase (3.4.4.–). Their product was identical in pH and reaction temperature optima with the soluble enzyme but much more stable. Activity was retained for several months at 4°C storage. The soluble preparation lost about 25% of its ac-tivity per day when stored under identical conditions. These authors ob-served that pronase, as had frequently been noted earlier for other proteo-lytic enzymes coupled to matrices, was more efficient in the catalysis of low molecular weight substrates than of high molecular weight molecules. They proposed an expression relating the molecular weight of the substrate to the soluble enzyme: the reaction rate with insoluble enzyme/reaction rate with soluble enzyme equals K + the quantity $k \times$ log of the molecular weight of the substrate.

Bar-Eli and Katchalski (1960) prepared polytyrosyl derivatives of trypsin and p-amino-DL-phenylalanine–leucine copolymer derivatives of trypsin using diazotized matrix polymers.

Glazer et al. (1962) used the diazotized copolymer of p-aminophenyl-alanine–leucine to bind trypsin (3.4.4.4). The insoluble polytyrosyl trypsin (IPTT) was found to contain 20 to 28 additional residues of tyrosine per mole of trypsin. Roughly half of the ϵ-amino groups of trypsin were acylated by the n-carboxyanhydride. The supported preparation was more resistant to autolysis than the soluble enzyme, although it was apparently identical in the effect upon it of urea and of the specific trypsin inhibitor derived from soybean. These authors considered the substrate specificity of polytyrosyl trypsin as compared with the soluble trypsin (Table III) ; in their experiments, equal concentrations of trypsin and polytyrosyl trypsin were compared. Substrate copolymers with a relatively high percentage of lysine were cleaved by the supported trypsin more rapidly than by the free solution enzyme (Table IV). The copolymers with a high percentage of nonpolar residues, however, were digested by soluble trypsin more rapidly than the supported enzyme.

Alexander et al. (1965) bound thrombin to the diazotized polytyrosine. Rimon et al. (1966a) studied the effect of insoluble derivatives of trypsin upon prothrombin activity. Other proteolytic enzymes which have been studied using the diazo method include chymotrypsin (Mitz and Summaria, 1961; Lilly et al., 1965), papain (Grubhofer and Schleith, 1953; Cebra et al., 1961, 1962; Goldman et al., 1965, 1968; Jaquet and Cebra, 1965; Silman et al., 1966), and ficin (Lilly et al., 1965; Hornby et al., 1968).

Rimon et al. (1963, 1966b) bound streptokinase (3.4.4.14) to a diazotized copolymer p-amino-DL-phenylalanine and L-leucine. The insoluble streptokinase was used in studies (Gutman and Rimon, 1964) of the so-called activator system of the blood-clotting phenomenon.

Diazo coupling methods were also used for studies of urease (3.5.1.5) (Riesel and Katchalski, 1964; Weetall and Hersh, 1969) and of apyrase (3.6.1.5) (Brown et al., 1968a).

E. OTHER METHODS

1. Cyanuric Chloride and Dichloro-s-triazine Methods

Dichloro-s-triazines containing solubilizing groups, such as carboxymethoxy or carboxymethylamino—the so-called Procion Brilliant dyes—

242   HARRY D. BROWN AND FRANK X. HASSELBERGER

readily react with cellulose in aqueous solution to give good yield of mono-chloro-s-triazinyl–cellulose complex (Kay and Crook, 1967). Stronger nucleophiles such as the free amino functions of proteins, will displace the remaining chlorine atom forming a carbon-nitrogen bond between the tri-azine ring and the protein molecule. Each triazine moiety attached to the insoluble matrix is potentially capable of binding a protein molecule. Con-ditions for reaction of various chloro-s-triazines with cellulose and of the product with chymotrypsin are described by Kay and Crook (1967). (See Figs. 16 and 17.)

The monochloro-s-triazinyl–cellulose complex described above is un-

TABLE III

COMPARISON OF INITIAL RATES OF HYDROLYSIS OF VARIOUS SUBSTRATES BY WATER-INSOLUBLE TRYPSIN AND TRYPSIN[a]

Substrate	Substrate concentration (mg/ml)	pH	Concentration of water-insoluble trypsin[b] mg/ml	Activity ratio of water-insoluble trypsin to trypsin
Benzoyl-L-arginine ethyl ester	7.8	7.8	0.06	0.14
L-Arginine methyl ester	4.3	6.2	0.10	0.16
L-Lysine methyl ester	3.2	6.2	0.10	0.14
Poly-L-lysine	6.0	8.2	0.25	0.01
Copolymer L-lysine-L-tyrosine (5:1)	6.0	8.2	0.25	0.014
Copolymer L-lysine-L-aspartate (3:1)	5.4	8.2	0.25	0.012
Casein	10.0	7.6	0.12–0.25	0.017

[a] From Glazer et al. J. Biol. Chem. 237, 1832–1838 (1962), by permission.
[b] Expressed in terms of bound protein.

charged when suspended in neutral or basic solutions. A modification of this technique has been developed by Kay et al. (1968) and Sharp et al. (1969) which can result in the formation of a positively charged matrix. This affords an electrostatic advantage since many enzymes are negatively charged under the alkaline conditions usually used to bond them to a ma-trix.

In this modification, cellulose is treated with normal sodium hydroxide, the excess removed by filtration, and the product added to an acetone solu-tion of cyanuric chloride. Water is added at once, and the acid generated in the resulting "hydrolysis" of cyanuric chloride reduces the pH of the reaction mixture rapidly enough to prevent the hydrolysis of a second chlorine. The reaction is allowed to proceed for 10 to 15 seconds. The mix-ture is quenched in 20% aqueous acetic acid and washed with cold acetone–

water mixtures. The resulting dichloro-s-triazine–cellulose complex can be reacted with protein and is capable of bonding two protein molecules for each dichloro-s-triazine moiety present. Alternatively, the complex may be reacted briefly with an amine, such as N-(3-aminopropyl)diethanolamine. This results in the formation of a positively charged matrix because of quaternization of the tertiary amine.

The actual intermediates formed and their subsequent reactions have not been determined, but a number of possibilties should be considered.

TABLE IV

COMPARISON OF INITIAL RATES OF HYDROLYSIS OF SOME AMINO ACID POLYMERS BY POLYTYROSYL TRYPSIN AND TRYPSIN[a]

Substrate[b]	Substrate concentration (mg/ml)	Activity ratio of polytyrosyl trypsin I to trypsin
Poly-L-lysine	7.2	1.4
	6.0	2.5
	5.5	2.4
Copolymer L-lysine-L-tyrosine (5:1)	6.0	1.4
Copolymer L-lysine–L-glutamate (5:1)	5.4	1.3
	2.7	1.5
Copolymer L-lysine–L-aspartate (3:1)	5.4	1.0
Copolymer L-lysine–L-glutamate (1:5)	7.5	1.0
Copolymer L-lysine–L-alanine (2:1)	7.0	1.0
	4.6	0.95
Copolymer L-lysine–L-alanine–L-tyrosine (2:1:1)	5.8	0.5
	2.9	0.4
Copolymer L-lysine–L-leucine–L-tyrosine (2:1:1)	5.8	0.35

[a] From Glazer et al., J. Biol. Chem. 237, 1832–1838 (1962), by permission.
[b] The number average of degree of polymerization ranged from 30 to 65.

The dichloro compound may be mono- or bisquaternized. The positively charged matrix exerts a "come-hither effect" on negatively charged proteins, and once attracted to the vicinity of the matrix, the nucleophilic moieties of the protein can displace "leaving groups" from the triazine ring. In the case of monoquaternization, a chloride ion or the quaternizing amine may be displaced. (If the amine is displaced, it can attack the remaining chlorine, restoring the positive charge.) It is possible that both the chlorine and the amine would be displaced by protein nucleophiles. If bisquaternization occurs during the amine–triazine reaction, 1 or 2 amine groups can be displaced. The final product could have 2 protein molecules, 1 protein molecule and 1 quaternary ammonium group, or 1 protein mole-

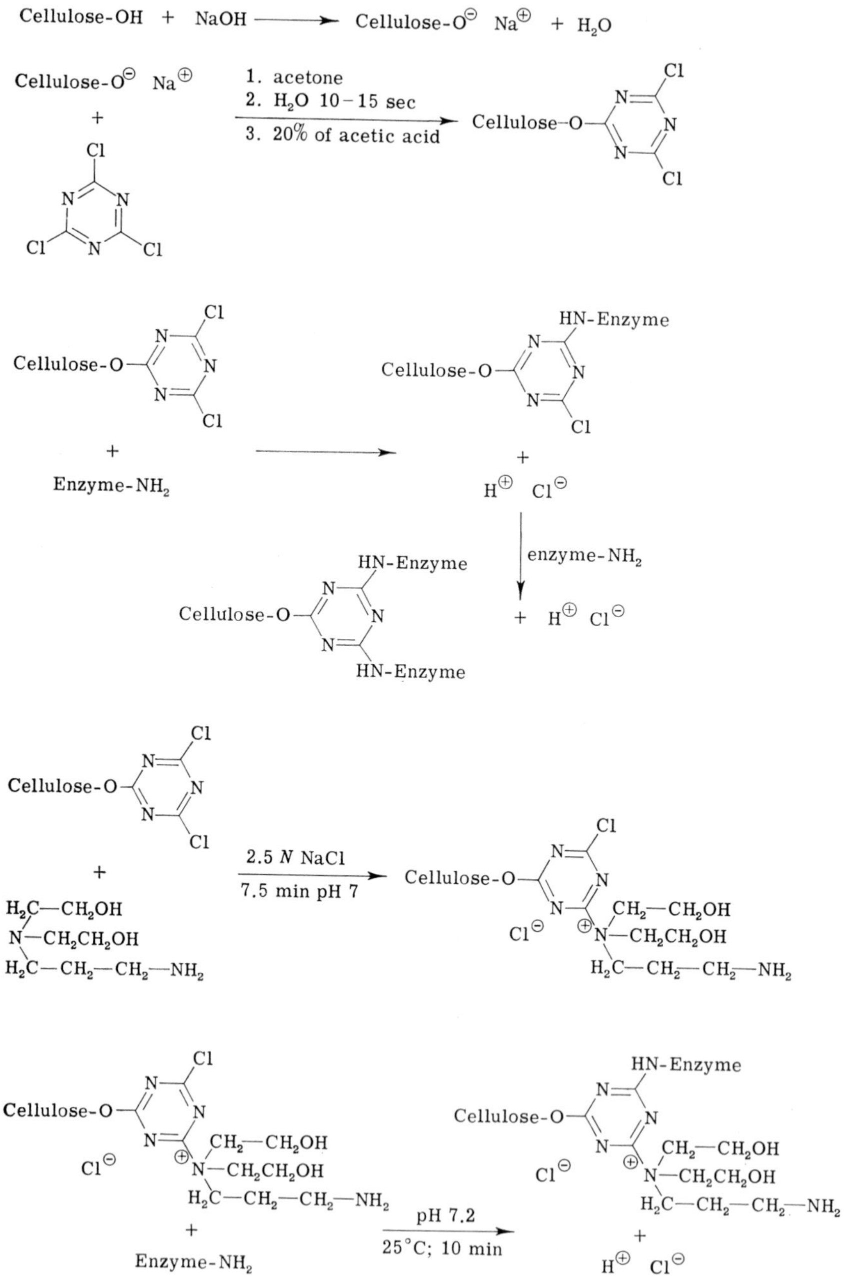

Fig. 16. Reactions representative of the cyanuric chloride method. Reactions of NH₂ groups are used for illustration, but other nucleophilic moieties (essentially N, O, and S functions) may also react. (From Kay *et al.*, 1968; Sharp *et al.*, 1969.)

cule and 1 chlorine atom attached to the triazine ring. The last possibility seems to be least likely because the chloride is a better "leaving group" than the amine. The most likely structure of the final product would seem to be that in which 1 protein molecule and 1 quaternary ammonium group are attached to the triazine ring.

If this is the correct structure, the positively charged support should attract negatively charged substrates and repel positively charged ones. This should be reflected in a lower value of K'_m for substrates that are negatively charged under assay conditions and an increased K'_m value for

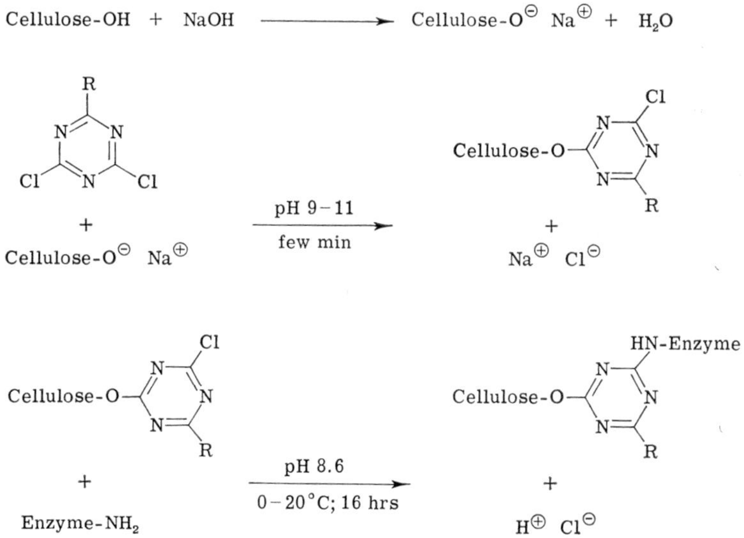

Cellulose-OH + NaOH $\longrightarrow$ Cellulose-O$^{\ominus}$ Na$^{\oplus}$ + H$_2$O

R
N N
Cl N Cl

+

Cellulose-O$^{\ominus}$ Na$^{\oplus}$

$\xrightarrow[\text{few min}]{\text{pH 9–11}}$

Cl
N
Cellulose-O N
N
R

+

Na$^{\oplus}$ Cl$^{\ominus}$

Cl
N
Cellulose-O N
N
R

+

Enzyme-NH$_2$

$\xrightarrow[\text{0–20°C; 16 hrs}]{\text{pH 8.6}}$

HN-Enzyme
N
Cellulose-O N
N
R

+

H$^{\oplus}$ Cl$^{\ominus}$

FIG. 17. Reactions representative of the dichloro-s-triazine method. (From Kay and Crook, 1967.)

positively charged substrates. Neutral substrates should not be affected (Hornby et al., 1966, 1968).

2. Cyanogen Bromide Method

Little is known about the chemistry of this useful method. A reactive intermediate is formed by treating a polysaccharide for a short period with an alkaline aqueous solution of cyanogen bromide. After washing the activated matrix to remove excess cyanogen bromide, base, and decomposition products, the reactive intermediate is coupled to the protein in neutral or preferably slightly alkaline solution (Axen et al., 1967).

Polymers containing primary amino groups can also be activated by the cyanogen halide method. The amino groups are believed to be converted into cyanamide intermediates which react with proteins to form guanidino derivatives.

The chemical nature of the intermediate in the polysaccharide reaction is unknown. Axen *et al.* (1967) reported that the intermediate contains nitrogen but no bromide and that ammonia is released in the coupling process. Imino carbonic acid esters, carbamic esters, and cyanate esters have been suggested as possible mediators in the reaction. The coupling process almost certainly involves a nucleophilic attack on the activated matrix.

The reaction conditions are very mild and the protein–matrix bond seems to be a strong one. Other cyanogen halides can be used, but the bromide has been used most commonly.

3. Isothiocyanate Method

The reactions of isothiocyanates which are of interest in solid-support chemistry are bimolecular nucleophilic additions to the carbon atom of the functional group. Many canonical forms can be written for this group, particularly when bonded to an aromatic ring, e.g.,

(Rao and Venkataraghavan, 1962; R. Arnold *et al.*, 1957). In these forms, nitrogen can be neutral, positively, or negatively charged, and sulfur can be neutral or negatively charged. Carbon can be neutral or positively charged; no reasonable structure can be written with a negative charge on carbon. This simple resonance model predicts that least electron density will be found on the carbon atom. This prediction is supported by the nature of the products formed when isothiocyanates react with nucleophilic groups, such as the amino, hydroxyl, and sulfhydryl groups of proteins.

A generalized equation for the reaction of isothiocyanates with these groups can be written:

where X is NH in primary amino groups, OH in hydroxy groups, and SH in sulfhydryl groups.

The reactions of aromatic isothiocyanates with various amino acids, thioglycolic acid, and free hydroxyl ion in weakly alkaline (pH 9.8) solution have been studied by Drobnica and Augustin (1965a,b,c). They found that sulfhydryl groups dissociated before addition and that amino groups were reactive only in the nonprotonated form. Rate constants for these nucleophilic additions were increased by electron-withdrawing groups on the aromatic ring of the isothiocyanate and were decreased by electron-releasing groups, as expected. Rate constants for the reaction of arylisothiocyanates with thioglycolate were reported to be higher by two orders of magnitude than the rate constants for the corresponding reactions with glycine or hydroxyl ion under the same experimental conditions. This result is of interest for two reasons. First, it deviates from the usual order of

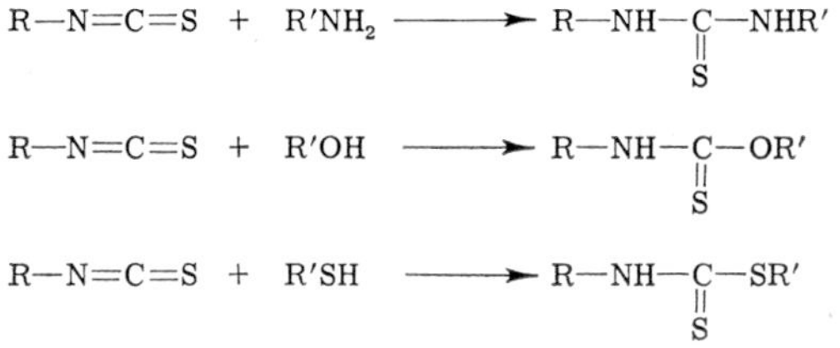

$$R-N{=}C{=}S \ + \ R'NH_2 \longrightarrow R-NH-\underset{\underset{S}{\|}}{C}-NHR'$$

$$R-N{=}C{=}S \ + \ R'OH \longrightarrow R-NH-\underset{\underset{S}{\|}}{C}-OR'$$

$$R-N{=}C{=}S \ + \ R'SH \longrightarrow R-NH-\underset{\underset{S}{\|}}{C}-SR'$$

Fig. 18. Reactions representative of the isothiocyanate method.

nucleophilicity, $RNH_2 > R-OH > RSH$, which has been reported for reaction of isocyanates by R. Arnold et al. (1957). Second, it suggests the necessity of protecting essential sulfhydryl groups when isothiocyanates are used to derivatize enzymes. Reactions representative of isothiocyanates are illustrated in Fig. 18.

Axen and Porath (1966) described the preparation of isothiocyanophenoxyhydroxypropyl dextrans using Sephadex G-25, G-100, and G-200 derivatives. These matrices were used to support β-amylase (3.2.1.2), chymotrypsin, and trypsin. Coupling was through the isothiocyano group of the matrix and amino group of the protein. The main product is bonded essentially as a thiourea. Histidine was also bonded to this support. Axen and Porath compared the catalytic activities of the histidine-containing proteins to those which Katchalski and associates (Bar-Eli and Katchalski, 1960, 1963; Katchalski et al., 1960) had studied bonded to polypeptide matrices. Characteristics of these enzymes in the two studies were quite similar, and the authors concluded that the nature of the polymer with reference to these specific examples did not influence the catalytic action displayed by the fixed groups. It is interesting also that similar catalytic constants for chymotrypsin were measured on Sephadex G-25 and on the much finer G-200, presumably indicating that the chymotrypsin buried in the interior of the matrix (G-200) is not functionally different from the surface-fixed enzyme (G-25).

Storage life of dried chymotrypsin polymer was unaltered after 5 weeks at −10°C, although it was labile at room temperatures when stored over long periods. The lower catalytic efficiency of the supported enzyme toward high molecular weight substrates referred to by most of the authors who have studied bound proteolytic enzymes was demonstrated with these dextran-supported proteins as well. Soluble starch, for example, was not hydrolyzed by α-amylase, whereas smaller molecular weight substrates were hydrolyzed efficiently. Presumably here too the phenomenon is explicable in terms of hindrance by the matrix.

Kay et al. (1968) pointed out that many of the methods available for coupling enzymes to insoluble supports result in a support with a net negative charge and that this, in turn, results in a preparation with relatively low activity toward many substrates, since the substrate frequently has a net negative charge under the assay conditions. If it is an advantage for the net charge on the support to be of sign opposite that carried by the enzyme at the pH of the binding reaction, in many practical stiuations this means that the support should carry a positive charge. Hence, studying LDH (1.1.1.27), Kay et al. used dichloro-s-triazinyl cellulose and N-(3-aminopropyl)diethanolamine enabling them to have either a neutral matrix or a positively charged one as the situation required. The same enzyme was bound (Wilson et al., 1968a,b) to anion exchange particles using Procion Brilliant Orange (PBO). The stability and kinetic properties of these preparations were described and equations developed which served to summarize the relationships involved in the passage of substrate through columns composed of bound enzyme. These authors (Kay et al., 1968) used the same LDH preparation together with sheets similarly prepared with pyruvate kinase (2.7.1.40) in a two-enzyme reactor system. In this series of studies creatine kinase (2.7.3.2), β-galactosidase (3.2.1.23), and chymotrypsin (3.4.4.5) were also used in various models.

Acetylcholinesterase (3.1.1.7) bound by the PBO technique had greatly enhanced stability even at elevated temperatures (A. B. Patel et al., 1969a; Stasiw et al., 1970).

VII. Enzymes Insolubilized by the Use of Bridging Compounds

A. Insoluble Maleic Acid–Ethylene Trypsin and Insoluble Polytyrosyl Trypsin

The concept of a reagent which serves in a reaction sequence to bind an enzyme to a polymeric support and of which some "part" remains as a bridge between the enzyme and the support has been involved in several of the techniques described above. The principle is, however, worthy of further consideration because it may provide a way to construct solid-supported enzymes with useful properties built into the complex. If one recognizes that at present the number of polymers for solid support of proteins is relatively limited and that a mild reaction condition is a further undesirable restriction upon the design of appropriate chemistry, the attachment of an enzyme to a third moiety offers a greater variety of approaches. More important, perhaps, rather specific control of charge distributions, conformation, and even the avoidance of steric hindrance might be more easily achieved by bridging than by more direct techniques.

A number of studies have been concerned with the insoluble maleic acid–ethylene trypsin (IMET) and insoluble polytyrosyl trypsin (IPTT) complexes (Glazer et al., 1962; Bar-Eli and Katchalski, 1963; Goldstein et al., 1964, 1967b; Levin et al., 1964; Alexander et al., 1965; Ong et al., 1966; Rimon et al., 1966a; Lowey et al., 1967). Insoluble maleic acid ethylene trypsin has been formed by coupling enzyme to EMA, cross-linking with hexamethylenediamine, and then hydrolyzing the unreacted anhydride groups. By a similar technique, IPTT was used to couple polytyrosyl trypsin to a water-soluble diazonium salt derived from a copolymer of p-amino-DL-phenylalanine and L-leucine.

The IPTT preparations used by Glazer et al. (1962) (see Section VI, D) had 20 to 28 additional residues of tyrosine per mole compared to the free enzyme (trypsin) bound with an average chain length of 2.5. In contrast to trypsin, the IPTT was sparingly soluble between pH 5 and 9, though it was considerably more resistant to autolysis than was the unmodified enzyme. No difference was observed in the behavior of the derivative and free enzyme toward denaturing agents such as urea and toward the trypsin inhibitor isolated from soybean.

Bar-Eli and Katchalski (1963) found that the copolymer-bound polytyrosyl trypsin retained 30% of its initial esterase activity after 12 months at 4°C. There was, however, a smaller residual proteolytic activity. Levin et al. (1964) used the IMET preparation in their study; activities ranged from 40 to 70% of that of the free enzyme. All of their preparations were considerably more stable than the free trypsin in the range pH 7.0–10.7. They studied the effect of an electrostatic field in the molecular domain of the enzyme and considered its effect upon properties and mode of action of the catalyst.

Some of the IMET preparations were active in 8 M urea and in the presence of soybean trypsin inhibitor which inactivated both the native trypsin and IPTT preparations. However, IMET was inhibited by a trypsin inhibitor derived from pancreas. Stability of the IMET preparations can be related to (1) prevention of autodigestion, (2) blocking of lysyl residues, and (3) change in pH between the domain of the polyelectrolyte gel and the bulk of the solution. The IMET preparations retained most of their activity after lyophilization or drying; IPTT did not. Stability of IMET may result from the hydrophilic nature of the polymer. Goldstein *et al.* (1964) sought to elucidate the factors responsible for properties of IMET. The pH activity profile at low ionic strength was displaced approximately 2.5 pH units toward alkaline. At higher ionic strength, the activity curve shifted toward acid values. The apparent Michaelis constant (esterase activity) was approximately 30 times lower than that for trypsin in a low ionic strength environment, although it approached that of free solution trypsin at high ionic strength when measured at the trypsin optimal pH (pH 7.5). Effects of the polyanionic carrier in the IMET preparation were explained by the contribution of the electrostatic potential of the carrier to the local concentration of hydrogen ions and the presence of positively charged substrate molecules in the microenvironment of the bound enzyme molecules.

Goldstein *et al.* (1967b) studied the action of IMET on myosin. A single first-order reaction class was found—a fact which was related to the formation of myosin subfragments. Trypsin, on the contrary, yielded two reaction classes. It is presumed that this dichotomy is explained by the fact that approximately half as many peptide bonds are available to IMET as to trypsin in solution.

Lowey *et al.* (1967) used IMET and IPTT as catalysts for the hydrolysis of myosin. As for soluble trypsin, their kinetic data were interpreted in terms of two first-order reaction classes. The IMET differed from free trypsin in that the fast reaction did not parallel the formation of meromyosins and the total number of peptide bonds available for enzymic hydrolysis was reduced by about 50%. Ten to twenty bonds were split in the fast reaction by IMET; 150 in the slow reaction. The fast rate is about ten times as rapid as the slower and both are about one-fiftieth the rate exhibited in systems containing native trypsin. Only 22% of the total available bonds of myosin are hydrolyzed by supported trypsin (IMET) in contrast to 45% by solution trypsin. The authors believed that supported enzyme offered the advantage of cleaving heavy meromyosin under relatively mild conditions and with little loss of nonprotein nitrogen.

Alexander *et al.* (1965) used trypsin, IMET, and IPTT to look at fibrinogen and the clotting phenomenon. Both IMET and IPTT promptly and progresively compromised the clottability of fibrinogen by thrombin as did native trypsin. The IMET was relatively more potent than IPTT. Once insoluble derivatives are removed, the elevated clotting time achieved by the IMET- and IPTT-catalyzed events remains constant. As with native trypsin, an increased amount of insoluble derivative caused a corresponding increase in clotting time. Both soybean and pancreatic trypsin inhibitors inhibited the effects of derivatives. Hence the increase in clotting time is a function of proteolysis by trypsin and trypsin derivatives which converts fibrinogen into nonclottable material. The action of IMET and trypsin on pepsinogen has also been compared (Ong et al., 1966). Trypsin hydrolyzed fifteen peptide bonds in pepsinogen, whereas the insoluble derivatives never hydrolyzed more than ten bonds. Ong et al. (1966) concluded that water-insoluble trypsin is more selective in specificity toward high molecular weight substrates than is the native enzyme. Rimon et al. (1966a) reported on the interrelationship of IMET and IPTT with prothrombin and found that marked differences occurred. These authors put forward the suggestion that if "identical catalytic sites in these enzymes are assumed, one has to postulate that the ultimate specificity of enzymes toward their corresponding protein substrate is predicated upon a high affinity between distinct noncatalytic sites of the enzymes and well defined regions of the high molecular weight substrates."

B. OTHER BRIDGES

Peroxidase (1.11.1.7) and acid phosphatase were conjugated to antibodies with bifunctional reagents in a study of enzyme-labeled antibodies for electron-microscopic localization of tissue antigens by Nakane and Pierce (1967). Their technique utilized the product of the enzyme-catalyzed reaction as the marker in immunohistochemical localization of tissue antigens. Peroxidase–antibody conjugates were stable after storage for several months at 4°C or indefinitely when stored at lower temperatures.

Peroxidase, glucose oxidase (1.1.3.4), and several "blameless proteins" were coupled to erythrocytes by the use of glutaraldehyde, cyanuric chloride, and tetraazotide O-dyanisadine. In the study (Avrameas et al., 1969), rabbit antisera to these proteins were prepared, and the suspension of the protein–erythrocyte complexes were then used for the passive hemoagglutination test. Lactate dehydrogenase (1.1.1.27) was derivatized in a study of antibody–enzyme conjugates (Casu and Avrameas, 1969).

Davis et al. (1969) evaluated the immunosuppressive properties of RNase (2.7.7.–)–human serum albumin complexes prepared with bifunctional reagents. A technical procedure was used earlier by Mowbray and Scholand (1966) with human γ-globulin, human ovalbumin, and bovine serum albumin as matrix for the RNase.

Use of N-ethyl-5-phenylisoxazolium-3'-sulfonate (Woodward's reagent K) as a bridging agent has been described for acetylcholinesterase (3.1.1.8) and apyrase by A. B. Patel et al. (1969b).

Green and Crutchfield (1969) coupled rennin (3.4.4.5) to aminoethyl cellulose by using glutaraldehyde as a bridging agent. The derivative was 5.9% protein with a specific activity of 0.014% of the soluble enzyme. Their study indicated that about 20% of the activity of the derivative preparation resided in protein which was adsorbed rather than covalently bound to the aminoethyl cellulose matrix.

Hexamethylenediamine was used to bind apyrase (3.6.1.5) to ethylene maleic acid copolymer by Brown et al. (1968b). Lysozyme (3.2.1.17) as well as chymotrypsin and peroxidase were linked to protein matrices by reaction with ethyl chloroformate (Avrameas and Ternynck, 1967).

VIII. Conclusions

The polymeric matrix adsorbed to, intertwined about, or covalently bound to a protein presents a basis for construction of systems which may have useful properties in modeling certain biological entities. Fortunately for the development of the "art," the same techniques lend themselves to the construction of useful conversion systems. However (and despite the length of our Table I), the variety of the applications and, more remarkably, the varieties of chemistries employed, scarcely represent a beginning. Most limiting has been the, usually covert, occasionally overt, assumption that reactions must be carried forward under extremely mild conditions and in aqueous solution—an assumption that is not necessarily valid. One expects that the promise of great reward in industrial application will stimulate development of chemical techniques offering a great variety in protein-binding sites and in effect upon the active proteins conformation and local environment.

From the standpoint of the biological model, two immediate needs present themselves. The first is the availability of a sufficient armamentarium of chemical methods for selecting those conditions that will serve the model system being constructed. The second, perhaps more difficult, is methodology for description of the polymer–enzyme complex in terms of its chemistry and conformation and the physical interaction of the elements of the model once it is formed. Ultimately it is hoped that our knowledge of the chemistry will be such as to allow us to predict the properties of the reaction product—our biological model.

REFERENCES

Alexander, B., Rimon, A., and Katchalski, E. (1965). Fed. Proc., Fed. Amer. Soc. Exp. Biol. 24, 804.
Arnold, R. Nelson, J. A., and Verbanc, J. J. (1957). Chem. Rev. 57, 47.

Arnold, W. N. (1966). *Arch. Biochem. Biophys.* **113**, 451.
Arsenis, C., and McCormick, D. B. (1964). *J. Biol. Chem.* **239**, 3093.
Arsenis, C., and McCormick, D. B. (1966). *J. Biol. Chem.* **241**, 330.
Ascherson. (1840). *Arch. Anat. Physiol.* **7**, 44.
Augenstine, E. S., Augenstine, L. G., and Lippincott, E. R. (1960). *J. Phys. Chem.* **64**, 1211.
Augenstine, L. G., Ghiron, C. A., and Nims, L. F. (1958). *J. Phys. Chem.* **62**, 1231.
Avrameas, S., and Ternynck, T. (1967). *J. Biol. Chem.* **242**, 1651.
Avrameas, S., and Ternynck, T. (1969). *Immunochemistry* **6**, 53.
Avrameas, S., Taudou, B., and Chuilon, S. (1969). *Immunochemistry* **6**, 67.
Axen, R., and Porath, J. (1966). *Nature (London)* **210**, 367.
Axen, R., Porath, J., and Ernback, S. (1967). *Nature (London)* **214**, 1302.
Axen, R., Heilbronn, E., and Winter, A. (1969). *Biochim. Biophys. Acta* **191**, 478.
Banks, T. E., Blossey, B. K., and Shafer, J. A. (1969). *J. Biol. Chem.* **244**, 6323.
Bar-Eli, A., and Katchalski, E. (1960). *Nature (London)* **188**, 856.
Bar-Eli, A., and Katchalski, E. (1963). *J. Biol. Chem.* **238**, 1690.
Barnett, L. B., and Bull, H. B. (1959a). *J. Amer. Chem. Soc.* **81**, 5133.
Barnett, L. B., and Bull, H. B. (1956b). *Biochim. Biophys. Acta* **36**, 244.
Bauman, E. K., Goodson, L. H., Guilbault, G. G., and Kramer, D. N. (1965). *Anal. Chem.* **37**, 1378.
Becker, W., and Pfeil, E. (1965). Ger. Pat. Appl. No. P 3673, IVb/120.
Becker, W., and Pfeil, E. (1966). *J. Amer. Chem. Soc.* **88**, 4299.
Becker, W., Freund, H., and Pfeil, E. (1965). *Angew. Chem., Int. Ed. Engl.* **4**, 1079; *Angew. Chem.* **77**, 1139 (1965).
Bernfeld, P., and Bieber, R. E. (1969). *Arch. Biochem. Biophys.* **131**, 587.
Bernfeld, P., and Wan, J. (1963). *Science* **142**, 678.
Bernfeld, P., Bieber, R. E., and MacDonnell, P. C. (1968). *Arch. Biochem. Biophys.* **127**, 779.
Bernfeld, P., Bieber, R. E., and Watson, D. M. (1969). *Biochim. Biophys. Acta* **191**, 570.
Bishop, W. H., Quiocho, F. A., and Richards, F. M. (1966). *Biochemistry* **5**, 4077.
Boyd, W. C., and Hooker, S. B. (1934). *J. Biol. Chem.* **104**, 329.
Boyd, W. C., and Mover, P. (1935). *J. Biol. Chem.* **110**, 457.
Brederoo, P., Th. Daems, W. Van Duijn, P., and Van der Ploeg, M. (1968a). *Electron Microsc. Proc. Euro. Reg. Conf., 4th, 1968* p. 79.
Brederoo, P., Th. Daems, W., Van Duijn, P., and Van der Ploeg, M. (1968b). *Proc. Roy Microsc. Soc.* **3**, 153 (abstr.).
Broomfield, C. A., and Scheraga, H. A. (1961). *Abstr., 140th Meet. Amer. Chem. Soc.* p. 14C.
Broun, G., Selegny, E., Avrameas, S., and Thomas, D. (1969). *Biochim. Biophys. Acta* **185**, 260.
Brown, H. D., Chattopadhyay, S. K., and Patel, A. (1966a). *J. Cell Biol.* **31**, 17a.
Brown, H. D., Chattopadhyay, S. K., and Patel, A. (1966b). *Biochem. Biophys. Res. Commun.* **25**, 304.
Brown, H. D., Patel, A., and Chattopadhyay, S. K. (1966c). *Plant Physiol.* **41**, Supp., lxvi.
Brown, H. D., Chattopadhyay, S. K., and Patel, A. (1967). *Enzymologia* **31**, 205.
Brown, H. D., Patel, A. B., and Chattopadhyay, S. K. (1968a). *J. Biomed. Mater. Res.* **2**, 231.
Brown, H. D., Patel, A. B., and Chattopadhyay, S. K. (1968b). *Amer. J. Bot.* **55**, 729.

Brown, H. D., Patel, A. B., Chattopadhyay, S. K., and Pennington, S. N. (1968c). *Enzymologia* **35**, 215.
Brown, H. D., Patel, A. B., Chattopadhyay, S. K., and Pennington, S. N. (1968d). *Enzymologia* **35**, 233.
Brown, H. D., Patel, A. B., and Chattopadhyay, S. K. (1968e). *J. Chromatogr.* **35**, 103.
Bull, H. B. (1947). *Advan. Protein Chem.* **3**, 95.
Busch, M., Patrascanu, N., and Weber, W. (1934). *J. Prakt. Chem.* [3] **140**, 117.
Campbell, D. H., Luescher, E., and Lerman, L. S. (1951). *Proc. Nat. Acad. Sci. U.S.* **37**, 575.
Carraway, K. J., and Koshland, D. E., Jr. (1968). *Biochim. Biophys. Acta* **160**, 274.
Casu, A., and Avrameas, S. (1969). *Ital. J. Biochem.* **18**, 166.
Cebra, J. J. (1964). *J. Immunol.* **92**, 977.
Cebra, J. J., Givol, D., Silman, I. H., and Katchalski, E. (1961). *J. Biol. Chem.* **236**, 1720.
Cebra, J. J., Givol, D., and Katchalski, E. (1962). *J. Biol. Chem.* **237**, 751.
Cheesman, D. F., and Davies, J. T. (1954). *Advan. Protein Chem.* **9**, 439.
Chibata, I., and Tosa, T. (1966). *Tampakushitsu Kakusan, Koso* **11**, 23.
Chung, S., Hamano, M., Aida, K., and Uemura, T. (1968). *Agr. Biol. Chem.* **32**, 1287.
Cresswell, P., and Sanderson, A. R. (1969). *Biochem. J.* (Agenda paper of the 500th meet., Biochem. Soc., No. 43, p. 40).
Crook, E. M. (1968). *Biochem. J.* **107**, 1p.
Cuatrecasas, P., Wilchek, M., and Anfinsen, C. B. (1968). *Proc. Nat. Acad. Sci. U.S.* **61**, 643.
Davis, R. C., Copperband, S. R., and Mannick, J. A. (1969). *J. Immunol.* **103**, 1029.
DeTar, D. F., and Silverstein, R. (1966). *J. Amer. Chem. Soc.* **88**, 1013.
DeTar, D. F., Silverstein, R., and Rogers, F. F., Jr. (1966). *J. Amer. Chem. Soc.* **88**, 1024.
Devaux, M. H. (1903). *Proc.-Verb. Seances Soc. Sci. Phys. Natur. Bordeaux*, p. 9.
Doyle, R. J., Bello, J., and Roholt, O. A. (1968). *Biochim. Biophys. Acta* **160**, 274.
Drobnica, L., and Augustin, J. (1965a). *Collect. Czech. Chem. Commun.* **30**, 99.
Drobnica, L., and Augustin, J. (1965b). *Collect. Czech. Chem. Commun.* **30**, 1221.
Drobnica, L., and Augustin, J. (1965c). *Collect. Czech. Chem. Commun.* **30**, 1618.
Engel, A., and Alexander, B. (1965). *Fed. Proc., Fed. Amer. Soc. Exp. Biol.* **24**, 512.
Epstein, C. J., and Anfinsen, C. B. (1962). *J. Biol. Chem.* **237**, 2175.
Erlanger, B. F. (1958). *Biochim. Biophys. Acta* **27**, 646.
Estermann, E. F., Peterson, G. H., and McLaren, A. D. (1959). *Soil Sci. Soc. Amer. Proc.* **23**, 31.
Fraenkel-Conrat, H. (1959). *In* "The Enzymes" (P. D. Boyer, H. Lardy, and K. Myrbäck, eds.), Vol. 1, p. 589. Academic Press, New York.
Fraser, M. J. (1957). *J. Pharm. Pharmacol.* **9**, 479.
Fraser, M. J., Kaplan, J. G., and Schulman, J. H. (1955). *Discuss. Faraday Soc.* **20**, 44.
Gelewitz, E. W., Riedeman, W. L., and Klotz, I. M. (1954). *Arch. Biochem. Biophys.* **53**, 411.
Geschwind, I. I., and Li, C. H. (1957). *Biochim. Biophys. Acta* **25**, 171.
Ghosh, S., and Bull, H. B. (1962). *Arch. Biochem. Biophys.* **99**, 121.
Glazer, A. N., Bar-Eli, A., and Katchalski, E. (1962). *J. Biol. Chem.* **237**, 1832.
Goldman, R., Silman, I. H., Caplan, S. R., Kedem, O., and Katchalski, E. (1965). *Science* **150**, 758.

Goldman, R., Kedem, O., Silman, I. H., Caplan, S. R., and Katchalski, E. (1968). *Biochemistry* **7**, 486.
Goldstein, L., Levin, Y., and Katchalski, E. (1964). *Biochemistry* **3**, 1913.
Goldstein, L., Levin, Y., Pecht, M., and Katchalski, E. (1967a). *Isr. J. Chem.* **5**, 90p.
Goldstein, L., Lowey, S., Cohen, C., and Luck, S. (1967b). *Isr. J. Chem.* **5**, 91p.
Green, M. L., and Crutchfield, G. (1969). *Biochem. J.* **115**, 183.
Grubhofer, N., and Schleith, L. (1953). *Naturwissenschaften* **40**, 508; *Chem. Abstr.* **48**, 7983 (1954).
Grubhofer, N., and Schleith, L. (1954). *Hoppe-Seyler's Z. Physiol. Chem.* **297**, 108.
Guilbault, G. G. (1966). *Anal. Chem.* **38**, 527R.
Guilbault, G. G. (1968). *Anal. Chem.* **40**, 459.
Guilbault, G. G., and Kramer, D. N. (1965). *Anal. Chem.* **37**, 1675.
Guilbault, G. G., and Montalvo, J. G., Jr. (1969). *J. Amer. Chem. Soc.* **91**, 2164.
Gutman, M., and Rimon, A. (1964). *Can. J. Biochem.* **42**, 1339.
Habeeb, A. F. S. A. (1967). *Arch. Biochem. Biophys.* **119**, 264.
Hartman, R. S., Bateman, J. B., and Edelhoch, H. E. (1953). *J. Amer. Chem. Soc.* **75**, 5748.
Hasselberger, F. X., Brown, H. D., Chattopadhyay, S. K., Mather, A., Stasiw, R. O., Patel, A. B., and Pennington, S. N. (1970a). *Cancer Res.* **30**, 2736.
Hasselberger, F. X., Brown, H. D., Osborn, K., and Stasiw, R. (1970b). *Mo. Acad. Sci.* **4**, 161.
Haynes, R., and Walsh, K. A. (1969). *Biochem. Biophys. Res. Commun.* **36**, 235.
Helfferich, F. (1962). "Ion Exchange." McGraw-Hill, New York.
Herriot, R. M. (1947). *Advan. Protein Chem.* **3**, 174.
Herzig, D. J., Rees, A. W., and Day, R. A. (1962). *Fed. Proc., Fed. Amer. Soc. Exp. Biol.* **21**, 410e.
Herzig, D. J., Rees, A. W., and Day, R. A. (1964). *Biopolymers* **2**, 349.
Hicks, G. P., and Updike, S. J. (1966). *Anal. Chem.* **38**, 726.
Higgins, H. G., and Fraser, D. (1952). *Aust. J. Sci. Res. Ser. A* **5**, 736.
Higgins, H. G., and Harrington, K. J. (1959). *Arch. Biochem. Biophys.* **85**, 409.
Hiremath, C. B., and Day, R. A. (1964). *J. Amer. Chem. Soc.* **86**, 5027.
Hoare, D. G., and Koshland, D. E., Jr., (1967). *J. Biol. Chem.* **242**, 2447.
Hornby, W. E., Lilly, M. D., and Crook, E. M. (1966). *Biocheim. J.* **98**, 420.
Hornby, W. E., Lilly, M. D., and Crook, E. M. (1968). *Biochem. J.* **107**, 669.
Howard, A. N., and Wild, F. (1957). *Biochem. J.* **65**, 651.
Isliker, H. C. (1953). *Ann. N. Y. Acad. Sci.* **57**, 225.
James, L. K., and Augenstein, L. G. (1966). *Advan. Enzymol.* **28**, 1.
James, L. K., and Hilborn, D. A. (1968). *Biochim. Biophys. Acta* **151**, 279.
Jansen, E. F., and Olson, A. C. (1969). *Arch. Biochem. Biophys.* **129**, 221.
Jaquet, H., and Cebra, J. J. (1965). *Biochemistry* **4**, 954.
Katchalski, E. (1962a). *Pont. Accad. Sci., Scripta Vria* p. 97.
Katchalski, E. (1962b). *Proc. Int. Symp., 1962* Pap. No. 26, p. 283.
Katchalski, E. (1968). *Proc. 7th Int. Congr. Biochem., 1967* Symp. III, 1, 2, p. 147.
Katchalski, E., Fasman, G. D., Simons, E., Blout, E. R., Gurd, F. R. N., and Koltun, W. L. (1960). *Arch. Biochem. Biophys.* **88**, 361.
Kauzmann, W. (1954). *In* "The Mechanism of Enzyme Action" (W. D. McElroy and H. B. Glass, eds.), p. 70. Johns Hopkins Press, Baltimore, Maryland.
Kauzmann, W. (1956). *J. Cell. Comp. Physiol.* **47**, Suppl. 1, 113.
Kay, G., and Crook, E. M. (1967). *Nature (London)* **216**, 514.
Kay, G., and Lilly, M. D. (1970). *Biochim. Biophys. Acta* **198**, 276.

Kay, G., Lilly, M. D., Sharp, A. K., and Wilson, R. J. H. (1968). *Nature (London)* **217**, 641.

Khorana, H. G. (1953). *Chem. Rev.* **53**, 145.

Kobamoto, N., Lofroth, G., Camp, P., VanAmburg, G., and Augenstein, L. (1966). *Biochem. Biophys. Res. Commun.* **24**, 622.

Kurzer, F., and Douraghi-Zadeh, K. (1967). *Chem. Rev.* **67**, 107.

Langmuir, I., and Schaefer, V. J. (1938). *J. Amer. Chem. Soc.* **60**, 1351.

Lerman, L. S. (1953). *Proc. Nat. Acad. Sci. U.S.* **39**, 232.

Levin, Y., Pecht, M., Goldstein, L., and Katchalski, E. (1964). *Biochemistry* **3**, 1905.

Lilly, M. D., Money, C., Hornby, W., and Crook, E. M. (1965). *Biochem. J.* **95**, 45.

Lilly, M. D., Hornby, W. E., and Crook, E. M. (1966). *Biochem. J.* **100**, 718.

Lilly, M. D., and Sharp, A. K. (1968). *Chem. Eng. (London)* **215**, 12–18.

Lilly, M. D., Kay, G., Sharp, A. K., and Wilson, R. J. H. (1968). *Biochem. J.* **107**, 5p.

Lofroth, G., and Augenstein, L. (1967). *Arch. Biochem. Biophys.* **118**, 73.

Lowey, S., Goldstein, L., Cohen, C., and Luck, S. M. (1967). *J. Mol. Biol.* **23**, 287.

Lynn, J., and Falb, R. (1969). *Abstr. Div. Biol. Chem., 158th ACS Nat. Meet.* Pap. No. 298.

McCormick, D. B. (1965). *Anal. Biochem.* **13**, 194.

McLaren, A. D. (1954a). *J. Phys. Chem.* **58**, 129.

McLaren, A. D. (1954b). *Soil Sci. Soc. Amer., Proc.* **18**, 170.

McLaren, A. D. (1957). *Science* **125**, 697.

McLaren, A. D. (1963). *In* "Cell Interface Reactions" (H. D. Brown, ed.), pp. 1–31. Scholar's Library, New York.

McLaren, A. D., and Babcock, K. L. (1959). *In* "Subcellular Particles" (T. Hayashi, ed.), pp. 23–35. Ronald Press, New York.

McLaren, A. D., and Estermann, E. F. (1956). *Arch. Biochem. Biophys.* **61**, 158.

McLaren, A. D., and Estermann, E. F. (1957). *Arch. Biochem. Biophys.* **68**, 157.

Manecke, G. (1962a). *Pure Appl. Chem.* **4**, 507.

Manecke, G. (1962b). *Brit. Pat.* 912,897 (1962): *Chem. Abstr.* **58**, 1413h (1963).

Manecke, G. (1964). *Naturwissenschaften* **51**, 25.

Manecke, G., and Foerster, H. J. (1966). *Makromol. Chem.* **91**, 136.

Manecke, G., and Gunzel, G. (1962). *Makromol. Chem.* **41**, 199.

Manecke, G. and Gunzel, G. (1967). *Naturwissenschaften* **20**, 531.

Manecke, G., and Singer, S. (1960a). *Makromol. Chem.* **37**, 119.

Manecke, G., and Singer, S. (1960b). *Makromol. Chem.* **39**, 13; *Chem. Abstr.* **55**, 1064 (1961).

Marfey, P. S., Nowak, H., Uziel, M., and Yphantis, D. A. (1965). *J. Biol. Chem.* **240**, 3264.

Messing, R. A. (1969). *Enzymologia* **38**, 39.

Micheel, F., and Ewers, J. (1949). *Makromol. Chem.* **3**, 200.

Mitchell, P. (1966a). "Chemiosmotic Coupling in Oxidative and Photosynthetic Phosphorylation," Glynn Research Ltd., Bodwin, England.

Mitchell, P. (1966b). *Biol. Rev.* **41**, 445.

Mitz, M. A. (1956). *Science* **123**, 1076.

Mitz, M. A., and Schlueter, R. J. (1959). *J. Amer. Chem. Soc.* **81**, 4024.

Mitz, M. A., and Summaria, L. J. (1961). *Nature (London)* **189**, 576.

Mitz, M. A., and Yanari, S. (1956). *J. Amer. Chem. Soc.* **78**, 2649.

Mkrtumova, N. A., and Deborin, G. A. (1962). *Dokl. Akad. Nauk SSSR* **146**, 1434.

Mosbach, K., and Mosbach, R. (1966). *Acta Chem. Scand.* **20**, 2807.

Mowbray, J. F., and Scholand, J. (1966). *Immunology* **11**, 421.

Nakane, P. K., and Pierce, G. B., Jr. (1967). *J. Cell Biol.* **33,** 307.

Nezlin, R. S. (1961). *Usp. Sovrem. Biol.* **52,** 19.

Nikolaev, A. Ya. (1962). *Biokhimiya* **27,** 843; see *Biochemistry* (*USSR*) **27,** 713.

Nikolaev, A. Ya., and Mardeshev, S. R. (1961). *Biokhimiya* **26,** 641; see *Biochemistry* (*USSR*) **27,** 565.

Ogata, K., Ottesen, M., and Svendsen, I. (1968). *Biochim. Biophys. Acta* **159,** 403.

Ong, E. B., Tsang, Y., and Perlmann, G. E. (1966). *J. Biol. Chem.* **241,** 5661.

Patel, A. B., Pennington, S. N., and Brown, H. D. (1969a). *Proc. Cell Biol. Biophys. Midwest* **SB-2,** 10.

Patel, A. B., Pennington, S. N., and Brown, H. D. (1969b). *Biochim. Biophys. Acta* **178,** 626.

Patel, A. B., Hasselberger, F. X., Ghiron, C. A., and Brown, H. D. (1971). In preparation.

Patel, R. P., and Price, S. (1967). *Biopolymers* **5,** 583.

Patel, R. P., Lopiekes, D. V., Brown, S. P., and Price, S. (1967). *Biopolymers* **5,** 577.

Patramani, I., Katsiri, K., Pistevou, E., Kalogerakos, T., Pavlatos, M., and Evangelopoulous, A. (1969). *Eur. J. Biochem.* **11,** 28.

Pauly, H. (1915). *Hoppe-Seyler's Z. Physiol. Chem.* **94,** 284.

Pennington, S. N., Brown, H. D., Patel, A. B., and Knowles, C. O. (1968a). *Biochim. Biophys. Acta* **167,** 479.

Pennington, S. N., Brown, H. D., Patel, A. B., and Chattopadhyay, S. K. (1968b). *J. Biomed. Mater. Res.* **2,** 443.

Pennington, S. N., Patel, A., Chattopadhyay, S. K., and Brown, H. D. (1968c). *Plant Physiol.* **43,** S-47.

Porath, J., Axen, R., and Ernback, S. (1967). *Nature* (*London*) **215,** 1491.

Quiocho, F. A., and Richards, F. M. (1964). *Proc. Nat. Acad. Sci. U.S.* **52,** 833.

Quiocho, F. A., and Richards, F. M. (1966). *Biochemistry* **5,** 4062.

Quiocho, F. A., Bishop, W. H., and Richards, F. M. (1967). *Proc. Nat. Acad. Sci. U.S.* **57,** 525.

Ramsden, W. (1903). *Proc. Roy. Soc., Ser. B* **72,** 156.

Rao, C. N. R., and Venkataraghavan, R. (1962). *Tetrahedron* **18,** 531.

Reese, E. T., and Mandels, M. (1958). *J. Amer. Chem. Soc.* **80,** 4625.

Riehm, J. P., and Scheraga, H. A. (1966). *Biochemistry* **5,** 99.

Riesel, E., and Katchalski, E. (1964). *J. Biol. Chem.* **239,** 1521.

Rimon, A., Gutman, M., and Rimon, S. (1963). *Biochim. Biophys. Acta* **73,** 301.

Rimon, A., Alexander, B., and Katchalski, E. (1966a). *Biochemistry* **5,** 792.

Rimon, S., Stupp, Y., and Rimon, A. (1966b). *Can. J. Biochem.* **44,** 415.

Rothen, A. (1947). *Advan. Protein Chem.* **3,** 123.

Selegny, E., Avrameas, S., Broun, G., and Thomas, D. (1968). *C. R. Acad. Sci., Ser. C* **266,** 1431.

Sharp, A. K., Kay, G., and Lilly, M. D. (1969). *Biotechnol. Bioeng.* **11,** 363.

Sheehan, J. C., and Hess, G. P. (1955). *J. Amer. Chem. Soc.* **77,** 1067.

Sheehan, J. C., and Hlavka, J. J. (1956). *J. Org. Chem.* **21,** 439.

Sheehan, J. C., and Hlavka, J. J. (1957). *J. Amer. Chem. Soc.* **79,** 4528.

Sheehan, J. C., Cruickshank, P. A., and Boshart, G. L. (1961). *J. Org. Chem.* **26,** 2525.

Silman, I. H., and Katchalski, E. (1966). *Annu. Rev. Biochem.* **35,** 873.

Silman, I. H., Wellner, D., and Katchalski, E. (1963). *Isr. J. Chem.* **1,** 65.

Silman, I. H., Albu-Weissenberg, M., and Katchalski, E. (1966). *Biopolymers* **4,** 441.

Skou, J. C. (1957). *Biochim. Biophys. Acta* **23,** 394.

Sri Ram, J., Bier, M., and Maurer, P. H. (1962). *Advan. Enzymol.* **24,** 105.

Stasiw, R. O., Brown, H. D., and Hasselberger, F. X. (1970). *Can. J. Biochem.* **48,** 1314.

Stevens, C. O., and Long, J. L. (1969). *Proc. Soc. Exp. Biol. Med.* **131,** 1312.

Sundaram, P. V., and Crook, E. M. (1967). *Indian J. Biochem.* **4,** 33.

Surinov, B. P., and Manoilov, S. E. (1966). *Biokhimiya* **31,** 337.

Suzuki, H., Ozawa, Y., and Maeda, H. (1966). *Agr. Biol. Chem.* **30,** 807.

Tabachnick, M., and Sobotka, H. (1960). *J. Biol. Chem.* **235,** 1051.

Takami, T. (1968). *Seikagaku* **40,** 749; *Chem. Abstr.* **70,** 74662j (1969).

Tosa, T., Mori, T., Fuse, N., and Chibata, I. (1966a). *Enzymologia* **31,** 214.

Tosa, T., Mori, T., Fuse, N., and Chibata, I. (1966b). *Enzymologia* **31,** 225.

Tosa, T., Mori, T., Fuse, N., and Chibata, I. (1967a). *Enzymologia* **32,** 153.

Tosa, T., Mori, T., Fuse, N., and Chibata, I. (1967b). *Biotechnol. Bioeng.* **9,** 603.

Tosa, T., Mori, T., Fuse, N., and Chibata, I. (1969a). *Agr. Biol. Chem.* **33,** 1047.

Tosa, T., Mori, T., and Chibata, I. (1969b). *Agr. Biol. Chem.* **33,** 1053.

Updike, S. J., and Hicks, G. P. (1967). *Nature (London)* **214,** 986–988.

Usami, S., and Taketomi, N. (1965). *Hakko Kyokaishi* **23,** 267; *Chem. Abstr.* **63,** 11939 (1965).

Van der Ploeg, M., and Van Duijn, P. (1964). *J. Roy. Microsc. Soc.* [3] **83,** 415.

Van der Ploeg, M., and Van Duijn, P. (1968). *J. Histochem. Cytochem.* **16,** 693.

Van Duijn, P., Pascoe, E., and Van der Ploeg, M. (1967). *J. Histochem. Cytochem.* **15,** 631.

Wagner, T., Hsu, C. J., and Kelleher, G. (1968). *Biochem. J.* **108,** 892.

Wang, J. H., and Tu, J. (1969). *Biochemistry* **8,** 4403.

Weetall, H. H. (1969a). *Nature (London)* **223,** 959.

Weetall, H. H. (1969b). *Science* **166,** 615.

Weetall, H. H., and Hersh, L. S. (1969). *Biochim. Biophys. Acta* **185,** 464.

Weetall, H. H., and Weliky, N. (1966). *Anal. Biochem.* **14,** 160.

Weliky, N., and Weetall, H. H. (1965). *Immunochemistry* **2,** 293.

Weliky, N., Brown, F. S., and Dale, E. C. (1969). *Arch. Biochem. Biophys.* **131,** 1.

Westman, T. L. (1969). *Biochem. Biophys. Res. Commun.* **35,** 313.

Wharton, C. W., Crook, E. M., and Brocklehurst, K. (1968a). *Eur. J. Biochem.* **6,** 565.

Wharton, C. W., Crook, E. M., and Brocklehurst, K. (1968b). *Eur. J. Biochem.* **6,** 572.

Wheeler, K. P., Edwards, B. A., and Wittam, R. (1969). *Biochim. Biophys. Acta* **191,** 187.

Whittam, R., Edwards, B. A., and Wheeler, K. P. (1968). *Biochem. J.* **107,** 3p.

Wilson, R. J. H., Kay, G., and Lilly, M. D. (1968a). *Biochem. J.* **108,** 845.

Wilson, R. J. H., Kay, G., and Lilly, M. D. (1968b). *Biochem. J.* **109,** 137.

Wold, F. (1961). *J. Biol. Chem.* **236,** 106.

Woodward, R. B., and Olofson, R. A. (1961). *J Amer. Chem. Soc.* **83,** 1007.

Woodward, R. B., Olofson, R. A., and Mayer, H. (1961). *J. Amer. Chem. Soc.* **83,** 1010.

Wu, H., and Ling, S. M. (1927). *Chin. J. Physiol.* **1,** 407.

Zahn, H., and Meienhofer, J. (1958). *Makromol. Chem.* **26,** 126.

Zahn, H., and Waschka, O. (1955). *Makromol. Chem.* **18,** 201.

Zahn, H., Zuber, H., Ditscher, W., Wegerle, D., and Meienhofer, J. (1956). *Chem. Ber.* **89,** 407.

CHAPTER VIII

Changes in Protein Conformation Associated with Chemical Modification

A. F. S. A. HABEEB

I. Introduction

Chemical modification of proteins is applicable in (a) studying the functional groups responsible for biological activity, e.g., enzymes, hormones, toxins, and antibodies; (b) altering the activity of proteins to render them suitable for medical or industrial use, e.g., using formaldehyde to inactivate bacteria or toxins for the preparation of vaccines and toxoids (French and Edsall, 1945); (c) altering the physical properties of proteins, e.g., increasing their solubility as in polyalanylation (Epstein et al., 1962; Fuchs and Sela, 1965; Freedman and Sela, 1966); (d) preparing insoluble derivatives of enzymes with considerable enzymic activity by introduction of intermolecular cross-links (Quiocho and Richards, 1964; Habeeb, 1967c) or by coupling the enzymes to an insoluble support (Grubhofer and Schleith, 1954; Mitz and Summaria, 1961; Glazer et al., 1962; Levin et al., 1964; Habeeb, 1967c); and (e) preparing derivatives which are suitable for sequence studies, e.g., blocking free amino groups or arginine residues in order to cleave the polypeptide chain specifically at arginine or lysine, respectively, by trypsin. Alternatively both arginine and lysine residues may be blocked, and by modifying cysteine specifically to aminoethylcysteine, trypsin can cleave at cysteinyl residues (Slobin and Singer, 1968).

The study of structure–function relationship has been an impetus for developing reagents with narrow specificities for many protein groups (for review, see Cohen, 1968). One important question to be answered before an unequivocal conclusion as to the essential nature of a given functional group for the biological activity of a protein is whether conformational changes have taken place and what their magnitude is.

Native proteins have an ordered conformation in physiological conditions. The molecular conformation of a protein is determined by the sequence of amino acid residues which constitute the primary structure and by the stabilizing effect of the noncovalent bonds (Scheraga, 1963; Schellman and Schellman, 1964) in the direction to minimize the total free energy of the system.

The thermodynamic hypothesis for the formation of the unique three-dimensional structure of native proteins has been inferred from experimental data. These data were obtained by converting native enzymes into an unfolded polypeptide chain devoid of secondary and tertiary structure by reduction of disulfide bonds. It was possible by oxidation under the optimum conditions (Epstein et al., 1963) (protein concentration, pH, and temperature) to recover the original protein with its native conformation, shown by various criteria (e.g., specific activity, optical rotation, viscosity, ultraviolet spectrum, immunochemistry, peptide map, chromatography, sedimentation, and electrophoresis). Recovery of enzymic activity after refolding of the reduced protein by reoxidation was much higher than what would be expected from random pairing of the sulfhydryl groups to form disulfide bonds. The recovered activity of ribonuclease was 95 to 100% (White, 1961; Anfinsen and Haber, 1961); of lysozyme 50 to 80% (Epstein and Goldberger, 1963; Goldberger and Epstein, 1963); of insulin 60 to 80% (Katsoyannis and Tometsko, 1966); of Taka amylase 48% (Isemura et al., 1963); of poly-DL-alanyl trypsin 8% (Epstein and Anfinsen, 1962); of alkaline phosphatase 80% (Levinthal et al., 1962); and of pepsinogen, 50% Frattali et al., 1963). Polyalanylated rabbit antibodies to bovine serum albumin recovered 68% of the antigenic determinants upon reoxidation of completely reduced polyalanyl rabbit antibody (Freedman and Sela, 1966).

With sperm whale myoglobin, the presence of the heme group was essential for the protein to regain its native conformation (Harrison and Blount, 1965; Breslow *et al.*, 1965).

It is expected that chemical modification by covalent introduction of new groups (without rupture of peptide bonds) will generate a new protein with a new amino acid composition and sequence. As a result the noncovalent interactions will undergo a "reorganization," and the new polypeptide chain will assume the three-dimensional arrangement that is compatible with the lowest conformational free energy—thermodynamically the most stable.

The magnitude of this reorganization will depend on the nature of the groups introduced and the nature and location of the groups substituted. The conformational reorganization associated with a particular chemical modification of a protein is characteristic of that protein. Great care should be exercised in extrapolating the results obtained with one protein.

Only through X-ray crystallography can the three-dimensional structure of a protein molecule in atomic detail be determined. Phillips (1966) showed that the binding of tri-*N*-acetylglucosamine to lysozyme resulted in conformational changes of the enzyme. These changes were largely restricted to the part of enzyme structure left of the cleft, which appeared to tilt as a whole to close the cleft slightly. As a result, the side chain of tryptophan-62 moved about 0.75 Å toward the position of the second sugar residue. Information on conformation can be derived from studies of "shape properties" and "short-range properties" (Kauzmann, 1959). Shape properties are those that depend on the overall shape of the molecule rather than specific spacial relationships between any particular atom and its immediate neighbors. Short-range properties in contrast are those influenced by changes in the immediate vicinity of individual groups in the molecule. Table I gives a summary of the type of information associated with the various physical properties of protein molecules.

Both classes of properties must be studied to understand the nature and magnitude of conformational changes. Properties that are sensitive to different kinds of conformational changes need to be studied. For example, optical rotation is particularly useful in studying the amount of α-helix present in a protein (Yang and Doty, 1957). Ultraviolet absorption spectra in the vicinity of 280 mμ serve to reflect local changes in the vicinity of the chromophoric groups.

Several methods used in probing the conformational changes of protein were reviewed. Schellman and Schellman (1964) discussed optical techniques: ultraviolet spectra, optical rotatory dispersion and circular dichroism, and infrared spectra and hydrogen exchange in proteins and polypeptides. Other reviews for evaluating conformational changes in proteins

TABLE I

SHAPE PROPERTIES OF PROTEINS[a]

Hydrodynamic properties
  A. Frictional ratio (from sedimentation diffusion)
  B. Viscosity increment
  C. Rotary diffusion constant (flow birefringence, dielectric relaxation, fluorescence depolarization)
Radiation scattering (angular dependence)
  A. Light scattering
  B. Small angle X-ray scattering
Long-range electrostatic effects on titration curves (Linderstrøm-Lang's w)
Electron microscopy
Second virial coefficients at moderate salt concentrations
Surface properties
  A. Force area curves at moderate pressures
  B. Surface dipole moment (?)
  C. Area of solid film
Dipole moment and/or Kirkwood-Shumaker effect
Diffusion through membranes with controlled pore size (Craig)

Short-range properties of proteins

Thermodynamic properties
  A. Energy and heat capacity
  B. Entropy
  C. (Free energy)
  D. Volume, compressibility, and coefficient of expansion
  E. Solubility, activity, distribution between solvents
Optical properties
  A. Optical rotation and dispersion
  B. Infrared absorption
  C. Visible and ultraviolet absorption
  D. Wide-angle X-ray diffraction
  E. Index of refraction (polarizability, anisotropy)
  F. Depolarization of fluorescence (in some cases)
Chemical properties
  A. Reactivity of groups
  B. Intrinsic pK values of acidic and basic groups
  C. H-D or H-T exchange
  D. Binding of small molecules, dyes, ions, etc.
  E. Immunochemical properties
  F. Digestibility by proteolytic enzymes
  G. Biological activity
  H. Electrophoresis (isoelectric point, zeta potential)
Nuclear and electronic magnetic resonance
Surface phenomena
  A. Spreadability
  B. Surface viscosity
  C. (Surface dipole moment)

[a] From Kauzmann (1959).

have appeared recently. These include examination of titration behavior (Nazaki and Tanford, 1967); hydrogen exchange DiSabato and Ottesen, 1967); difference spectroscopy (Herskovitz, 1967); fluorescence measurements (Brand and Witholt, 1967); environmentally sensitive groups attached to proteins (Horton and Koshland, 1967); susceptibility to attack by proteolytic enzyme (Rupley, 1967); and microcomplement fixation (Levine and Van Vunakis, 1967). The present review deals with four methods useful in evaluating the conformational changes which accompany chemical modification of proteins. These methods include determination of shape changes, availability of disulfide bridges to reduction or sulfitolysis, susceptibility of the modified protein to proteolytic digestion, and precipitability of modified protein with antibodies directed against native protein.

II. Shape Properties

A. CHANGES IN VISCOSITY

The viscosity of a liquid is a measure of its resistance to flow when subjected to a shearing force. When macromolecules are dissolved in the liquid, they disturb the flow pattern of the liquid, and the system behaves like a pure liquid with higher viscosity. For spherical, rigid, uncharged solute molecules, Einstein (1906, 1911) derived the equation.

$$\left(\frac{\eta}{\eta_0} - 1 \right) = 2.5\Phi \tag{1}$$

where η = viscosity of solution, η_0 = viscosity of solvent, and Φ = volume fraction occupied by the particles; the equation is valid only as long as $\Phi < 0.03$. Equation (1) for spheres may be written

$$\nu = 2.5 \tag{2}$$

where ν = viscosity increment = $1/\Phi[(\eta/\eta_0) - 1]$. For proteins the value of ν is always greater than 2.5 and can be explained partly by hydration and by deviation of the protein molecule from a sphere. Asymmetric particles, such as ellipsoidal molecules, are subjected to Brownian movement; as a result of the rotation of the particles, the effective volume occupied by the particles is much larger than their true geometric volume. This random orientation of asymmetric macromolecules causes a greater increase in viscosity than spherical particles of the same volume and is proportional to the axial ratio of the ellipsoidal molecules.

Viscosity measurements of proteins can be conveniently determined by

Oswald viscometer from the relative flow time of different concentrations of protein solutions and the solvent. Care must be taken to carry out the determination at a constant temperature and in the presence of neutral salt (0.1–0.2 M) to overcome charge effects (Kragh, 1961).

To evaluate conformational changes that affect the shape of protein on modification, the determination of the intrinsic viscosity offers valuable information. However the information is only meaningful when one molecular species exists since the presence of aggregates would greatly influence the interpretation.

Modification of the free amino groups of bovine serum albumin, human γ-globulin, and β-lactoglobulin by succinylation resulted in modified proteins which showed marked increase in their intrinsic viscosity compared to the respective native proteins (Habeeb et al., 1958), indicative of unfolding or swelling of the modified proteins. These conformational changes, produced as a result of succinylation of antibodies to bovine serum albumin, gave rise to nonspecific loss of the ability of modified antibodies to precipitate with bovine serum albumin (Habeeb et al., 1958). The shape changes accompanying various modifications of bovine serum albumin were studied by Habeeb (1966a). The results (Fig. 1) show that guanidination and amidination did not affect the viscosity, indicating no detectable shape changes on modification. Nitroguanidinated and acetylated bovine serum albumin showed an increase in their viscosity, the increase being greater with acetyl bovine serum albumin than with nitroguanyl bovine serum albumin (Habeeb, 1966a). That these conformational changes were not due simply to increased net negative charge was indicated by the fact that although nitroguanyl bovine serum albumin had an electrophoretic mobility which was greater than succinyl bovine serum albumin (-10.9 compared to -9.5 cm^2 V second $\times 10^{-5}$) (Habeeb, 1964), its limiting viscosity number was less (5.8 compared to 33). This indicated that, in nitroguanyl bovine serum albumin, some intramolecular noncovalent interactions were of a magnitude to counteract the electrostatic repulsive force and limit shape changes. With succinyl bovine serum albumin, electrostatic repulsive forces were operative and caused the dramatic increase in asymmetry of the molecule and, hence, the increased viscosity. The nature of the protein is important in determining the extent of shape changes on chemical modification. No significant change in viscosity of succinylated subtilopeptidase was observed compared to the native enzyme (Gounaris and Ottesen, 1965). Removal of sialic acid from a glycoprotein isolated from sheep plasma resulted in a decrease in viscosity indicating folding and a relatively close packing of the molecule (Anantha Samy, 1967). Removal of heme from whale metmyoglobin resulted in an increase in the intrinsic viscosity which was consistent with an overall

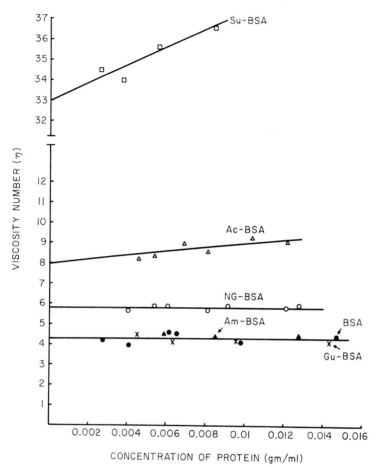

FIG. 1. Viscosity of bovine serum albumin and modified bovine serum albumin in phosphate buffer ionic strength 0.1, pH 7.5 at 26.2°C. (●——●) Bovine serum albumin (BSA); (▲——▲) amidinated bovine serum albumin (Am-BSA); (×——×) guanidinated bovine serum albumin (Gu-BSA); (○——○) nitroguanyl bovine serum albumin (NG-BSA); (△——△) acetyl bovine serum albumin (Ac-BSA); and (□——□) succinyl bovine serum albumin (Su-BSA). (From Habeeb, 1966a.)

swelling of the protein moiety or increased asymmetry of the molecule (Crumpton and Polson, 1965).

B. CHANGES IN SEDIMENTATION COEFFICIENT

Sedimentation velocity studies in addition to providing a method for investigating the homogeneity of the macromolecule offer useful information on molecular size and shape of the molecule. The macromolecules are allowed to move under the influence of a high centrifugal field, and the rate of movement is a function of the molecular weight of the solute and the frictional resistance which the molecules experience moving through the solvent. The equation describing the relationship between the sedimentation rate and the properties of the molecule is

$$s = \frac{M(1 - \bar{V}\rho)}{Nf} \tag{3}$$

where s = sedimentation coefficient, M = molecular weight of macromolecules, $\bar{V}$ = partial specific volume of macromolecule, ρ = density of the solution, N = Avogadro's number (6.02×10^{23}) and f = molecular frictional coefficient which is a function of size, shape, and hydration of the sedimenting molecules. Values of $f/f_0 > 1$ (where f_0 is the frictional coefficient of a sphere of the same molecular weight) indicate that the macromolecule deviates from an anhydrous sphere, possibly due to asymmetry or to hydration of the molecule or, most probably, to a combination of these two factors. Protein molecules which carry 0.3 gm water of hydration per gram of protein give a frictional ratio of 1.13. A number of proteins have a frictional ratio lower than 1.13 (ribonuclease, pepsin), and these molecules may conceivably be spherical. Proteins with frictional ratios of 1.5 or higher (zein and antipneumonococcus serum globulin) are expected to deviate very markedly from spherical shape (Cohn and Edsall, 1943). Perrin (1936) has calculated the relation between f/f_0 and the axial ratio $p = a/b$ for ellipsoid of rotation (a = semiaxis of revolution and b = equatorial semiaxis of ellipsoid)—a model frequently used to describe the shape of protein molecule. Scheraga and Mandelkern (1953) proposed using an effective hydrodynamic ellipsoid to describe protein molecules. However the dimensions of the macromolecule derived from hydrodynamic data are still semiquantitative.

Sedimentation coefficients of proteins are in some cases concentration dependent (Schachman, 1959). If the macromolecule is highly charged, it will show a sedimentation potential (Pedersen, 1958) that will cause a reduction in sedimentation coefficient of the protein. However this sedimentation potential is eliminated by working with proteins in buffers of high ionic strength or by extrapolating the sedimentation coefficient to zero protein concentration (Kronman and Foster, 1957; Habeeb, 1967b). A decrease in the sedimentation coefficient of succinylated and acetylated bovine serum albumin was observed (Habeeb et al., 1958; Habeeb, 1967b) which indicated shape changes as demonstrated by increased asymmetry of the molecules. No shape changes were observed with succinylated subtilopeptidase (Gounaris and Ottesen, 1965). Guanidination of bovine serum albumin (Habeeb, 1959, 1960), ovalbumin (Habeeb, 1961), ribonuclease (Klee and Richards, 1957), and mercuripapain (Shields *et al.*, 1959) did not produce significant changes in their sedimentation coefficients compared to the respective native proteins. However, guanidinated β-lactoglobulin showed a decrease in sedimentation coefficient indicating increased asymmetry of the molecule (Habeeb, 1960). Amidination of bovine serum albumin, antibenzene arsonate antibodies (Wofsy and Singer, 1963), α-lactalbumin (Robbins *et al.*, 1965) and nitroguanidination of bovine serum albumin (Saroff and Evans, 1959; Habeeb, 1964) did not produce shape changes as determined by similarity of sedimentation coefficients of native and modified proteins.

The observed loss of the ability of several antibodies after acetylation to precipitate with the homologous antigen as opposed to guanidinated antibodies (Habeeb *et al.*, 1959) was due to nonspecific loss resulting from conformational changes. Participation of an amino group at the active site of antibody was thus unconfirmed. Removal of heme from sperm whale metmyoglobin resulted in a decrease in sedimentation coefficient of the apomyoglobin indicating increased asymmetry of the molecule (Crumpton and Polson, 1965; Breslow, 1964).

C. CHANGES IN THE STOKES RADIUS OF THE MACROMOLECULE

Gel filtration on Sephadex columns has extensive application for separating macromolecules based on their molecular size (Porath and Flodin, 1959; Pedersen, 1962; Hjerten and Mosbach, 1962; Porath, 1962). Some workers found that a linear relationship existed between the logarithm of molecular weight of a protein and the ratio of elution volume to the column void volume (Whitaker, 1963; Roubal and Tappel, 1964; Leach and O'Shea, 1964) or between logarithm of molecular weight and elution volume (Andrews, 1964), hemoglobin being an exception. Laurent and Killander (1964), Ackers (1964), and Siegel and Monty (1965, 1966) have found that the elution volume correlated with the Stokes radius. The reason for this apparent inconsistency was that all proteins used as calibrating standards were globular, possessing closely similar frictional ratios and partial specific volumes. For such proteins a correlation of molecular weight with elution volume is virtually indistinguishable from a correlation with Stokes radius as is evident from Eq. (4):

$$f/f_0 = a \left/ \left(\frac{3\bar{V}M}{4\Pi N}\right)^{1/3}\right. \tag{4}$$

where f/f_0 = frictional ratio, a = Stokes radius, $\bar{V}$ = partial specific volume, M = molecular weight, and N = Avogadro's number (6.02×10^{23}).

Ackers (1964) postulated a restricted diffusion mechanism for the operation of Sephadex columns, in which K_D is a kinetic parameter representing steric and frictional hindrance to diffusion,

$$K_D = \frac{V_e - V_0}{V_i} \tag{5}$$

where K_D = distribution coefficient, V_e = effluent peak volume of the macromolecule, V_0 = void volume of the column and is the elution volume of an excluded molecule, e.g., blue dextran, V_i = volume of unbound solvent internal to the gel phase = total volume of gel bed − (V_0 + volume of gel grains).

During flow a macromolecule will only penetrate the gel if its Stokes radius is smaller than the pore radius r. If a molecule does enter the gel pore, it encounters increased hydrodynamic frictional resistance to motion and, therefore, has a lower diffusion coefficient than in free solution.

Ackers derived an equation to relate the distribution coefficient of a macromolecule from Sephadex column to the Stokes radius a and effective pore radius of the gel r:

$$\frac{V_e - V_0}{V_i} = \left(1 - \frac{a}{r}\right)^2 \left[1 - 2.104 \frac{a}{r} + 2.09 \left(\frac{a}{r}\right)^3 - 0.95 \left(\frac{a}{r}\right)^5\right] \tag{6}$$

For particles of known Stokes radius a, the effective pore radius r within the gel is determined from the Sephadex column data. The validity of the derivation of Eq. (6) was well demonstrated by the good agreement of the values of r obtained by chromatography of proteins with known Stokes radius a. Once r is known for a given column, it is possible to calculate the Stokes radius of a macromolecule from its elution position (or distribution coefficient) on a calibrated column. Ackers has constructed a table to relate $(V_e - V_0)/V_i$ to a/r which aids in calculating the Stokes radius of an unknown macromolecule from its behavior on a calibrated Sephadex column.

Siegel and Monty (1965, 1966) found that fibrinogen (mol wt = 330,000 and a = 107 Å) eluted earlier than ferritin (mol wt = 1,300,000 and a = 79 Å) on a Sephadex G-200 column. This finding provided further evidence that the elution volume of a macromolecule correlated best with its Stokes radius. The discrepancy observed with hemoglobin disappears when elution volume is correlated with the Stokes radius instead of molecular weight. Based on these observations, Habeeb (1966b, 1967b) utilized the sensitivity of the elution volume of a protein molecule on a column of Sephadex to its Stokes radius to evaluate the conformational changes associated with chemical modification of bovine serum albumin and human γ-globulin.

The Stokes radius provides a workable dimension for comparison of the shape of a protein molecule and its modified counterpart. The usefulness of Sephadex chromatography for the determination of the Stokes radius can be attributed to (a) the sensitivity of the elution volume of a macromolecule to the Stokes radius, (b) the well-documented insensitivity of the elution volume to the concentration of the macromolecule in the range of 0.1 to 50 mg/ml (Ackers, 1964; Whitaker, 1963; Atassi and Caruso, 1968; Andrews, 1964), and (c) the Stokes radius calculated from the diffusion coefficient [Eq. (7)] being in good agreement with that obtained from a

TABLE II

STOKES RADII OF VARIOUS MODIFIED BOVINE SERUM ALBUMINS[a]

Protein	Number of amino groups modified	Stokes radius (mμ)	f/f_0
Bovine serum albumin	0	3.7	1.36
Succinylated bovine serum albumin	59	11.2 and 7.75[b]	2.78[b]
Acetylated bovine serum albumin	60	8.4 and 5.5[b]	2.0[b]
Nitroguanyl bovine serum albumin	54	4.5	1.62
Guanidinated bovine serum albumin	54	3.78	1.37
Amidinated bovine serum albumin	58	3.88	1.41

[a] From Habeeb (1966b).
[b] Monomer species.

calibrated Sephadex column (Habeeb, 1967b):

$$a = \frac{RT}{6\Pi\eta ND} \tag{7}$$

where R = gas constant (8.3×10^7), T = absolute temperature, η = viscosity of solvent, and D = diffusion coefficient. Sephadex chromatography on a calibrated column permits the determination of a Stokes radius independent of concentration. By contrast, sedimentation coefficients are generally concentration dependent and invariably insensitive to minor conformational changes (Habeeb, 1967b; Atassi and Caruso, 1968).

Conformational changes that affect the shape of various modified bovine serum albumins were evaluated from the Stokes radii (Habeeb, 1966b), and a summary of the results is shown in Table II.

These results indicated that succinylation and acetylation resulted in the formation of two molecular species. The one with the largest Stokes radius was an aggregate as shown from sedimentation data (Habeeb, 1967b). An increase in the Stokes radius and frictional coefficient is the result of an

270 A. F. S. A. HABEEB

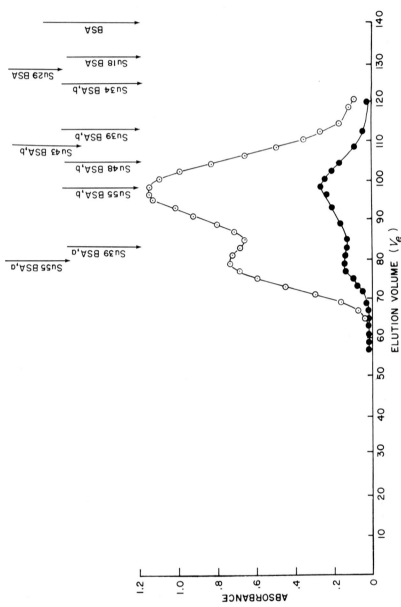

Fig. 2. The elution pattern of succinyl bovine serum albumin with 55 modified amino groups (Su 55 BSA) on a calibrated Sephadex G-200 column (1.5 × 120 cm). The elution volumes of the various succinylated bovine serum albumins are indicated by an arrow. (a) Rapidly eluting component and (b) slowly eluting component. (○——○)

increase in the asymmetry of the molecule due to swelling or unfolding (providing no aggregation occurs). It is evident (Table II) that succinylation produced the greatest shape changes followed by acetylation and nitroguanidination. Amidination and guanidination of the amino groups in bovine serum albumin produced no significant shape changes. These results were in agreement with those obtained from viscosity measurements previously mentioned (Habeeb, 1966a).

To study the conformational changes that affect the overall shape of the molecule accompanying progressive succinylation of the bovine serum albumin and human γ-globulin, the Stokes radii of the modified proteins were determined (Habeeb 1967b). Figure 2 shows the elution pattern of a preparation of bovine serum albumin with 55 amino groups succinylated together with the elution volume of native bovine serum albumin and other succinylated bovine serum albumins. Preparations with more than 35 amino groups succinylated showed evidence of heterogeneity due to appearance of a rapidly eluting component. The proportion of the rapidly eluting component (found to be an aggregate from the sedimentation coefficient) increased with progressive succinylation. Elution volume of the monomeric species of succinylated bovine serum albumins decreased with progressive succinylation indicating an increase in the Stokes radii of the derivatives. The relationship between the Stokes radii of succinylated bovine serum albumins and the number of amino groups succinylated is shown in Fig. 3. There is a gradual increase in the Stokes radius with succinylation of up to 34 amino groups (about 0.322 mμ/10 amino groups succinylated). Then a steep increase in the Stokes radius follows (1.44 mμ/ 10 amino groups succinylated). These results show that a gradual increase in asymmetry of the molecule occurs up to succinylation of 34 amino groups which is followed by extensive asymmetry with further modification. Human γ-globulin gave a similar pattern on succinylation, but the shape changes were less pronounced than with the derivatives of bovine serum albumin (Habeeb, 1967b).

Atassi and Caruso (1968) found that modification of tryptophan-7 (with 2-hydroxy-5-nitrobenzyl bromide) in sperm whale metmyoglobin did not change the Stokes radius of the molecule. Modification of tryptophan-7 and -14, resulted in a change of the Stokes radius from 18.5 Å, for the native molecule, to 34 Å, for the modified protein, indicating swelling or unfolding. In agreement with previous results obtained with succinylated bovine serum albumin (Habeeb, 1967b), the sedimentation coefficients of metmyoglobin modified at tryptophan-7 or at tryptophan-7 and -14 were similar, and shape changes were not detectable. However, metmyoglobin nitrated at all three tyrosines exhibited a Stokes radius identical to that of native protein (Atassi, 1968), indicating no conformational changes associated with this modification. Differences in conformation between human hemoglobin and various artificial hemoglobins were established by Atassi and Skalski (1969) from values of Stokes radii. The Stokes radius of a zinc

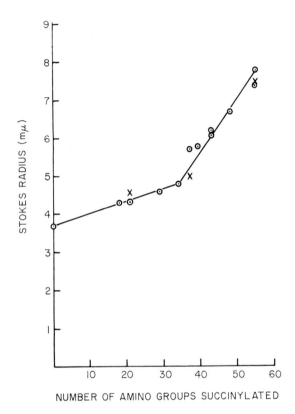

NUMBER OF AMINO GROUPS SUCCINYLATED

Fig. 3. The relationship between the Stokes radius and the number of amino groups succinylated in bovine serum albumin. ($\bigcirc$) Values obtained from a calibrated Sephadex G-200 column and ($\times$) values calculated from diffusion coefficient determined with the ultracentrifuge. (From Habeeb, 1967b.)

metalloporphyrin–globin complex was 34.4 Å compared to 25.5 Å for ferrihemoglobin, but the two proteins behaved identically on sedimentation.

Sephadex chromatography of the products of partial sulfitolysis of bovine serum albumin (Habeeb and Borella, 1966) revealed a heterogenous preparation which consisted of three species: a monomer of Stokes radius of 3.7 mμ and $s_{20,w}^0$ of 4.6 S; an expanded reduced molecule with Stokes radius of 20.2 mμ and $s_{20,w}^0$ of 1.85 S (by virtue of the large Stokes radius and decreased sedimentation); and an aggregate reduced species having a Stokes radius of 20.2 mμ and $s_{20,w}^0$ of 15.9 S. The relative proportion of the expanded and the compact monomer increased with increase in urea concentration at which sulfitolysis was performed. These results were considered evidence of microheterogeneity of bovine serum albumin where some molecules were more susceptible than others to sulfitolysis. Similar behavior was also observed with human serum albumin when reduced with β-mercaptoethanol in 3 M urea for 6 hours (Habeeb, 1968). Two components were obtained, one with Stokes radius of 3.7 mμ and $s_{20,w}^0$ of 4.45 S and an expanded form with Stokes radius of 21 mμ and $s_{20,w}^0$ of 1.15 S. The relative amounts of both components varied in populations obtained by fractionating human serum albumin with 3 M KCl. The results indicate that human serum albumin consists of several populations which vary in their susceptibility to reduction due to differences in pairing the disulfide groups or to other factors that impart internal stabilization to the molecule. Evidence of intermolecular and intramolecular cross-links formed by reacting glutaraldehyde with bovine serum albumin, ovalbumin, and human γ-globulin has been established from Sephadex chromatography and sedimentation coefficient (Habeeb and Hiramoto, 1968). The reaction product showed two peaks—an intermolecularly cross-linked aggregate which eluted at the void volume and an intramolecularly cross-linked monomer.

III. Short-Range Properties

A. Susceptibility of Disulfide Bonds to Reaction

Disulfide bridges play an important role in stabilizing the three-dimensional structure of protein molecules. They bring distant parts of the polypeptide chain in close proximity and thus contribute to the topographical stability of the native molecule. In native proteins, disulfide bonds show limited reactivity but become susceptible to reaction with reducing or oxidizing reagents as the molecule unfolds on denaturation. It was reasoned that if chemical modification caused some conformational reorganization of a protein molecule, changes in the atomic environment of the disulfide groups would result and thus render them accessible for reaction. A chemical method based on this reasoning was devised (Habeeb, 1965, 1966a) to detect conformational changes that accompanied exhaustive modification of the free amino groups of bovine serum albumin and human γ-globulin. Modification was performed by use of a variety of reagents which resulted in modified proteins with a wide distribution in their net electri-

274 A. F. S. A. HABEEB

cal charge. Reagents for the disulfide groups were β-mercaptoethanol, sodium sulfite, and peracetic acid. Figure 4 shows the susceptibility of the disulfide groups to reduction with β-mercaptoethanol in modified bovine serum albumins together with reducible disulfides in unmodified bovine serum albumin as a function of urea concentration. This does not imply

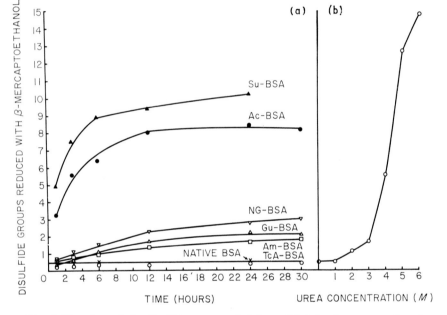

FIG. 4. Reaction of the disulfide groups of (a) modified bovine serum albumins with 0.05 M β-mercaptoethanol in 0.05 M acetate buffer pH 6, at room temperature. (×——×) Native bovine serum albumin (BSA); (○——○) trichloroacetic acid-treated bovine serum albumin (TcA-BSA); (□——□) amidinated bovine serum albumin (Am-BSA); (△——△) Guanidinated bovine serum albumin (Gu-BSA); (▽——▽) nitroguanidinated bovine serum albumin (NG-BSA); (●——●) acetyl bovine serum albumin (Ac-BSA); and (▲——▲) succinyl bovine albumin (Su-BSA). (b) Bovine serum albumin in the presence of urea. (From Habeeb, 1966a.)

that the location of reducible disulfides is the same in modified and urea-treated bovine serum albumin. Conformational reorganization was detected on guanidination and amidination of bovine serum albumin since in these modified proteins 2 disulfide groups were susceptible to reduction. Nitroguanyl bovine serum albumin showed 3, acetyl bovine serum albumin 8, and succinyl bovine serum albumin 10 reducible disulfide groups (Habeeb, 1966a). Although there was some variation in the number of disulfide groups reacting with sodium sulfite or peracetic acid as compared to β-mercaptoethanol, the order was similar. The order of susceptible disulfides was as follows: succinyl bovine serum albumin > acetyl bovine serum albumin > nitroguanyl albumin > amidinated and guanidimated albumin > native bovine serum albumin or trichloroacetic acid-precipitated albumin. This chemical method was sensitive to conformational changes which were not readily detectable from viscosity, sedimentation,

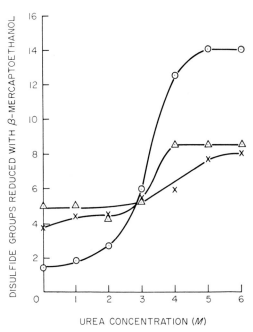

UREA CONCENTRATION (*M*)

Fig. 5. The susceptibility of the disulfide bonds in native and modified bovine serum albumin to reduction with β-mercaptoethanol in absence and presence of urea. (O——O) Bovine serum albumin; (×——×) formaldehyde-treated albumin; and (△——△) glutaraldehyde-treated albumin. (From Habeeb, 1969.)

or the Stokes radius measurements (Section II) as these occur with guanidinated and amidinated bovine serum albumin. Increased susceptibility of the disulfide groups to reduction in exhaustively succinylated human γ-globulin has also been observed (Habeeb, 1965, 1966a; Lenard and Singer, 1966). The ability of reagents to effect intramolecular cross-links was demonstrated in the case of formaldehyde or glutaraldehyde. Modification with these reagents (Habeeb, 1969) resulted in molecular relaxation shown by increased susceptibility of disulfide groups to reduction (Fig. 5), and the modified proteins showed resistance to the unfolding action of urea.

Meaningful information can be derived by observing the effect of chem-

ical modification of a protein on both its overall shape and the reactivity
of the disulfide bonds. An assessment of the interrelation of one parameter
with the other becomes available. The changes in Stokes radii and susceptibility of the disulfide groups as a result of succinylation of bovine serum

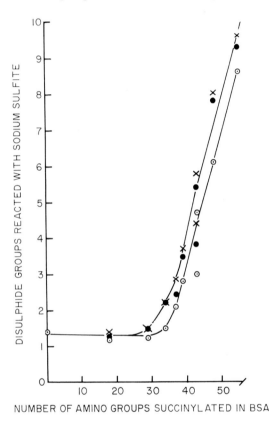

FIG. 6. The number of disulfide groups susceptible to sulfitolysis in bovine serum
albumin as a function of amino groups succinylated. (○) Sulfitolysis for 1 hour; (●)
sulfitolysis for 3 hours; and (×) sulfitolysis for 6 hours. Reaction was performed in
0.1 M sodium phosphate buffer pH 7 containing 0.5 mg ethylenediaminetetraacetate/
ml in 0.05 M sodium sulfite. (From Habeeb, 1967b.)

albumin and human γ-globulin have been examined (Habeeb, 1967b).
Figure 6 shows that succinylation of the first 30 amino groups in bovine
serum albumin does not affect the reactivity of the disulfide groups despite
the observed slight shape changes manifested by an increase in the Stokes
radius (see Fig. 3). However succinylation of the remaining amino groups
resulted in a dramatic increase in the number of the disulfide groups susceptible to sulfitolysis. Based on this evidence, it is reasonable to assume

that the first 30 to 34 amino groups of bovine serum albumin are situated on the surface of the molecule in such a way that succinylation causes only a small increase in the Stokes radius. This limited relaxation of conformation does not affect the accessibility of the disulfide groups. More extensive succinylation results in a dramatic increase in the Stokes radius which

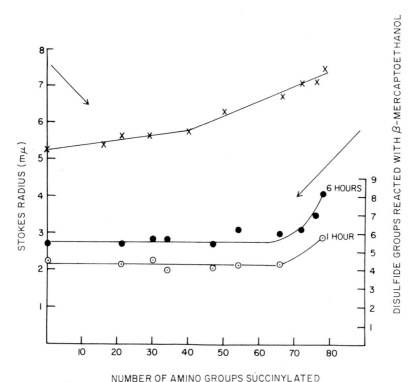

NUMBER OF AMINO GROUPS SUCCINYLATED

FIG. 7. The relationship between the Stokes radius and the number of reducible disulfide groups (reduction in 0.1 M sodium phosphate pH 7 containing 0.5 mg ethylenediaminetetraacetate/ml with 0.05 M β-mercaptoethanol) as a function of amino groups succinylated in γ-globulin. (From Habeeb, 1967b.)

causes increased reactivity of the disulfide groups. Electrostatic repulsions due to the high net negative charge exert a dominant effect and cause unfolding. It is, therefore, possible by limited succinylation to produce minor shape changes despite the increase in net negative charge of the molecule. These modified derivatives were useful to investigate the effect of shape changes on the antigen–antibody precipitates (Habeeb, 1967a). Succinylation of human γ-globulin resulted in conformational reorganization, the magnitude of which was different from that observed in succinylated bovine serum albumin (Fig. 7). Such behavior may reflect characteristic structural features inherent in human γ-globulin. Progressive increase in the Stokes radius accompanied increased succinylation of human γ-globulin. The increase was small with succinylation of up to 40 amino groups; then it was followed by a greater change on further succinylation. On the other hand, succinylation of up to 65 amino groups did not affect the reducibility of the disulfide groups with β-mercaptoethanol. Despite the increase in the net negative electrical charge and the unfolding that took place as a result of succinylating 65 amino groups, no additional disulfides were reducible. These disulfide groups seemed to occupy positions in the core of the γ-globulin molecule, in contrast to what was observed with bovine serum albumin. Lysozyme represents a very stable molecule. Neither addition of detergent nor prolonged heating at 50°C in the presence of detergent significantly affected its optical rotatory properties (Jirgensons, 1963). Moreover, an important degree of residual structure survived 9 M urea (Steiner, 1964). The disulfide groups in exhaustively succinylated lysozyme showed very limted reactivity (Habeeb, 1967b) despite the fact that there is localization of arginine and lysine in the neighborhood of cystine bridges (Canfield, 1963; Jauregui-Adell et al., 1965).

Modification of tyrosine-20 and -23 by tetranitromethane which introduces one nitro group, ortho to the phenolic hydroxyl group (Atassi and Habeeb, 1969), produced conformational changes detectable from the reducibility of one disulfide group by β-mercaptoethanol compared with 0.03 group in the native enzyme. It is remarkable that an increase in molecular weight by 90 due to introduction of 2 nitro groups resulted in perceptible conformational changes. The nature of the added groups was very important, since addition of 7 succinyl groups did not produce as pronounced a conformational change as occurred by introduction of the 2 nitro groups. Tyrosine modification was associated with a decrease of enzymic activity of lysozyme against Micrococcus lysodeikticus. This decrease was interpreted to result from the conformational changes since tryosine-20 and -23 are far removed from the cleft which forms the active site. All the tryptophan residues of lysozyme were modified by reaction with 2-nitrophenyl sulfenyl chloride with concomitant conformational change, as shown by reducibility of 2.6 disulfide groups (Habeeb and Atassi, 1969). There was complete loss of the lytic activity of the modified enzyme. It is not possible to conclude unequivocally whether the loss of activity was due to participation of a tryptophan residue at the active site or was nonspecific resulting from conformational changes. Previous evidence (Hayashi et al., 1965; Hartdegen and Rupley, 1967) showed that tryptophan-62 and -108 were present at the binding site of the enzyme. No attempt has been made in the aforementioned studies to localize the disulfide bonds which become available for reduction in modified proteins. Such studies are feasible when proteins with known amino acid sequence and three-dimensional structure are used. Valuable information can be derived about the region of the molecule where conformational reorganization is favored.

Not only can the chemical method detect conformational changes as a result of chemical modification, but it is also capable of revealing confor-

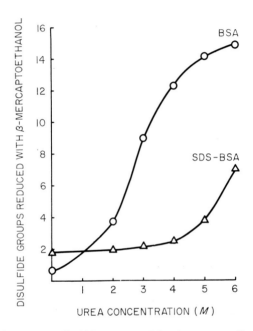

FIG. 8. Reduction of the disulfide groups of bovine serum albumin and sodium dodecyl sulfate–bovine serum albumin (SDS-BSA) complex with 0.05 M β-mercaptoethanol in the presence of different urea concentrations. Reduction in 0.05 M acetate buffer pH 7 containing 0.5 mg ethylenediaminetetraacetate/ml. (From Habeeb, 1966a.)

mational changes resulting in increased rigidity of the molecule. Thus, ligand-induced conformational changes of bovine serum albumin produced by sodium dodecyl sulfate were detected (Habeeb, 1966a). The ligand–protein complex exhibited a considerable amount of stability toward the denaturing action of urea (Fig. 8). Stabilization against denaturation resulted from the formation of detergent bridges between the ϵ-amino groups (Habeeb, 1966c) and hydrophobic regions on the protein. Markus et al. (1964) demonstrated this protective action of detergent on bovine serum albumin from optical rotation measurements.

B. Susceptibility to Proteolytic Digestion

Native proteins occur as compact structures with highly organized folding. Their susceptibility to proteolytic digestion is dependent on the number of amino acid residues in the polypeptide chain accessible to the enzyme and compatible with the specificity of the enzyme. A change in the protein conformation may result in a more compact molecule with decreased rate of proteolysis or a relaxed conformation with a concomitant increase in rate of proteolysis. Both effects were observed in the binding of ligands to proteins and in chemically modified proteins when compared with their native counterpart. Markus (1965) showed that the binding of methyl orange to human serum albumin (1 mole/mole protein) caused a significant decrease in the rate of digestion by chymotrypsin, trypsin, papain, pronase, and subtilisin. Ligand binding to human serum albumin not only affected the rate of digestion but also changed the pathway of digestion as demonstrated from analysis of breakdown products obtained in the presence and absence of ligands (Markus *et al.*, 1967). A large fragment of molecular weight 51,000 which splits from serum albumin by trypsin in absence of methyl orange was found to be greatly reduced with methyl orange–serum albumin complex. The suppression of this formation depended on concentration of methyl orange. In addition, the presence of methyl orange modified the pathway of tryptic digestion as revealed by decrease of smaller fragments. Crystal violet (which belongs to a group of positively charged triphenylmethane dyes) as ligand increased the digestion of serum albumin by trypsin. Thus the binding of human serum albumin by several dyes, e.g., methyl orange, *p*-hydroxy analog of methyl orange, and crystal violet during digestion, besides modifying the digestion rate, had a profound influence on the pathway of digestion. These observations suggested that the modification of conformation by the ligands was dependent on the ligands due to adaptation of the contours of the binding site to the steric requirements of the ligand. This adaptation may involve more than the immediate vicinity of the bound ligand so as to include a conformational rearrangement of the entire protein molecule. Similarly the binding of the strong inhibitor, 2′-cytidylate, to the active center of ribonuclease significantly reduced the rate of formation of ribonuclease S (Markus *et al.*, 1968) by subtilisin. The bond 20–21 which is the site of action of subtilisin and is present in a region remote from the binding site became resistant to proteolytic attack. Analysis of amino terminal residues of trypsin- and chymotrypsin-treated ribonuclease and 2′-cytidylate-bound ribonuclease revealed changes in conformation of ribonuclease at nine defined places (Markus *et al.*, 1968), and none of the identified bonds cleaved by the proteases were in the immediate vicinity of the active center.

Aspartate transcarbamylase acquired a "relaxed" conformation (Mc-Clintock and Markus, 1968) in the presence of the substrate aspartate shown by increased rate of digestion. In the presence of adenosine 5'-triphosphate (an allosteric activator) or cytidine 5'-triphosphate (an inhibitor), a "constrained" conformation resulted which showed decreased digestibility by trypsin, subtilisin, pronase, and, to a smaller extent, by chymotrypsin.

Various chemically modified proteins have shown increased digestibility by proteolytic enzymes, depending on the enzyme and the type of modification. Removal of sialic acid from sheep plasma glycoprotein (Anantha Samy, 1967) and esterification of bovine serum albumin (Sri Ram and Maurer, 1959) resulted in an increased proteolysis with trypsin and chymotrypsin of the modified proteins compared to the corresponding native protein. Oxygenation of hemoglobin resulted in the increased digestibility with carboxypeptidase, and the effect was greater for the β than for the α chain (Zito et al., 1964). Modification of 11 free amino groups of the 19 amino groups in Takaamylase by 2-methoxy-5-nitrotropane resulted in increased digestion by subtilopeptidase A as compared to the native enzyme (Tamaoki et al., 1968). Removal of 5-nitrotroponyl residues with hydrazine resulted in recovery of the native conformation evidenced by the resistance of the recovered protein to proteolytic digestion (Tamaoki et al., 1968).

Knowledge of the primary structure of the protein, when combined with identification of the peptides produced by proteolysis (with enzymes of restricted specificities) of the modified protein, yields valuable information about the peptide bonds that became exposed as a result of the conformational changes on modification. Atassi and Habeeb (1969) showed that modification of tyrosine-20 and -23 in lysozyme with tetranitromethane resulted in release by trypsin of the following amino acids and peptides (Fig. 9) (lysozyme released only leucine-129) which were identified from fingerprint maps: lysine-1 and -97; arginine-14; leucine-129; asparaginyl–arginine 113–114; sequence 69–73; sequence 97–112; a spot with sequence 15–21, 1–5, or both, and finally a spot probably corresponding to sequence 98–112. Eight out of nine peptides released are located on one side of the lysozyme molecule relative to the active site of the enzyme. It is reasonable to expect that modification of the tyrosines would induce displacements of parts of the polypeptide chain with respect to one another, in the region of modification. It is noteworthy, however, to point out that the bonds Arg-68—Thr-69, Arg-73—Asn-74, which are accessible, are present on the opposite side of the molecule to the region of modification. Therefore the reorganization of the immediate region around the modified tyrosines apears to have induced a conformational change in a more distant region of the molecule.

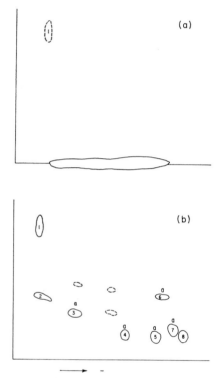

FIG. 9. Peptide maps of tryptic hydrolyzates of (a) lysozyme and (b) lysozyme modified at tyrosine-20 and -23. Tryptic hydrolysis was carried out on the proteins *without* reduction and carboxymethylation of the disulfide bonds. The spots corresponded to the following sequences. (1) position 129 (leucine); (2) possibly sequence 98–112; (3) sequence 97–112; (4) sequence 69–73; (5) sequence 113–114; (6) sequence 15–21 and/or sequence 1–5 which often overlapped; (7) position 14 (arginine); (8) position 1 or/and 97 (lysine). Spots denoted by (a) were positive with Sakaguchi stain for arginine. Spots outlined by broken line appeared only when papers were overloaded. (From Atassi and Habeeb, 1969.)

The loss of enzymic activity (50%) of the modified lysozyme was most likely the result of conformational changes. None of the tyrosyl residues in lysozyme has been found to be at the binding site of the inhibitor *N*-acetyl glucosamine or its dimer by X-ray crystallography (Blake *et al.*, 1965). However it is clear that the segment 99–114 underwent some conformational reorganization upon modification of tyrosine-20 and -23. This segment carries almost all of the helix between residues 108 through 115 and includes tryptophan-108 which is present in the binding site of the enzyme (Blake et al., 1965; Johnson and Phillips, 1965; Hayashi et al., 1965). Disturbance of the native conformation of segment 108–115 upon modification may contribute to loss of lytic activity.

All the tryptophan residues of lysozyme were modified by reaction with 2-nitrophenyl sulfenyl chloride (Habeeb and Atassi, 1969) with complete loss of activity. This modification was associated with conformational changes since several peptide bonds became available for tryptic digestion with release of the following amino acids and peptides: lysine-1 and/or lysine-97; arginine-14; leucine-129; sequences 113–114, 69–73; 15–21, and 34–45. The loss of activity on modifying tryptophan cannot be considered to imply the essentiality of tryptophan residue since considerable conformational changes have taken place and these in themselves would disrupt the binding site.

It is observed that modification of tryosine or tryptophan in lysozyme exposed some identical peptide bonds for proteolysis in addition to those that were characteristic for modification. X-ray crystallography of modified proteins is of importance in revealing the atomic details of these conformational changes.

The peptides released by tryptic digestion of the modified lysozymes may not necessarily reflect bonds that became simultaneously accessible due to conformational changes. There is the possibility that some bonds may be hydrolyzed in larger fragments that were initially released by cleavage of accessible bonds in the protein. The results are still helpful in revealing the general location of the most susceptible region in the derivative.

It is noteworthy that photooxidation of lysozyme sensitized by methylene blue (Jori et al., 1968) converted the 2 methionine residues into methionine sulfoxide, thus introducing a polar center in place of the thioether function. The modified lysozyme was found to lose about 95% of the lytic activity which was attributed to conformational changes. These conformational changes caused disruption of the native structure to expose nine peptide bonds to tryptic activity compared to one peptide bond in the native enzyme. However, the segment of the polypeptide chain involved in the conformational change may contain some tryptophyl residues indicated by the difference absorption spectrum. There was no analysis of the cleaved peptides by trypsin in methionine-modified lysozyme, and so the site of conformational changes could not be determined. Both the enzymic and native structure were recovered by reducing the methionine sulfoxide to methionine (Jori et al., 1968).

C. IMMUNOCHEMICAL PROPERTIES

The immunochemical behavior of globular proteins is strongly influenced by changes in the native conformation. Antibody response is mostly directed against the three-dimensional structure of the protein molecule. In globular proteins, antigenic determinants are formed of regions which may be distant in sequence but which are brought to close proximity by the folding of the polypeptide chain. Any change in the conformation of a protein would result in a decrease in the ability of the modified protein to react with antibodies to the native protein. The reaction of antigen with the homolgous or heterologous antibodies is conveniently determined from quantitative precipitin curves.

Quantitative precipitin tests are performed by adding increasing quantities of the antigen in a given volume (usually 200 μl–1 ml) to an equal volume of antiserum. The reaction mixture is incubated at 37°C for 30 minutes and then at 4°C for 2 to 3 days. The immune precipitate is washed three times with saline, and the precipitate is dissolved in 3 ml 0.1 N NaOH and the absorbance of the solution determined at 280 mμ.

Reduction or oxidation of the disulfide bonds of ribonuclease which produced disruption of the three-dimensional structure resulted in decreased ability of modified ribonuclease to react with antiribonuclease antiserum (Brown et al., 1959; Brown, 1962). Scission of seven disulfide groups in bovine serum albumin with sodium sulfite (Habeeb and Borella, 1966) resulted in 80% reduction of ability of modified bovine serum albumin to react with antibovine serum albumin antiserum. Modification of the amino groups of bovine serum albumin by succinylation, guanidination, nitroguanidination, and amidination (Habeeb, 1967a) resulted in a decrease in the ability of modified serum albumins to precipitate with antibovine serum albumin antiserum. The decrease in precipitated antibody varied between 9 and 15% with amidinated, guanidinated, and nitroguanidinated bovine serum albumins. These modifications resulted in minimal conformational changes (Sections II, C and III, A). Changes in the net electrical charge of the modified albumin did not seem to be important if conformational changes were minimal. A correlation existed between the Stokes radii of succinylated bovine serum albumins and the amount of antibody precipitated at equivalence (Fig. 10). Maximum precipitation of antibody occurred with the homologous system and was followed by a slow decrease in the amount of precipitated antibody with an increase or decrease in the Stokes radius of the antigen to a point after which a dramatic decrease in the amount of precipitated antibody was observed. This decrease may result from changes in the topography of the antigen molecule which would inhibit or partly destroy some of the sites responsible for the complementarity with the antibody.

Modification of tyrosine-20 and -23 in lysozyme (Atassi and Habeeb,

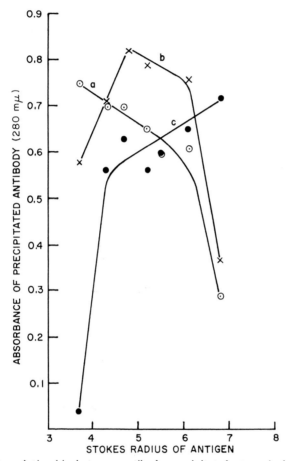

FIG. 10. The relationship between antibody precipitated at equivalence and the Stokes radius in millimicrons of different preparations of succinylated bovine serum albumin by using (a) antibovine serum albumin antiserum, (b) antisuccinyl (32) bovine serum albumin, and (c) antisuccinyl (57) bovine serum albumin. The numbers in parentheses refer to the number of amino groups succinylated. (From Habeeb, 1967a.)

1969) resulted in a loss of 10 to 23% of the ability of the modified lysozyme to precipitate with antilysozyme antiserum as compared to native lysozyme. This loss of precipitability was attributed in part to conformational changes since these were manifested by proteolysis and susceptibility of the disulfide bonds to reduction. Extensive conformational changes of lysozyme modified at tryptophan (Habeeb and Atassi, 1969) resulted in an 82% decrease of ability of modified lysozyme to react with antilysozyme antiserum. In sperm whale myoglobin, where the heme group is not part of an antigenic site, derivatives were prepared (Atassi, 1967) with various modified porphyrins and metalloporphyrins. Therefore, in such models the modification clearly did not involve any antigenic reactive regions of the protein, and conformational alterations were imposed as a result of the different coordination tendencies of the various metals or the modification of the side chains of heme. The presence of these conformational changes was confirmed by studying the antigenic reactivities of these derivatives (Atassi, 1967).

The influence of conformational changes on antigenic reactivity can be better understood now that the first antigenic structure of a globular protein (i.e., myoglobin) has become available (Atassi and Saplin, 1968; Atassi and Thomas, 1969). It was found by these authors that the antigenic regions occupied some discrete surface segments on corners between helical portions of the molecule. No reactive regions were found in the interior of the polypeptide chain. Atassi and Thomas (1969), therefore, suggested that the antibody response is directed primarily against the native three-dimensional structure of the protein. It is quite likely that this is applicable to other proteins. Changes in conformation that will give rise to distortion of the native shape of some reactive regions or to dislocation of reactive regions with respect to one another will influence the antigenic reactivity. The native shape of each reactive region and the mode of approach of these regions in the three-dimensional structure are essential for appropriate fit onto the antibody-combining site.

IV. Conclusions

The dependence of the biological activity of proteins on the presence of certain functional groups to carry out the catalytic function as well as on the native conformation has been recognized. In cases where correlation of the nature of functional groups (by chemical modification) with the biological activity is desired, it is essential to establish whether conformational changes have taken place. This problem can be approached by determining (a) changes in overall shape and (b) changes in short-range properties. Among the methods used to determine shape changes, viscosity measurements are only useful when the protein is homogeneous and is represented by one molecular species. The presence of more than one molecular species has been indicated in the reaction product of succinylation of bovine serum albumin, reduction of human serum albumin, and sulfitolysis of bovine serum albumin. Sedimentation measurement is less sensitive and may not be able to detect minor conformational changes. The most sensitive method to detect shape changes is determination of the Stokes radius on a calibrated Sephadex column. It requires readily available material and is capable of detecting, separating, and determining the Stokes radius of each molecular species in cases where molecular heterogeneity was present. This has been shown in succinylated bovine serum albumin, sulfitolysis of bovine serum albumin, and reduction of human serum albumin. Moveover it is very sensitive and is capable of detecting shape changes that were not detected from sedimentation measurements, e.g., metmyoglobin modified at trypophan-7 and -14. The Stokes radius measurement should prove valuable in following changes in overall shape of a modified macromolecule.

Short-range properties have been found to detect in some cases conformational changes in modified proteins that were not detectable from changes in shape properties, e.g., conformational changes in amidinated and guanidinated bovine serum albumin. The reactivity of the disulfide group was found to be a valuable tool in probing the conformational changes associated with various modifications of several proteins and also changes associated with binding of sodium dodecyl sulfate to bovine serum albumin. The use of this method to detect conformational changes associated with the binding of substrates and inhibitors to enzymes or haptens with antibodies has not been explored, but it should prove useful. It would be useful to identify the susceptible disulfide groups (due to conformational changes) by working with proteins of known primary structure and analyzing the peptide maps to localize the site of conformational reorganization.

Susceptibility to proteolysis of modified proteins of known amino acid sequence accompanied by identification of the peptides liberated has been found useful to evaluate and localize the site of conformational changes. Some valuable information has been obtained by the technique. The addition of two nitro groups to tyrosine residues 20 and 23 of lysozyme resulted in conformational reorganization which exposed peptide bonds distant from the site of modification. This method is sensitive and simple to perform.

The value of these various methods to evaluate conformational changes associated with chemical modification has been ascertained either separately or jointly and should be useful in determining the essential nature of a particular functional group to biological activity.

REFERENCES

Ackers, G. K. (1964). *Biochemistry* **3**, 723.
Anantha Samy, T. S. (1967). *Arch. Biochem. Biophys.* **121**, 703.
Andrews, P. (1964). *Biochem. J.* **71**, 222.
Anfinsen, C. B., and Haber, E. (1961). *J. Biol. Chem.* **236**, 1361.
Atassi, M. Z. (1967). *Biochem. J.* **103**, 29.
Atassi, M. Z. (1968). *Biochemistry* **7**, 3078.
Atassi, M. Z., and Caruso, D. R. (1968). *Biochemistry* **7**, 699.
Atassi, M. Z., and Habeeb, A. F. S. A. (1969). *Biochemistry* **8**, 1385.
Atassi, M. Z., and Saplin, B. J. (1968). *Biochemistry* **7**, 688.
Atassi, M. Z., and Skalski, D. J. (1969). *Immunochemistry* **6**, 25.
Atassi, M. Z., and Thomas, A. V. (1969). *Biochemistry* **8**, 3385.
Blake, C. C. F., Koenig, D. F., Mair, G. A., North, A. C. T., Phillips, D. C., and
    Sarma, V. R. (1965). *Nature (London)* **206**, 757.
Brand, L., and Witholt, D. E. (1967). *Methods Enzymol.* **11**, 715.
Breslow, E. (1964). *J. Biol. Chem.* **239**, 486.
Breslow, E., Beychock, S., Hardman, K. D., and Gurd, F. R. N. (1965). *J. Biol.
    Chem.* **240**, 304.
Brown, R. K. (1962). *J. Biol. Chem.* **237**, 1162.
Brown, R. K., Duriex, J., Delany, R., Leikhim, E., and Clark, B. J. (1959). *Ann. N. Y.
    Acad. Sci.* **81**, 524.
Canfield, R. E. (1963). *J. Biol. Chem.* **238**, 2698.
Cohen, L. A. (1968). *Annu. Rev. Biochem.* **37**, 695.
Cohn, E. J., and Edsall, J. T. (1943). "Proteins, Amino Acids and Peptides," p. 428.
    Reinhold, New York.
Crumpton, M. J., and Polson, A. (1965). *J. Mol. Biol.* **11**, 722.
Di Sabato, G., and Ottesen, M. (1967). *Methods Enzymol.* **11**, 734.
Einstein, A. (1906). *Ann. Physik.* [4] **19**, 289.
Einstein, A. (1911). *Ann. Physik.* [4] **34**, 591.
Epstein, C. J., and Anfinsen, C. B. (1962). *J. Biol. Chem.* **237**, 3464.
Epstein, C. J., and Goldberger, R. F. (1963). *J. Biol. Chem.* **238**, 1380.
Epstein, C. J., Anfinsen, C. B., and Sela, M. (1962). *J. Biol. Chem.* **237**, 3458.
Epstein, C. J., Goldberger, R. F., and Anfinsen, C. B. (1963). *Cold Spring Harbor
    Sym. Quant. Biol.* **28**, 439.
Frattali, V., Steiner, R. F., Millar, D. B. S., and Edelhoch, H. (1963). *Nature (London)*
    199, 1186.
Freedman, M. H., and Sela, M. (1966). *J. Biol. Chem.* **241**, 5225.
French, D., and Edsall, J. T. (1945). *Advan. Protein Chem.* **2**, 277.
Fuchs, S., and Sela, M. (1965). *J. Biol. Chem.* **240**, 3558.
Glazer, A. N., Bar-Eli, A., and Katchalski, E. (1962). *J. Biol. Chem.* **237**, 1832.
Goldberger, R. F., and Epstein, C. J. (1963). *J. Biol. Chem.* **238**, 2988.
Gounaris, A., and Ottesen, M. (1965). *C. R. Trav. Lab. Carlsberg* **35**, 37.
Grubhofer, N., and Schleith, L. (1954). *Hoppe-Seyler's Z. Physiol. Chem.* **297**, 108.
Habeeb, A. F. S. A. (1959). *Biochim. Biophys. Acta* **34**, 274.
Habeeb, A. F. S. A. (1960). *Can. J. Biochem. Physiol.* **38**, 493.
Habeeb, A. F. S. A. (1961). *Can. J. Biochem. Physiol.* **39**, 729.
Habeeb, A. F. S. A. (1964). *Biochim. Biophys. Acta* **93**, 533.
Habeeb, A. F. S. A. (1965). *Fed. Proc., Fed. Amer. Soc. Exp. Biol.* **24**, 224 (abstr.).
Habeeb, A. F. S. A. (1966a). *Biochim. Biophys. Acta* **115**, 440.
Habeeb, A. F. S. A. (1966b). *Biochim. Biophys. Acta* **121**, 21.

Habeeb, A. F. S. A. (1966c). *Anal. Biochem.* **14**, 328.

Habeeb, A. F. S. A. (1967a). *J. Immunol.* **99**, 1264.

Habeeb, A. F. S. A. (1967b). *Arch. Biochem. Biophys.* **121**, 652.

Habeeb, A. F. S. A. (1967c). *Arch. Biochem. Biophys.* **119**, 264.

Habeeb, A. F. S. A. (1968). *Can. J. Biochem.* **46**, 789.

Habeeb, A. F. S. A. (1969). *J. Immunol.* **102**, 457.

Habeeb, A. F. S. A., and Atassi, M. Z. (1969). *Immunochemistry* **6**, 555.

Habeeb, A. F. S. A., and Borella, L. (1966). *J. Immunol.* **97**, 951.

Habeeb, A. F. S. A., and Hiramoto, R. (1968). *Arch. Biochem. Biophys.* **126**, 16.

Hebeeb, A. F. S. A., Cassidy, H. G., and Singer, S. J. (1958). *Biochim. Biophys. Acta* **29**, 587.

Habeeb, A. F. S. A., Cassidy, H. G., Stelos, P., and Singer, S. J. (1959). *Biochim. Biophys. Acta* **34**, 439.

Harrison, S. C., and Blout, E. R. (1965). *J. Biol. Chem.* **240**, 299.

Hartdegen, F. J., and Rupley, J. A. (1967). *J. Amer. Chem. Soc.* **89**, 1743.

Hayashi, K., Imoto, T., Fanatsu, G., and Fanatsu, M. (1965). *J. Biochem.* (Tokyo) **58**, 227.

Herskovitz, T. T. (1967). *Methods Enzymol.* **11**, 748.

Hjerten, S., and Mosbach, R. (1962). *Anal. Biochem.* **3**, 109.

Horton, H. R., and Koshland, D. E., Jr. (1967). *Methods Enzymol.* **11**, 856.

Isemura, T. T., Takagi, V., Maeda, V., and Yutani, K. (1963). *J. Biochem.* (*Tokyo*) **53**, 155.

Jauregui-Adell, J., Jolles, J., and Jolles, P. (1965). *Biochim. Biophys. Acta* **107**, 97.

Jirgensons, B. (1963). *J. Biol. Chem.* **238**, 2716.

Johnson, L. N., and Phillips, D. C. (1965). *Nature* (*London*) **206**, 761.

Jori, G., Galiazzo, G., Marzotto, A., and Scoffone, E. (1968). *J. Biol. Chem.* **234**, 4272.

Katsoyannis, P. G., and Tometsko, A. (1966). *Proc. Nat. Acad. Sci. U.S.* **55**, 1554.

Kauzmann, W. (1959). *Advan. Protein Chem.* **14**, 1.

Klee, W. A., and Richards, F. M. (1957). *J. Biol. Chem.* **229**, 489.

Kragh, A. M. (1961). *In* "A Laboratory Manual of Analytical Methods of Protein Chemistry" (P. Alexander and R. J. Block, eds.), Vol 3., p. 179. Pergamon Press, Oxford.

Kronman, M. J., and Foster, J. F. (1957). *Arch. Biochem. Biophys.* **72**, 205.

Laurent, T. C., and Killander, S. (1964). *J. Chromatogr.* **14**, 317.

Leach, A. A., and O'Shea, P. C. (1964). *J. Chromatogr.* **17**, 211.

Lenard, J., and Singer, S. J. (1966). *Nature* (*London*) **210**, 536.

Levin, Y., Pecht, M., Goldstein, L., and Katchalski, E. (1964). *Biochemistry* **3**, 1905.

Levine, L., and Van Vunakis, H. (1967). *Methods Enzymol.* **11**, 928.

Levinthal, C., Singer, E. R., and Fetherolf, K. (1962). *Proc. Nat. Acad. Sci. U.S.* **48**, 1230.

McClintock, D. K., and Markus, G. (1968). *J. Biol. Chem.* **243**, 2855.

Markus, G. (1965). *Proc. Nat. Acad. Sci. U. S.* **54**, 253.

Markus, G., Love, R. L., and Wissler, F. (1964). *J. Biol. Chem.* **239**, 3687.

Markus, G., McClintock, D. K., and Castellani, B. A. (1967). *J. Biol. Chem.* **242**, 4402.

Markus, G., Barnard, E. A., Castellani, B. A., and Saunders, D. (1968). *J. Biol. Chem.* **243**, 4070.

Mitz, M. A., and Summaria, L. J. (1961). *Nature* (*London*) **189**, 576.

Nazaki, Y., and Tanford, C. (1967). *Methods Enzymol.* **11**, 715.

Pedersen, K. O. (1958). *J. Phys. Chem.* **62**, 1282.

Pedersen, K. O. (1962). *Arch. Biochem. Biophys.* Suppl. 1, 157.

290	A. F. S. A. HABEEB

3
Perrin, G. (1936). *J. Phys. Radium* **7**, 1.
Phillips, D. C. (1966). *Sci. Amer.* **215**, 78.
Porath, J. (1962). *Advan. Protein Chem.* **17**, 209.
Porath, J., and Flodin, P. (1959). *Nature (London)* **183**, 1657.
Quiocho, F. A., and Richards, F. M. (1964). *Proc. Nat. Acad. Sci. U. S.* **52**, 833.
Robbins, F. M., Kronman, M. J., and Andreotti, R. E. (1965). *Biochim. Biophys. Acta* **109**, 223.
Roubal, W. T., and Tappel, A. L. (1964). *Anal. Biochem.* **9**, 211.
Rupley, J. A. (1967). *Methods Enzymol.* **11**, 905.
Saroff, H. A., and Evans, R. L. (1959). *Biochim. Biophys. Acta* **36**, 511.
Schachman, H. K. (1959). "Ultracentrifugation in Biochemistry," p. 90. Academic Press, New York.
Schellman, J. A., and Schellman, C. (1964). *In* "The Proteins" (H. Neurath, ed.), 2nd ed., Vol. 2, p. 1. Academic Press, New York.
Scheraga, H. A. (1963). *In* "The Proteins" (H. Neurath ed.), 2nd ed., Vol. 1, p. 477. Academic Press, New York.
Scheraga, H. A., and Mandelkern, L. (1953). *J. Amer. Chem. Soc.* **75**, 179.
Shields, G. S., Hill, R., and Smith, E. L. (1959). *J. Biol. Chem.* **234**, 1747.
Siegel, L. M., and Monty, K. J. (1965). *Biochem. Biophys. Res. Commun.* **19**, 494.
Siegel, L. M., and Monty, K. J. (1966). *Biochim. Biophys. Acta* **112**, 346.
Slobin, L. I., and Singer, S. J. (1968). *J. Biol. Chem.* **243**, 1777.
Sri Ram, J., and Maurer, P. H. (1959). *Arch. Biochem. Biophys.* **85**, 512.
Steiner, R. F. (1964). *Biochim. Biophys. Acta* **79**, 51.
Tamaoki, H., Murase, Y., Minato, S., and Nakanishi, K. (1967). *J. Biochem. (Tokyo)* **62**, 7.
Whitaker, J. R. (1963). *Anal. Chem.* **35**, 1950.
White, F. H., Jr. (1961). *J. Biol. Chem.* **236**, 1353.
Winzer, D. J., and Nichol, L. W. (1965). *Biochim. Biophys. Acta* **104**, 1.
Wofsy, L., and Singer, S. J. (1963). *Biochemistry* **1**, 104.
Yang, J. T., and Doty, P. (1957). *J. Amer. Chem. Soc.* **79**, 761.
Zito, R., Antonini, E., and Wyman, J. (1964). *J. Biol. Chem.* **239**, 1804.

Author Index

Numbers in italics refer to the pages on which the complete references are listed.

A

Abdulla, Y. H., 166, *171*
Ablett, S., *183*
Abramson, M. B., 85, 86, *171*
Ackers, G. K., 79, *176*, 267, 268, 269, *288*
Adlfinger, K. H., 23, 24, 25, 26, 159, *180*
Adolph, E. F., *183*
Ager, D. V., 137, *172*
Agnihotri, V. P., 120, *172*
Aida, K., 190, 206, *254*
Aidanova, O. S., 38, 124, 157, *179*
Albu-Weissenberg, M., 192, 195, 221, 232, 241, *257*
Alexander, B., 192, 195, 196, 241, 249, 250, 251, *252*, *254*, *257*
Almeida, J. P., *183*
Anantha Samy, T. S., 264, 281, *288*
Andjus, R. K., 146, *172*
Andreotti, R. E., 267, *290*
Andrewartha, H. G., 121, *172*
Andrews, P., 267, 269, *288*
Anfinsen, C. B., 189, 193, 233, 234, *254*, 260, *288*
Angelakos, E. T., 148, *172*
Antonini, E., 281, *290*
Apffel, C. A., 148, 158, *172*
Arnett, E. M., 18, 72, *172*
Arnold, R., 246, 247, *252*
Arnold, W. N., 190, 201, *253*
Arsenis, C., 189, *253*
Ascherson, *253*
Asunmaa, S. K., 93, *180*
Atassi, M. Z., 269, 271, 272, 278, 281, 282, 283, 285, 286, *288*, *289*
Augenstein, L. G., 186, 190, 202, *255*, *256*
Augustine, E. S., 190, *253*
Augustin, J., 197, 247, *254*
Avrameas, S., 191, 192, 194, 217, 238, 240, 251, 252, *253*, *254*, *257*
Axen, R., 194, 195, 245, 246, 247, *253*, *257*

B

Babcock, K. L., 201, *256*
Bach, S. A., 143, 162, 163, *172*
Baldwin, J., 132, 133, *172*
Baldwin, J. J., 80, *172*
Bangham, A. D., 28, 100, 101, *172*, *177*
Banks, T. E., 193, 198, 225, *253*
Bar-Eli, A., 195, 241, 242, 243, 247, 249, *253*, *254*, 259, *288*
Bargoot, F. G., *181*
Barker, R., 77, 78, *178*
Barnard, E. A., 280, *289*
Barnett, L. B., 188, 190, 199, *253*
Barrall, E. M., 84, *172*
Barry, P. H., 92, *172*
Bateman, J. B., 190, *255*
Bauman, E. K., 190, 206, 211, *253*
Bean, R., 86, *172*
Becker, G., 117, *174*
Becker, M., 95, *182*
Becker, W., 190, 206, *253*
Belding, H. S., 154, *172*
Bello, J., 192, *254*
Ben-Naim, A., 11, 18, 41, 52, 56, *172*
Berendsen, H. J. C., 35, 51, 52, 53, *172*
Bergqvist, E., *10*
Berlin, E., *183*
Bernal, J. D., 6, 48, *172*
Bernfeld, P., 191, 209, 210, 213, 214, *253*
Bieber, R. E., 191, 210, 213, 214, *253*
Bier, M., 223, *257*
Biltonen, R., 66, 69, 73, 82, *178*
Binet, 148, *172*
Birch, L. C., 121, *172*
Bishop, W. H., 192, 219, *253*, *257*
Blake, C. C. F., 282, 283, *288*
Blanchard, K. C., *183*
Blandamer, M. J., 11, 18, 45, *172*, *173*, *181*
Blei, 89, *179*

291

Cussler, E. L., 105, *182*
Cuthbert, A. W., *183*
Cyr, T. J. R., 84, *174*

D

Dale, E. C., 193, 227, *258*
Dalton, T., 88, *174*
Damadian, R., 39, *174*
Damaschke, von K., 117, *174*
Danford, M. D., *6, 7*
Danielli, J. F., 85, 87, 88, 92, *174, 176*
Das Gupta, 166, *174*
Davey, C. B., 120, 121, *174*
David, G., 149, *173*
Davies, J. T., 186, *254*
Davis, C. M., Jr., *7*
Davis, R. C., 192, 194, 251, *254*
Davson, H., 85, 87, 88, *174*
Dawe, A. R., 149, *178*
Dawson, R. M. C., 86, *174*
Day, A. F., *183*
Day, R. A., 192, 215, 216, 222, *255*
Deborin, G. A., 190, 201, *256*
De Bruijne, A. W., 43, *174*
de Greiff, H. J., 102, *176*
DeHaven, J. C., 44, 114, 148, *174*
Delany, R., 284, *288*
Del Bene, J., 5, 59, *174*
Derjaguin, B. V., 23, 38, 124, *174*
Desnoyers, J. E., 9, *174*
De Sylva, D. P., 133, 164, *174*
DeTar, D. F., 226, *254*
Deutsch, I. S., 148, *172*
Devaux, M. H., 186, *254*
Diamond, J. M., 90, 91, 92, *174, 182*
Diana, A. L., 86, *182*
Dick, D. A. T., 39, *174*
Di Sabato, G., 263, *288*
Ditscher, W., 215, *258*
Dixon, M., 75, *174*
Dodt, E., 109, *174*
Doty, P., 261, *290*
Douraghi-Zadeh, K., 226, *256*
Doyle, R. J., 192, *254*
Dreyer, G., 25, 26, 27, *174*
Drobnica, L., 197, 247, *254*
Drost-Hansen, W., 9, 11, 23, 24, 25, 28, 30, 31, 34, 41, 49, 51, 81, 85, 89, 100, 102, 103, 105, 106, 112, 113, 115, 119, 120, 122, 124, 125, 126, 127, 133, 146, 155, 164, 165, *174, 175, 178, 179, 180*

Duby, P., 89, 100, *180*
Dunant, Y., *183*
Dunnell, B. A., 84, *174*
Duriex, J., 284, *288*

E

Eaks, I. L., *123*
Edelhoch, H., 260, *288*
Edelhoch, H. E., 190, *255*
Edner, O. J., *181*
Edsall, J. T., 259, 266, *288*
Edwards, B. A., 193, 194, 237, *258*
Egelstaff, P. A., 20, *175*
Ehrenberg, A., 39, *175*
Einstein, A., 263, *288*
Eisenberg, D., 3, 31, 34, 138, *175*
Eisentraut, M., 149, *175*
Ekwall, P., 83, *178*
Elsner, J., 155, *182*
Engel, A., 196, *254*
Epstein, C. J., 193, 233, 234, *254*, 259, 260, *288*
Erlanger, B. F., 190, 197, 203, *254*
Ernback, S., 194, 245, 246, *253, 257*
Erpenbeck, J., 25, 26, 27, *174*
Estermann, E. F., 190, 201, 202, *254, 256*
Evangelopoulous, A., 192, *257*
Evans, M. W., 3, *10, 175*
Evans, R. L., 267, *290*
Ewers, J., 193, 230, *256*
Eyring, H., *7*, 75, 76, 82, 83, *177*

F

Falb, R., 191, 192, *256*
Falk, M., 75, 95, *175*
Fanatsu, G., 278, 283, *289*
Fanatsu, M., 278, 283, *289*
Farrell, E. F., 9, *182*
Farrell, J., 121, 136, *175*
Fasman, G. D., 247, *255*
Fetherolf, K., 260, *289*
Ficalbi, A., 25, 28, *173*
Fischer, E. H., 80, *178, 180*
Flautt, T. J., 84, *175*
Fleischer, S., 86, *180*
Flodin, P., 267, *290*
Foerster, H. J., 193, 196, 235, *256*
Fogg, G. E., 134, *175*
Fontell, K., 83, *178*
Forslind, E., *6, 10*, 24, 25, 104, *175*

M

Maaløe, O., 132, *177*
McClintock, D. K., 280, 281, *289*
McCormick, D. B., 189, *253*, *256*
MacDonnell, P. C., 191, 213, *253*
McElhaney, R. N., *181*
McKelvey, D. R., 18, 72, *172*
McLaren, A. D., 190, 201, 202, 203, 204, *254*, *256*
McLauchlan, K. A., 39, *173*
McWilliam, J. R., 126, *178*
Maeda, H., 190, 202, *258*
Maeda, V., 260, *289*
Mair, G. A., 282, 283, *288*
Mak, H. D., 21, *176*
Mak, T. C. W., 48, *178*
Malik, S. K., 18, *182*
Malmstrom, B. G., 39, *175*
Mandelkern, L., 266, *290*
Mandell, L., 83, *178*
Mandels, M., 190, 201, *257*
Manecke, G., 189, 190, 193, 196, 197, 223, 225, *256*
Mannick, J. A., 192, 194, 251, *254*
Manoilov, S. E., 194, 195, *258*
Mardeshev, S. R., 190, 205, *257*
Marfey, P. S., 192, 216, 218, *256*
Margottini, M., 155, *173*
Markovitz, A., 80, *178*
Markus. G., 279, 280, 281, *289*
Martin-Löf, S., 85, *179*
Marzotto, A., 283, *289*
Massey, V., 78, *179*
Mather, A., 193, 194, 195, 229, 237, *255*
Maurer, P. H., 223, *257*, 281, *290*
Mayer, H., 223, *258*
Mazur, P., 150, 151, 152, 153, *179*
Meinhofer, J., 215, *258*
Meryman, H. T., 69, 153, *179*
Messing, R. A., 190, 205, *256*
Metsik, M. S., 38, 124, 157, *179*
Meyer, G. H., 29, *182*
Meyer, H. H., 80, *179*
Micheel, F., 193, 230, *256*
Migchelsen, C., 52, 53, *172*
Mikhailov, V. A., 11, *179*
Miller, D. B. S., 260, *288*
Miller, J. A., Jr., 146, *179*
Miller, R. J., 104, 120, 121, *174*

Miller, S. L., 85, 138, *179*
Millero, F. J., 9, 28, *178*, *179*
Minanikawa, T., *123*
Minato, S., 281, *290*
Mishra, R. K., 90, *173*
Mitchell, H. K., 120, *179*
Mitchell, P., 222, *256*
Mitz, M. A., 190, 193, 194, 195, 199, 202, 230, 235, 238, 240, 241, *256*, 259, *289*
Mkrtumova, N. A., 190, 201, *256*
Moelwyn-Hughes, E. A., 85, 103, *177*
Mondovi, B., 155, *173*
Money, C., 193, 194, 195, 234, 240, 241, *256*
Montalvo, J. G., Jr., 190, 191, 213, *255*
Monty, K. J., 267, 268, *290*
Morcom, K. W., *173*
Morgan, J., *6*
Mori, T., 190, 205, *258*
Moricca, G., 155, *173*
Morris, L. L., *123*
Mosbach, K., 191, 212, *256*
Mosbach, R., 191, 212, *256*, 267, *289*
Moskowitz, J. W., 9, *176*
Mover, P., 238, *253*
Mowbray, J. F., 192, 216, 238, 251, *256*
Murase, Y., 281, *290*
Murata, T., *123*

N

Nachmansohn, D., 87, *173*
Nakane, P. K., 192, 193, 251, *257*
Nakanishi, K., 281, *290*
Narten, A. H., *6*
Nazaki, Y., 263, *289*
Neihof, R. A., 94, 95, *180*
Nelson, G. J., 86, 89, *179*, *180*
Nelson, J. A., 246, 247, *252*
Nemethy, G., 7, 56, *179*
Neuberger, A., *179*
Nezlin, R. S., 198, *257*
Nichol, L. W., *290*
Nicholls, J. G., 109, *178*
Nichols, B. L., 36, 111, 113, *176*
Nichols, C. T., 149, *173*
Nikolaev, A. Ya., 190, 205, *257*
Nims, L. F., 190, 202, *253*
Nishiyama, I., 123, *179*
North, A. C. T., 282, 283, *288*
Nosnova, T. A., *10*
Nowak, H., 192, 216, 218, *256*

O

O'Brien, R. C., 150, *178*
Odeblad, E., 39, *179*
Ogata, K., 192, *257*
Ohki, S., 85, 92, *176, 179*
Ohsaka, A., 78, *181*
Olmstead, E. G., 155, *179*
Olofson, R. A., 223, *258*
Olson, A. C., 192, 221, *255*
Ong, E. B., 192, 196, 249, 251, *257*
Oppenheimer, C. H., 120, 122, 124, 126, 127, *179*
Osborn, K., 193, 194, 195, 196, 225, 240, *255*
O'Shea, P. C., 267, *289*
Otori, T., 92, *176*
Ottesen, M., 192, *257*, 263, 264, 267, *288*
Ozawa, Y., 190, 202, *258*

P

Page, D. I., 20, *175*
Pak, C. Y. C., 44, *179*
Pallansch, M. J., *183*
Pankhurst, K. G. A., 87, *174*
Papahadjopoulos, D., 100, *172*
Park, R. B., 87, *173*
Pascoe, E., 191, 207, 212, *258*
Patel, A. B., 191, 192, 193, 194, 195, 196, 206, 207, 208, 209, 210, 212, 213, 222, 228, 230, 231, 234, 235, 237, 241, 248, 251, *253, 257*
Patel, R. P., 195, 198, 223, 224, 225, *257*
Patramani, I., 192, *257*
Patrascanu, N., 239, *254*
Patsatsiya, K. M., 11, *178*
Pauling, L., 7, 41, 138, *179*
Pauly, H., 238, *257*
Pavlatos, M., 192, *257*
Peachey, L. D., 89, *179*
Pecht, M., 192, 196, 227, 229, 249, *255, 256*, 259, *289*
Pedersen, K. O., 266, 267, *289*
Pennington, S. N., 192, 193, 194, 195, 196, 209, 212, 222, 230, 234, 237, 248, 251, *254*, *257*
Perlmann, G. E., 192, 196, 249, 251, *257*
Perrin, G., 266, *290*
Peschel, G., 23, 24, 25, 26, 159, *180*
Peters, J. H., 148, 158, *172*

Peterson, G. H., 190, 201, *254*
Pethica, B. A., 19, 55, 159, 160, *173, 177, 180*
Pfeil, E., 190, 206, *253*
Phillips, D. C., 261, 282, 283, *288, 289, 290*
Phillips, M. C., 84, *180*
Piccardi, G., 117, *180*
Pierce, G. B., Jr., 192, 193, 251, *257*
Piguet, A., 80, *180*
Pistevou, E., 192, *257*
Platteeuw, J. C., 9, 163, *182*
Podolsky, R. J., 114, *178*
Poland, D., 70, *180*
Polissar, M. J., 75, 76, 82, 83, *177*
Polson, A., 265, 267, *288*
Poole, A. G., 75, *175*
Pople, J. A., 5, *6*, 59, *174*
Porath, J., 194, 195, 245, 246, 247, *253, 257*, 267, *290*
Porter, R. S., 85, *180*
Powell, H. M., 46, *180*
Prather, J. W., 92, *180, 182*
Precht, H., 116, *180*
Price, S., 195, 198, 223, 224, 225, *257*
Privalov, P. L., 67, *180*
Prokop'eva, E. M., 147, *180*
Prosser, C. L., 135, *180*
Pryce, N. G., *183*

Q

Quiocho, F. A., 192, 216, 218, 219, 221, *253, 257*, 259, *290*
Quist, A. S., 7, *175*

R

Rader, R. L., *181*
Rajender, S., 20, 68, 70, 71, 72, *178*
Ramanow, W. J., *6*
Ramiah, M. V., 16, 103, *180*
Ramsden, W., 186, *257*
Rand, R. P., 84, *178*
Rao, C. N. R., 246, *257*
Rath, N. S., 21, *176*
Ravenhill, J., 20, *175*
Redfearn, J. W. T., 109, *178*
Ree, T., *7*
Rees, A. W., 192, 215, 216, 222, *255*
Reese, E. T., 190, 201, *257*
Reinert, J. C., *181*
Reiss-Husson, F., 84, *178*

Reitnauer, P. G., 155, *182*
Reitsma, H. F., 80. *177*
Resing, H. A., 94, 95, *180*
Richards, F. M., 67, 180, 192, 216, 218, 219, 221, *253, 257,* 259, 267, *289, 290*
Riddiford, A. C., 87, *174*
Riedeman, W. L., 238, *254*
Rieger, F., 155, *182*
Riehm, J. P., 193, *257*
Riesel, E., 195, 241, *257*
Rigaud, J. L., *176, 183*
Rimon, A., 192, 195, 196, 241, 249, 250, 251, *252, 255, 257*
Rimon, S., 195, 241, *257*
Rivas, E., 84, *178*
Robbins, F. M., 267, *290*
Robertson, R. E., 16, *180*
Rogers, F. F., Jr., 226, *254*
Roholt, O. A., 192, *254*
Rorschach, H. E., 36, 113, *176*
Rosano, H. L., 89, 100, *180*
Rose, A. H., 86, 115, 116, 121, 126, 136, *175, 180*
Rose, S. M., *176*
Rosenberg, M. D., 87, *174*
Rossi-Fanelli, A., 155, *173*
Rothen, A., 186, *257*
Roubal, W. T., 267, *290*
Rouser, G., 86, *180*
Rupley, J. A., 263, 278, *289, 290*
Rushe, E. W., 28, *180*

S

Safford, G. J., 4, 8, 9, 10, 11, 13, 51, 71, *180*
Salsbury, N. J., 84, *173*
Samoilov, O. Ya., 3, *6,* 9, *10,* 11, 18, 22, *180*
Sanderson, A. R., 195, 240, *254*
Saplin, B. J., 286, *288*
Sarma, V. R., 282, 283, *288*
Saroff, H. A., 267, *290*
Sato, M., 110, *182*
Saunders, D., 280, *289*
Schachman, H. K., 266, *290*
Schaefer, V. J., 188, *256*
Schellman, C., 260, 261, *290*
Schellman, J. A., 260, 261, *290*
Scheraga, H. A., 56, 70, *179,* 193, 215, *253, 257,* 260, 266, *290*

Schleich, T., 67, 68, *180, 182*
Schleith, L., 194, 195, 238, 240, 241, *255,* 259, *288*
Schlueter, R. J., 190, 202, *256*
Schmidt, D., 155, *177*
Schmidt, M. G., 112, 120, 122, 124, 127, *180*
Schölgl, R., 87, *180*
Schoffeniels, E., 87, *180*
Scholand, J., 192, 216, 238, 251, *256*
Schreiner, H. R., 74, 139, *180*
Schröder, W., 140, 141, 142, *180*
Schulman, J. H., 44, 89, 100, *180,* 199, *254*
Schultz, R. D., 93, *180*
Schwan, H. P., 43, *180*
Scoffone, E., 283, *289*
Sealock, R. W., 81, *176*
Sears, D. F., 99, *182*
Seidell, 148, *180*
Sela, M., 259, 260, *288*
Selegny, E., 191, 194, 217, 238, 240, *253, 257*
Senghaphan, W., 28, *181*
Senior, W. A., *8,* 19, *173*
Shaafi, R. I., 97, 98, *182*
Shafer, J. A., 193, 198, 225, *253*
Shah, D. O., 83, *181*
Shapiro, N. Z., 44, 114, 148, *174*
Sharp, A. K., 193, 194, 197, 198, 242, 244, 248, *256, 257*
Sheehan, J. C., 198, 223, 224, 225, *257*
Sheraga, H. A., *7, 180*
Shields, G. S., 267, *290*
Siegel, L. M., 267, 268, *290*
Siegel, S. M., 89, *181*
Siepe, V., 21, *177*
Silman, I. H., 189, 192, 194, 195, 203, 221, 232, 238, 241, *254, 255, 257*
Silverstein, R., 226, *254*
Simon, G., 86, *180*
Simons, E., 247, *255*
Simons, R., 88, *174*
Singer, E. R., 260, *289*
Singer, S., 193, 196, 235, *256*
Singer, S. J., 260, 264, 266, 267, *289, 290*
Sitte, P., 106, *181*
Skalski, D. J., 272, *288*
Skinner, F. A., 135, *181*
Skou, J. C., 237, *257*
Slobin, L. I., 260, *290*

van der Waals, J. H., 9, 163, *182*
Van Duijn, P., 191, 206, 207, 212, *253, 258*
Vannel, F., 25, *173*
Vanngard, T., 39, *175*
Van Steveninck, J., 43, *174*
Van Vunakis, H., 263, *289*
Vaslow, F., *10*, 11, 114, *182*
Veeger, C., 80, *181*
Venkataraghavan, R., 246, *257*
Verbanc, J. J., 246, 247, *252*
Vieira, F. L., 97, 98, *182*
von Ardenne, M., 155, *182*
von Hippel, A., 9, *182*
von Hippel, P. H., 67, 68, *180, 182*

W

Wagner, T., 195, 228, *258*
Walker, J. M., 121, *182*
Wall, T. F., *6*
Wallach, D. F. H., 86, *173*
Walrafen, G. E., *8*
Walsh, K. A., 190, 192, 203, 221, *255*
Wan, J., 191, 209, 213, *253*
Wang, J. H., 81, 104, *176, 182*, 192, 218, *258*
Warner, D. T., 41, 46, *182*
Warren, B. E., *6*
Waschka, O., 215, *258*
Watkins, J. C., 100, *172*
Watson, D. M., 191, 210, *253*
Webb, E. C., 75, *174*
Weber, W., 239, *254*
Weetall, H. H., 193, 195, 225, 227, 238, 240, 241, *258*
Wegerle, D., 215, *258*
Weissmann, G., 100, *172*
Weliky, N., 193, 225, 227, *258*
Wellner, D., 194, 240, *257*
Wen, W. Y., 3, *7, 10, 175*
Wersuhn, G., 135, 136, *182*
Westman, T. L., 195, 227, 229, *258*
Wetlaufer, D. B., 18, 73, *182*
Wetzel, R., 95, *182*
Wharton, C. W., 193, 236, *258*
Wheeler, K. P., 193, 194, 237, *258*
Whitaker, J. R., 267, 269, *290*

White, F. H., Jr., 260, *290*
Whittam, R., 193, 237, *258*
Wilchek, M., 189, *254*
Wild, F., 198, 238, 239, *255*
Williams, E. J., *183*
Williams, L., 148, *172*
Wilson, R. J. H., 194, 197, 198, 242, 244, 248, *256, 258*
Winter, A., 194, *253*
Winzer, D. J., *290*
Wishnia, A., 73, *182*
Wissler, F., 279, *289*
Witholt, D. E., 263, *288*
Witt, I., 109, *177*
Wittam, R., 193, 194, *258*
Woessner, D. E., 29, 30, *182*
Wofsy, L., 267, *290*
Wold, F., 215, *258*
Wolken, J., 87, *181*
Woods, M., 155, *173*
Woodward, R. B., 223, *258*
Wotten, M. J., *173*
Wright, E. M., 90, 91, 92, *174, 180, 182*
Wu, H., *258*
Wyman, J., 281, *290*

Y

Yamashita, S., 110, *182*
Yanari, S., 190, *256*
Yang, J. T., 261, *290*
Yastremskii, P. S., 11, *182*
Yphantis, D. A., 192, 216, 218, *256*
Yu, N-T., *183*
Yutani, K., 260, *289*

Z

Zahn, H., 215, *258*
Zana, R., 16, *178*
Zhilenkov, A. P., 38, *182*
Zimmerman, G. O., 28, *181*
Zimmerman, Y., 29, *182*
Zirwer, D., 95, *182*
Zito, R., 281, *290*
Zotterman, Y., 109, *174, 182*
Zuber, H., 215, *258*
Zundel, G., 93, *182*

Subject Index

A

Acclimation, 132
Acetone, 21, 45, 49
  powder, 205
Acetonitrile, 228
Acetylation, 239, 267, 271
Acetylcholinesterase, 74, 133, 191, 192, 194–196, 212, 234, 248, 251
Acid phosphatase, 191, 193, 207, 212, 227, 251
Actinomycin, 51
Activation energy, 82, 214, 221, 228
  discrete distribution, 82
  of permeation, 91
Activation parameters, 26, 27, 76, 100
Activator system, 241
Active sites, 197, 230
Active transport, 35, 106, 107, 118, 237
  ATP, 106
Activity, 262
  biological, 259, 262
  catalytic, 186
  conformation relationship, 203
  gradients, 107
  profiles, 229
  proteolytic, 236, 249
  references, 236
  restored, 213
  specific, 260
Actomyosin, 111
Acylase, 202
Acylazide reaction, 197
Adaptation, 129, 130
  cellular, 130
  enzyme kinetics, 132
  thermal, 126, 130, 169
Additions
  bimolecular nucleophilic, 246
  nucleoplilic, 197
Adenosine monophosphate, 218
Adenosinetriphosphatase (ATPase), 106, 192, 193, 237
  detergent-solubilized, 237
  metal-sensitive, 222
  sarcoplasmic reticulum, 237
Adenosine triphosphate (ATP), 56, 106, 107
  synthesis, 222

Adenosine triphosphate creatine phosphotransferase, 230, 231, 240
Adenosine triphosphate deaminase, 190, 206
Adenosine triphosphate phosphotransferase (creatine kinase), 193
Adsorption, 186, 188, 190, 198, 204
  specific, 114
Aggregates, 264
Air–oil interface, 115
Air–solid interface, 203
Air–water interface, 29, 34, 50, 66, 139, 186, 201
Air–water surface, 115
Ajuga reptans, 126
Alanine, 138
L-Alanine, 122
  random copolymer, 228
Alaskan king crab, 132
Albumin, 215
  modified, 270
  reduced, 273
  Stokes radii, 269
  succinylated, 272
Alcohol–water mixtures, 11, 16
Alcohols
  aliphatic, 14, 45
  aqueous, 11, 16
  partial molar volumes, 15
Aldehyde lyase, 206
Aldehyde lyase D-oxynitrilase, 190
Aldolase, 78, 191, 192, 208, 210, 213, 214
  radioactive, 213
Algae
  flagellate, 135
  marine, 41
  thermophilic, 116
Alkaline phosphatase, 191, 193, 195, 207, 212, 227, 240, 260
Alkaloids, 45, 117
  plant survival, 117
Alkylamine hydrates, 58
Alkylating compounds, 156
All-or-none processes, 66
All-or-none responses, 41
All-or-none transitions, 66
Allosteric activator, 281
Amide bonds, 197, 227, 230

303

Amide groups, 223
Amide linkages, 223
Amidination, 264, 267, 271, 274
Amine-triazine reaction, 243
Amino acid(s), 45, 129
  acid, 223
  basic, 223
  composition, 261
  sequence, 66, 260
D-Amino acid oxidase, 79
Aminoacylase, 190, 205
  column, 205
p-Aminobenzyl cellulose, 230, 231, 238, 240
Aminoethyl cellulose, 252
  matrix, 252
Aminoethylcysteine, 260
Amino group(s), 238
  free, 223, 234
    blocking, 260
  terminal, 223
ε-Amino group, 238, 239
p-Aminophenylalanine-leucine copolymer, 240, 241, 249
Aminopropyl diethanolamine, 198
Amino oxidase, 209
Ammonium
  quaternary, 61
  salts, 97
Ammonium sulfate, 64
Amplitude, thermal, 34
Amylase, 210
α-Amylase, 191, 248
β-Amylase, 80, 191, 195, 247
  barley, 80
  wheat, 80
Anaerobic conditions, 146
Anesthesia, 46, 54, 100, 108, 138, 140, 145
Anhydride function, 227
Anion exchange particles, 248
Anisotropy, 262
Anteater, 118
Antibenzene arsonate antibodies, 267
Antibodies, 197, 251, 259, 263, 264, 267, 285, 287, see also specific types
  combining site, 286
  enzyme conjugates, 251
  modified, 264
  response, 284, 286
Anticholinergic agents, 211

Antigens, 197, 267, 284, see also specific types
  antibody precipitates, 277
  homologous, 267
  topography, 284
Antigenic determinants, 260, 284
Antigenic reactive regions, 286
Antigenic site, 286
Antigenic structure, 286
Antilysozyme antiserum, 285, 286
Antipneumonococcus serum globulin, 266
Antiribonuclease antiserum, 284
Antisera, 251, 284, see also specific types
Apatite, 51
Apoenzyme, 220, 221
  metal chelating agent complex, 221
Apomyoglobin, 267
Apyrase, 191–196, 237, 251, 252
  bound, 237
  water-insoluble, 222
Arabinose, 121
Arginine, 223, 238
Argon, 47, 140, 144, 161, 163
Armadillo, 118
Arrhenius activation parameters, 96
Arrhenius equation, 28, 100
Arrhenius nonlinearity, 129
Arrhenius plots, 81
  changes in, 76
Arsanilic acid, diazotized, 238
Arsenic, protein-bound, 239
Arylisothiocyanates, 247
Asparaginase, 190, 193–195, 205, 229
  matrix-supported, 237
  native vs. adsorbed, 205
Asparagine, 223
Aspartate transcarbamylase, 281
Aspartic acid, 120
Asphyxia, 154
Association–induction hypothesis, 27, 112
Asymmetry, 265
Atomic details, 283
Atomic environment, 273
ATP, see Adenosine triphosphate
ATPase, see Adenosinetriphosphatase
Autodigestion, 250
Autolysis, 204
  resistance to, 241
Avogadro's number, 268
Azide inhibition, 227

314 SUBJECT INDEX

D-α-Hydroxynitriles, 206
Hydroxyquinoline sulfonic acid, 220
Hyperglycemia, 148
Hyperthermia, 124, 154, 169
  therapy, 156
  of cancer, 39
Hypertropical belt, 137
Hypocalcemic, 148
Hypothermia, 82, 119, 138, 145–147, 149
  clinical, 149
  multistep, 155, 157
  patients, 148
  rabbit, 147
Hysteresis effects, 37, 95
  thermal, 161, 162, 170

I

Ice, 57, 68
  cluster, 35
  hexagonal, 9
  lattice, 162
  melting, 5
  liquid, 33
  polymorphism, 9
  polymorphs, 37, 43, 46, 153
    high pressure, 9, 31, 36, 57
  structure, 21
  surface, 113
Ice-likeness, 152
  hexagonal, 46
Ice-like model, 18
Ice-like state, 54
Ice-I, 7, 57
  structure, 6
Ice-Ih, 9, 22, 46, 75
  lattices, 5, 34, 46
Ice-Ih-like, 20
  structures, 45
Ice-II, 46
Ice-III, 7
Iceberg, 20, 68
  building, 10
  formation, 26
  microscopic, 10
Image forces, 159
IMET, 248, 249
Imidazole, 223, 238
Imino carbonic acid esters, 246
Imino group, 223
Immune precipitate, 284

Immunochemistry, 260
Immunohistochemical localization, 251
Immunology, 198
Index of refraction, 262
Indolyl group, 238
Inductive effects, 92
Infectious diseases, 154
Infrared absorption, 262
Infrared spectra, 261
Infrared spectroscopy, 68, 74, 75, 94
Infrared studies, 12, 13
Infrared techniques, 67
Inhibition constant, 220
Inhibitors, 220
Injury, thermal, 154
Insects, 110, 120
Insoluble, derivatives, 259, see also specific compounds
Insulin, 215, 260
  hypoglycemia, 97
Interactions
  chemical bond-type, 50
  homotropic, 218
  hydrophobic, 56, 67
  molecular, 84
Interfaces, 9, 22, 23, 25, 29, 30, 31, 34, 35, 37, 40, 43, 44, 71, 72, 77, 93, 112, 150, 156, see also specific types
  biological, 43, 163
  macromolecular, 115
  polar, 115
  solid, 44, 92–94, 167
Interfacial phenomena, 29
Interfacial reorganization, 86
Interfacial systems, rate data, 27
Interfacial tension, 50
Internal energy, 8
Interstitial fluid, 106
Invariant boundaries, 137
Invertase, 190, 193, 196, 201, 202, 235
Iodide–water interface, 29
Ion–dipole interactions, 50, 57, 113
Ion exchange, 190, 199
  system, 233
Ion hydration, 94
Ion pairing, 159
Ion transport, 222
Ion–water dipole interactions, 29
IPTT, 248, 249
Isobutane, 138

320 SUBJECT INDEX

Pines, 120
Pinus cembroides, 120
pK
  shift, 200
  values, 262
Plaice, 133
Plant–animal interaction, 117
Plants, 115
  alkaloids from, 117
  higher, 237
  physiology of, 102
Plasma membrane, 93
Plasminokinase, 195
Platelet(s), 161
  adhesion, 160
  aggregation, 160
Platelet–glass interface, 161
Pleuronectes platessa, 133
Poikilotherms, 120, 124, 154
Poiseuille flow, 97, 105
Polarizability, 262
Polarographic method, 209
Polyacrylamide, 191
  cross-linked, 209
  enzyme, 207
  films, 207
  gels, 206
Pollutants
  chemical, 164
  thermal, 130, 135, 154, 163–165, 170
Polyacrylamide–phosphatase model, 207
Polyacrylic acid, 223
Polyaminostyrene, diazotized, 240
Polyalanylation, 259
Polyalanyl rabbit antibodies, 260
Poly-DL-alanyl trypsin, 260
Polyanionic carrier, 250
Polyelectrolyte
  gel, 250
  membranes, 93
Poly (adenylic acid), 63
Poly(L-glutamic acid), 63, 223
Poly-L-glutamyl chymotrypsin, 229
Poly(L-lysine), 63, 64
Polymer(s)
  carboxyl groups of, 236
  charged, 64
  entanglement, 60
  hydrophobic, 235
  surfaces, 160

water-interacting, 65
  of water, 5
Polymer–polymer interactions, 64
Polymer–water interactions, 64
Poly(methacrylic acid), 63
Polyornithyl chymotrypsin, 229
Polypeptides, 45, 56, 61, 65, 68, 260
  chain, 74, 260, 280, 286
  synthetic, 238
  three-dimensional arrangement, 261
Polysaccharides, 61, 158
Polystyrene, 92
Polythermostat, 126
Polytyrosyl trypsin, 241, 243, 248
  copolymer-bound, 249
Poly-U, 64
Polyurethan pad, 211
Poly(uridylic acid), 63
Poly(vinyloxazolidinone), 63
Polyvinylpyrrolidone (PDP), 63, 151
Polyvinyl toluene–water interface, 29
Polywater, 54
Pores, 103–105, 168, 215, 262
  concept, 99
  diameters, 105
  discrete, 105
  narrow, 23
  radius, 268
  water-filled, 90, 99, 100
Porphyrins, modified, 286
Positron annihilation, 54
Positronium, 54
  formation, 54
Postnatal period, 114
Potassium
  permeability, 100
  uptake, 121
Potato
  apyrase, 237
  tubers, 126
Precipitability, 263
Precipitin curves, 284
Pressure
  anomalies, 141, 170
  disjoining, 23, 25
  hydrostatic, 40
  osmotic, 157, 159
  thermomolecular effect, 158
Procion Brilliant dyes, 241
Procion Brilliant Orange, 194, 248

lattice, 10, 47
  unperturbed, 68
layers, 44
long-range effects, 44, 62
long-range order, 7, 42, 60
metastable conditions, 44, 143
metastable cooperative state, 38
near-infrared spectrum, 71
nonsolvent, 42, 43
order in, 4
ordered, 32, 33, 111
puckered hexagonal rings, 7
in protein processes, 70
in protein solutions, 75
protoplasmic, 38
Raman measurements of, 8
rigid, 49
role of, in biological systems, 2
rotational modes, 103
shell of, 68
short-range order, 21, 42
short-range properties, 261, 262, 273, 286, 287
spectroscopic studies, 93
state in biological systems, 67
structurally modified, 42
structural regions, 6
structure, 2, 4, 39, 55, 112
  changes, 87
  effects, 168
  electrolyte effects, 67
  energetically favored, 24
  H-ion effect, 80
  of membranes, 56
  models, 4, 6
  stabilized, 48
supercooled, 45, 150
in tissues, 39
transfer to deuterium, 72
vicinal, 2, 45, 136
Water-air interface, 139
Water-alcohol solutions, 75
Water-cellulose interactions, 103
Water-cellulose system, 103
Water-host structure, 58
Water-hydrocarbon interface, 204
Water-ion interactions, 10

Water-lipid interactions, 84
Water-lipid interface, 100
Water-macromolecule stability, 61
Water-membrane interactions, 104
Water-nonpolar hydrocarbon interface, 21
Water-oil emulsions, 55
Water-polyhydric alcohol, 16
Water-pore system, 117
Water-rich regions, 12
Water-soap systems, 55
Water-solid interface, 66
Water-solid particle, 55
Water-solute interactions, 55
Water-swollen materials, 103
Water-tetrahydrofuran, 57
Water-urea clusters, 75
Water-water hydrogen bonding, 21
Water-water interactions, 57, 58
Wave reaction, 222
Wayne-Kerr conductance bridge, 88
Whale metmyoglobin, 264
Wheat, 121
Woodward's reagent K, 195, 197, 223, 226, 228, 229, 251

X

Xenon, 47, 54, 140, 161
X-ray, 12
  crystallography, 261, 283
  data, 10
  diffraction, 18
  studies, 41
  scattering, 20, 57
  small-angle, 262
  wide-angle, 262

Y

Yeast, 61, 151
  cells, 43
  invertase, 202

Z

Zein, 266
Zeta potential, 262
Zinc, 219, 220
  atom, 221